ANNUAL REVIEW OF NUMERICAL FLUID MECHANICS AND HEAT TRANSFER

Annual Review of Numerical Fluid Mechanics and Heat Transfer

ANNUAL REVIEW OF NUMERICAL FLUID MECHANICS AND HEAT TRANSFER

Volume 2

Edited by

C. L. Tien

A. Martin Berlin Professor of Mechanical Engineering
University of California
Berkeley, California

and

T. C. Chawla

Argonne National Laboratory
Argonne, Illinois

⬤ HEMISPHERE PUBLISHING CORPORATION
A member of the Taylor & Francis Group

New York   Washington   Philadelphia   London

ANNUAL REVIEW OF NUMERICAL FLUID MECHANICS AND HEAT TRANSFER, Volume 2

1 2 3 4 5 6 7 8 9 0   B C B C   8 9 8 7 6 5 4 3 2 1 0 9 8

This book was set in Times Roman by Edwards Brothers, Inc. The editors were Linda A. Dziobek and Brenda Brienza.

BookCrafters, Inc. was printer and binder.

ISBN 0-89116-740-4
ISSN 0892-6883

CONTENTS

CONTRIBUTORS

M. B. BUSH
Department of Mechanical
  Engineering
The University of Western
  Australia
Nedlands, Western Australia

S. G. CHUECH
Department of Aerospace
  Engineering
University of Michigan
Ann Arbor, Michigan

A. M. DAVIES
Proudman Oceanographic
  Laboratory
Bidston Observatory
Birkenhead, Merseyside
United Kingdom

W. J. M. DOUGLAS
Pulp and Paper Research Institute
  of Canada
Pointe Claire, Canada

D. J. EVANS
Department of Computer Studies
Loughborough University of
  Technology
Loughborough, Leicestershire
United Kingdom

G. M. FAETH
Department of Aerospace
  Engineering
University of Michigan
Ann Arbor, Michigan

J. P. GORE
Department of Aerospace
  Engineering
University of Michigan
Ann Arbor, Michigan

B. HENDERSON-SELLERS*
Department of Mathematics and
  Computer Science
University of Salford
Salford, United Kingdom

*Present address: School of Mathematics, University of New South Wales, P.O. Box 1, Kensington, New South Wales 2033, Australia.

B. HUANG
Department of Chemical
  Engineering
McGill University
Montreal, Canada

S.-M. JENG
Center of Laser Applications
University of Tennessee Space
  Institute
Tullahoma, Tennessee

EGON KRAUSE
Aerodynamisches Institut
Aachen, West Germany

A. S. MUJUMDAR
Department of Chemical
  Engineering
McGill University
Montreal, Canada

DOMINIQUE H. PELLETIER
Applied Mathematics
  Department
Ecole Polytechnique of
  Montreal
Montreal, Canada

N. PHAN-THIEN
Department of Mechanical
  Engineering
University of Sydney
Sydney, Australia

J. N. REDDY
Department of Engineering Science
  and Mechanics
Virginia Polytechnic Institute and
  State University
Blacksburg, Virginia

M. S. SAHIMI
Department of Computer Studies
Loughborough University of
  Technology
Loughborough, Leicestershire
United Kingdom

JOSEPH A. SCHETZ
Department of Aerospace and Ocean
  Engineering
Virginia Polytechnic Institute and
  State University
Blacksburg, Virginia

A. SHIMA
Institute of High Speed Mechanics
Tohoku University
Sendai, Japan

R. I. TANNER
Department of Mechanical
  Engineering
University of Sydney
Sydney, Australia

Y. TOMITA
Institute of High Speed Mechanics
Tohoku University
Sendai, Japan

PREFACE

Research and development in numerical fluid mechanics and heat transfer continue at an accelerating pace. The increasing use of parallel or vector processing and supercomputers makes the whole field even more exciting and challenging. Many complex physical problems, previously considered intractable, are now within numerical reach. These rapid developments have led to an evergrowing body of literature in a wide range of topics in fluid mechanics and heat transfer. Against this broad and dynamic background this book series was initiated to devote itself exclusively to expository, comprehensive, state-of-the-art reviews by specialists in various fields.

The favorable response from the science and engineering communities to the first volume of Annual Review of Numerical Fluid Mechanics and Heat Transfer confirms the view that this serial publication is playing a significant role in serving the scholarly community. The present volume contributes to the original objective of improved communication by presenting coordinated and unified chapters on important developments in numerical fluid mechanics and heat transfer. Each chapter starts from well-understood basic principles underlying the physics and computational aspects of the applications and then proceeds in a logical fashion to bring the reader up to the forefront of the topic. We express our deep appreciation to the authors of this volume for their excellent contributions.

During the preparation for this volume the founding editor, Dr. T. C. Chawla, became ill and required a lengthy period of recuperation. Due to this unexpected happening, Dr. Chang-Lin Tien, A. Martin Berlin Professor of Mechanical Engineering at the University of California at Berkeley, has now become the new editor of the series. The new editor looks forward to this most challenging endeavor and sincerely welcomes suggestions, cooperation, and support from all interested members of the scholarly community.

T. C. Chawla
C. L. Tien

ANNUAL REVIEW OF
NUMERICAL FLUID MECHANICS
AND HEAT TRANSFER

ONE

RADIATION FROM TURBULENT DIFFUSION FLAMES

G. M. Faeth, J. P. Gore, S. G. Chuech, and S.-M. Jeng

ABSTRACT

The thermal radiation properties of nonluminous and luminous turbulent diffusion flames are reviewed, considering: the scalar structure of flames, excluding soot; soot properties of luminous flames; and turbulence/radiation interactions. Progress has been greatest for nonluminous flames. Scalar structure needed to analyze radiation properties is generally estimated using the laminar flamelet concept, e.g., approximating turbulent flames as wrinkled laminar flames, noting that the scalar properties of laminar flames are nearly universal functions of fuel-equivalence ratio. Estimates of spectral radiation intensities and radiative heat fluxes based on these ideas are typically within 20–30 percent of measurements. Extension of the laminar flamelet concept to the properties of soot which influence radiation from luminous flames has been studied. There is some evidence to support the extension and radiation predictions based on this idea; however, additional evaluation of the concept is needed. Turbulence/radiation interactions have been studied using stochastic analysis. The effects of turbulence/radiation interactions are most significant for luminous flames, where turbulent fluctuations can increase heat fluxes up to 2–3 times higher than estimates based on mean properties.

NOMENCLATURE

a	acceleration of gravity
A	area
C_i	constants in turbulence model

1

d	burner exit diameter
$e_{\lambda,b}$	blackbody monochromatic emissive power
f	mixture fraction
f_v	soot volume fraction
g	square of mixture fraction fluctuations
h	square of total enthalpy fluctuations
H	total enthalpy
I_λ	spectral intensity
J	radiance
J_λ	spectral radiance
k	turbulent kinetic energy
K_λ	spectral absorption cross-section
L_e	dissipation length scale of eddy
n	real-part of complex refractive index of soot
$\bar{P}(f)$	time (Reynolds)-averaged probability density function of f
$\tilde{P}(f)$	mass weighted (Favre)-averaged probability density function of f
Q_r	radiative source term
r	radial distance
Re	burner Reynolds number
s	distance along optical path
Sc	laminar Schmidt number
S_ϕ	source term
T	temperature
u	axial velocity
v	radial velocity
x	height above burner
ϵ	rate of dissipation of turbulence kinetic energy
$\theta,\ \theta_A$	angle between path and surface normal
κ	imaginary part of the complex refractive index of soot
λ	wavelength
μ	laminar viscosity
μ_{eff}	effective viscosity
μ_t	turbulent viscosity
ρ	density
σ_i	turbulent Prandtl/Schmidt number
τ_λ	spectral transmittance
ϕ	generic property, fuel equivalence ratio
ω	solid angle

Subscripts

c	centerline value
s	soot property
o	burner exit condition
∞	ambient condition

Superscripts

$(\bar{\ })$ time-averaged quantity
$(\tilde{\ })$ Favre-averaged quantity
$(\bar{\ })'$ time-averaged fluctuating quantity
$(\tilde{\ })''$ Favre-averaged fluctuating quantity

1 INTRODUCTION

1.1 Overview

Flame radiation is an important aspect of heat transfer in furnaces, industrial flares, internal-combustion power and propulsion systems, and unwanted fires. This has motivated many studies of flame radiation. The objective of this paper is to review recent work in the field. Particular emphasis is placed on nonluminous and luminous turbulent diffusion flames, since they represent a large fraction of practical applications. At this limit, four major research issues are involved: (1) scalar flame structure, excluding flame-generated particulates (largely soot); (2) local properties of flame-generated particulates in luminous flames; (3) turbulence/radiation interactions; and (4) predicting radiative properties given the instantaneous scalar structure. The present paper emphasizes the first three issues, since the last currently causes fewest problems.

Measurements and predictions carried out by the authors will be emphasized, since both structure and radiation properties have been studied for the same flames. However, methods used draw on the work of many; therefore, their consideration provides a reasonable overview of the field. Only turbulent diffusion flames in free shear flows will be considered, in order to avoid the complications of low Reynolds number turbulence near surfaces and surface radiation properties.

The paper begins with a discussion of past work treating radiation from flames. Next, methods used to estimate the scalar structure of turbulent diffusion flames, and their evaluation with experiments, are considered. Analysis of spectral radiation intensities, based on predicted scalar properties, is then described and compared with measurements. These results are then combined to provide predictions of radiative heat fluxes which are also compared with measurements. The paper concludes with suggestions concerning current research issues.

1.2 Background

Several recent reviews treat aspects of flame radiation. Tien and Lee [1] provide a comprehensive summary of the radiation properties of nonhomogeneous and particle containing media typical of flame environments. deRis [2] surveys aspects of radiation from turbulent fires, where buoyancy influences the process. Eickhoff [3] discusses methods for predicting the scalar structure of turbulent diffusion flames. General background material on radiation in pure and particulate-laden gases is provided by Siegel and Howell [4], Ludwig et al. [5], Goody [6] and Hottel and Sarofim [7].

While radiation from turbulent flames is of greatest interest, studies of laminar flames are useful since the complexities of turbulent hydrodynamics are avoided. Lloyd and co-workers [8,9] report pioneering work in this area. Laminar methane diffusion flames, containing negligible soot, were analyzed using the boundary-layer approximations. Given the structure, radiation properties were found using the exponential wide-band model [10]. The results were promising, with greatest uncertainties associated with the structure predictions. Other recent studies of laminar diffusion flames have emphasized effects of soot, e.g., Grosshandler and Vantelon [11] and Markstein and co-workers [12–14]. Capabilities for rational predictions of soot concentrations in flames are very limited; therefore, current results rest heavily on empirical information.

In order to avoid the uncertainties of flame structure predictions, several studies have used measured scalar properties in flames to test predictions of flame radiation. Grosshandler and Sawyer [15] examined the nonluminous spectral radiation properties of methanol/air combustion products. Radiation properties were predicted employing the Goody [6] statistical narrow-band model, using the Curtis-Godson approximation for inhomogeneous gas paths, following Ludwig et al. [5]. Their measurements and predictions were in good agreement in the post-flame region. Grosshandler and co-workers [16,17] also developed a method to reduce the computations needed for radiation predictions.

Karman and Steward [18] report similar work for mixtures of propane, propylene and carbon particles (1–10 μm diameter range). Mie scattering from large particles was treated following Steward and Guruz [19]. The use of measured mean scalar properties provided good predictions of spectral radiation properties in the post-flame region.

Souil et al. [20,21] studied radiation from turbulent flame environments. They used measured mean temperatures, in conjunction with prescribed values of the absorption coefficient, finding good fits of measured radiative heat flux distributions.

The combined problem of predicting both turbulent diffusion flame structure and radiation properties has also been addressed. Wilcox [22] reports an early attempt. An integral model was used to predict structure while an exponential wide-band model (based on mean properties) was used to predict spectral radiation intensities. The predictions were compared with measurements for liquified natural gas pool fires having little continuum radiation from soot [23]. The structure model agreed with measurement of overall quantities like flame height. The spectral radiation model yielded good predictions of radiation intensities in the 4.3 μm band of carbon dioxide. The method contains substantial empiricism, however, and its generality has not been demonstrated.

Fishburne and Pergament [24] consider large-scale hydrogen flames burning in still air. Their predictions of flame structure were based on a mixing-length model of turbulence, while Arrhenius expressions were used to prescribe turbulent reaction rates. The radiation analysis was similar to Ludwig et al. [5]. Encouraging agreement was obtained between their predictions and existing measurements; however, flame widths were not predicted very well and the use of Arrhenius expressions for turbulent diffusion flames is questionable [3].

Tamanini [25] reports structure and radiation predictions for buoyant, round turbulent flames. A k-ϵ-g turbulence model, involving an algebraic stress model to treat buoyancy/turbulence interactions, was used to predict the flame structure. Flame radiation was taken to be a fixed fraction of the energy release of reaction, emitted at a rate proportional to the local rate of turbulent reaction. The predictions were compared with measurements reported by Markstein [26]. The comparison between predictions and measurements was encouraging, however, many empirical aspects of the structure model were not established very accurately. The generality of the radiation prediction method also requires further assessment.

Several recent experimental studies provide information on the radiative properties of turbulent flames. Pfenning [27] measured mean temperatures and heat fluxes for large natural gas flares, ca. 20–30 m high, which will be used in the following to test effects of flame scale. Claus [28], Emmerich [29] and Najjar and Goodger [30] report radiation measurements, including spectral radiation intensities, from turbulent soot-containing flames. Emmerich [29] suggests that internal absorption calculations, considering wavelength-dependent properties, are needed for accurate analysis of these flows. The results indicate that continuum radiation from soot dominates the spectrum for the combustor conditions considered in these studies. It is generally accepted that soot radiation dominates radiation from fires as well [2].

Mean properties are most often used to compute flame radiation. It is recognized, however, that turbulent fluctuations make this a questionable practice since radiation properties are nonlinear. Williams and Fuhs [31] treated turbulent fluctuations by assuming that a turbulent flame is a wrinkled laminar flame, but it is difficult to apply their methods to practical turbulent flames. Assuming a gray gas, Cox [32], finds that for fluctuating temperature intensities greater than 40 percent, which is not unusual for turbulent flames, radiance values allowing for fluctuations are more than twice those predicted using mean temperatures. While this suggests a strong effect of turbulence on radiation properties, the gray gas approximation is not very appropriate for turbulent flames, where gas bands and continuum radiation from soot cause a complex variation of radiation properties with wavelength.

Kabashnikov and Kmit [33] treat the combined effects of fluctuating absorption coefficient and temperature. They consider a linear variation of absorption coefficient in the Wien spectral region, where effects of fluctuations are largely due to the strong temperature dependence of the Planck function. When the radiation path is more than several times the characteristic turbulence length scale, and the optical depth of a single fluctuation is small, they show that the correct intensity can be found by using the average values of the product of the Planck function and the absorption coefficient. They state that using the conventional approach (multiplying the averages of both) results in lower estimates of intensities by factors of 2–3. However, while the radiation model they examine yields a convenient closed-form solution of the problem, it is not very representative of flame radiation properties.

Grosshandler and Joulain [34] treat effects of turbulence on flame radiation by assigning properties along a line-of-sight path using prescribed probability density functions for scalar properties. Both nonluminous and luminous conditions were considered. It was found that intensities could exceed predictions based on time-mean properties by as much as a factor of two. However, the actual intensity was *lower* than results based on time-mean properties for some strongly sooting situations. While the radiation analysis used by Grosshandler and Joulain is state of the art, their prescription for scalar fluctuations along the path is relatively ad hoc; therefore, whether turbulence/radiation interactions can yield effects of this type and magnitude must still be established.

There are few available measurements of radiance fluctuations for flames. An exception is Portscht [35] who reports direct measurements of the frequency spectrum of flame radiation, considering both nonluminous and luminous flames. The power spectral densities for various detectors and flames were measured. The results show interesting spikes characteristic of pulsations of the entire flame structure; however, no attempt was made to analyze the phenomena in terms of fundamental radiation or turbulence properties.

It is evident that capabilities for predicting radiation from turbulent diffusion flames rests on accurate scalar structure predictions. Furthermore, most flames involve flame-generated particulates (usually soot); therefore, scalar property predictions must include particulate properties which are relevant to radiation. The findings of Grosshandler and Sawyer [15] and Karman and Steward [18] suggest that the use of mean scalar properties, along with state-of-the-art narrow-band radiation models, provides good predictions of spectral radiation intensities in the post-flame region. However, several studies [31–34] find that turbulence/radiation interactions cause radiance levels in flames to be significantly different (by factors of two to three) than estimates based on mean properties. We conclude that existing narrow-band models provide an adequate framework for estimating flame radiation in nonluminous and luminous flames. However, capabilities to predict scalar structure, including particulates, and to treat radiation/turbulence interactions in turbulent flames are uncertain; therefore, work considering these aspects of flame radiation will be emphasized in the following.

2 ANALYSIS OF FLAME STRUCTURE

2.1 Description

The conserved-scalar formalism is currently popular for analysis of turbulent diffusion flames [3], and will be used here. The method parallels the use of Schvab-Zeldovich variables for analysis of laminar diffusion flames [36]. In both cases, the intent is to relate all scalar properties to one or more conserved scalars, e.g., quantities which can be found by solution of governing equations which do not involve source terms. Similar to Schvab-Zeldovich analysis, the turbulence formulation is considerably simplified when the turbulent exchange coefficients of all species and heat are the same (analogous to the unity Lewis number assumption of laminar flow); therefore, we make this assumption which is almost always adopted.

We limit present considerations to conditions where the simplest conserved-scalar formulation is achieved. This implies that the process only involves mixing between two streams; flow Mach numbers are low, so that viscous dissipation and kinetic energy can be ignored in the mean flow equations; effects of heat losses by radiation and convection to surfaces are small; and a classical diffusion flame structure can be assumed, e.g., reaction rates are sufficiently fast so that the rate of reaction is controlled by the rate of diffusion of reactants to an infinitely thin flame zone. The last assumption clearly implies a wrinkled laminar flame picture of turbulent flames. Under these approximations, all instantaneous scalar properties are only a function of a single conserved scalar, frequently taken to be the mixture fraction (the fraction of mass at a point which originated from the burner stream). In the following, the representation of scalar properties as a function of mixture fraction will be termed the state relationships. Appropriate state relationships are a critical issue; therefore, they are considered in some detail later. For the present, we assume that such relationships can be found and formulate the analysis in terms of mixture fraction.

The flows to be considered are steady, round, turbulent diffusion flames burning in still air; therefore, the boundary-layer approximations are adopted with little error. In order to consider turbulence/radiation interactions, information concerning turbulence properties is needed. This is obtained using the conserved-scalar formalism in conjunction with a k-ϵ-g turbulence model, along the lines of Lockwood and Naguib [37]. Lockwood and Naguib [37] employ a time (Reynolds)-averaged analysis. Bilger [38] points out, however, that use of mass-weighted (Favre) averages avoids neglecting many relatively large terms involving density fluctuations in the governing equations (which is conventional practice for Reynolds-averaged analysis). Thus, the convenience of a Favre-averaged formulation is adopted in the following.

The flames to be considered radiate an appreciable fraction (10–40 percent) of their energy release of combustion to the surroundings. Highly luminous flames also have significant radiant energy exchange within the flame itself [29]. Considering these phenomena would involve full coupling between analysis of structure and radiation. To begin, we avoid this complexity and treat radiative exchange as a perturbation, assuming that the flame is optically thin, e.g., it is implied that a fixed fraction of the chemical energy release of combustion has been radiated to the surroundings at all points in the flame. This involves initially analyzing the structure, while only correcting the energy release of reaction to account for heat loss from the flame by radiation, and then solving the equation of radiative transfer, using the structure predictions, in the second phase of analysis. Clearly, this approach can be iterated to complete a coupled flow-structure/

radiation solution, however, there are some fundamental problems with this procedure which will be discussed once the results of the simplified analysis have been studied. The present approximation is most accurate when the fraction of the chemical energy release radiated to the surroundings is small. This is appropriate for most nonluminous flames which are turbulent and for high-intensity combustors.

Many flames are subject to effects of buoyancy. Again, in an effort to simplify the analysis and to reduce empiricism, buoyancy is only considered in the mean flow equations, neglecting buoyancy/turbulence interactions. Jeng et al. [39] find that this simplified approach is adequate for predicting mean properties.

2.2 Governing Equations

Distances and velocities are denoted x and u in the streamwise direction, and r and v in the radial direction. Under present assumptions, the solution is closed by solving governing equations for conservation of mass, momentum and mixture fraction as well as additional modeled governing equations for turbulence kinetic energy, k, the rate of dissipation of turbulence kinetic energy, ϵ, and the square of the mixture fraction fluctuations, g. The governing equations for these quantities can be written in the following generalized form [38]:

$$\frac{\partial}{\partial x}(\bar{\rho}\tilde{u}\phi) + \frac{1}{r}\frac{\partial}{\partial r}(r\bar{\rho}\tilde{v}\phi) = \frac{1}{r}\frac{\partial}{\partial r}\left(r\mu_{\text{eff},\phi}\frac{\partial\phi}{\partial r}\right) + S_\phi \qquad (1)$$

where $\phi = 1, \tilde{u}, \tilde{f}, k, \epsilon$, or g. The formulation is based on Favre (mass weighted)-averaged quantities, $\tilde{\phi}$, e.g.,

$$\tilde{\phi} = \frac{\overline{\rho\phi}}{\bar{\rho}} \qquad (2)$$

where an overbar represents a conventional time average [38]. Expressions for $\mu_{\text{eff},\phi}$ and S_ϕ, appearing in Eq. (1), are summarized in Table 1. When written in Favre-averaged form, the governing equations are similar to those found for constant-density flows. However, a velocity/pressure gradient term, due to buoyancy, in the k equation, and an analogous term in the ϵ equation, have been ignored to minimize the number of empirical constants. Past work suggests that these terms are relatively unimportant for present flows [39]. The empirical constants appearing in the model, cf. Table 1, were established by matching predictions and measurements for constant density round jets and have not been subsequently changed [40]. They are not very different from the values originally proposed by Lockwood and Naguib [37].

2.3 Initial and Boundary Conditions

The boundary conditions for Eq. (1) are:

$$r = 0, \frac{\partial\phi}{\partial r} = 0 \qquad r \to \infty, \phi = 0 \qquad (3)$$

Table 1 Summary of turbulence model parameters

ϕ	$\mu_{\text{eff},\phi}$	S_ϕ
1	—	O
$\bar{u}$	$\mu + \mu_t$	$a(\bar{\rho} - \rho_\infty)$
$\tilde{f}$	$(\mu/Sc) + (\mu_t/\sigma_f)$	O
k	$\mu + (\mu_t/\sigma_k)$	$\mu_t(\partial\bar{u}/\partial r)^2 - \rho\epsilon$
ϵ	$\mu + (\mu_t/\sigma_\epsilon)$	$(C_{\epsilon1}\mu_t(\partial\bar{u}/\partial r)^2 - C_{\epsilon2}\bar{\rho}\epsilon)(\epsilon/k)$
g	$(\mu/Sc) + (\mu_t/\sigma_g)$	$C_{g1}\mu_t(\partial\tilde{f}/\partial r)^2 - C_{g2}\bar{\rho}g\epsilon/k$

C_μ	$C_{\epsilon t}$	$C_{g t}$	$C_{\epsilon 2} = C_{g 2}$	σ_k	σ_ϵ	$\sigma_f = \sigma_g$	Sc
0.09	1.44	2.8	1.87	1.0	1.3	0.7	0.7

$\mu_t = \bar{\rho}C_\mu k^2/\epsilon$

where the conditions at $r = 0$ are only applied beyond the end of the potential core.

Initial conditions must be specified at some axial station, often a burner exit. For some flames considered here $\bar{u}_0$ and k_0 were measured, while ϵ_0 was estimated by matching the decay of k in the potential core. By definition, $\tilde{f}_0 = 1$ and $g_0 = 0$. Quantities in the shear layer were then found from the governing equations, assuming linear profiles of mean quantities and neglecting convection and diffusion terms in the governing equations for turbulent quantities. Subsequent boundary conditions along the inner edge of the shear layer were found by solving the transport equations within the potential core. The initial conditions for some flows to be considered were not well defined; therefore, in these cases burner exit properties were estimated following past practice [39].

2.4 Scalar Properties

Under present assumptions, instantaneous scalar properties are only functions of the instantaneous mixture fraction—termed the state relationships $\phi(f)$. Specification of $\phi(f)$ is discussed subsequently; for the present, the $\phi(f)$ are assumed to be known.

Given the $\phi(f)$, the Favre-averaged mean and variance of scalar flow properties are found from the mass-averaged (Favre) probability density function (PDF) of f, $\tilde{P}(f)$, as follows [38]:

$$\tilde{\phi} = \frac{\overline{\rho\phi}}{\bar{\rho}} = \int_0^1 \phi(f)\tilde{P}(f)df \qquad (4)$$

$$\tilde{\phi}''^2 = \frac{\overline{\rho\phi''^2}}{\bar{\rho}} = \int_0^1 (\phi(f) - \tilde{\phi})^2\tilde{P}(f)df \qquad (5)$$

Noting that the time-averaged PDF of f, $\bar{P}(f)$, is given by [38]

$$\bar{P}(f) = \frac{\rho\tilde{P}(\tilde{f})}{\rho(f)} \qquad (6)$$

time-averaged scalar properties become

$$\bar{\phi} = \bar{\rho} \int_0^1 \left[\frac{\phi(f)}{\rho(f)} \tilde{P}(f) \right] df \tag{7}$$

$$\bar{\phi'^2} = \bar{\rho} \int_0^1 \left[(\phi(f) - \bar{\phi})^2 \frac{\tilde{P}(f)}{\rho(f)} \right] df \tag{8}$$

The value of $\bar{\rho}$ is needed to integrate Eq. (1); this is found by setting $\phi = 1$ in Eq. (7).

A functional form must be assumed for $\tilde{P}(f)$. Past work has shown that scalar properties are not strongly influenced by the specific form chosen for $\tilde{P}(f)$; therefore, the solution can be completed, for the present closure of turbulent quantities, using any two-parameter formula. The two unknown parameters are prescribed by noting the following identities:

$$\tilde{f} = \int_0^1 f \tilde{P}(f) df \tag{9}$$

$$g = \int_0^1 (f - \tilde{f})^2 \tilde{P}(f) df \tag{10}$$

Since both $\tilde{f}$ and g are known, from solving Eq. (1), Eqs. (9) and (10) provide two implicit equations to find the two parameters of the Favre PDF. Equations (4)–(8) are then integrated to yield the various moments of scalar properties. Similar to past work [37,39], a clipped Gaussian function is used for $\tilde{P}(f)$.

2.5 Numerical Solution

The equations are solved using a modified version of GENMIX [41]. Thirty cross-stream grid nodes were used for most computations, with forward step sizes limited to less than 1.5 percent of the flow width or an entrainment increase of 10 percent—whichever was smaller. Use of 90 cross-stream grid nodes and halving the limits of streamwise stepsize yielded similar results—within 1 percent; therefore, numerical closure was adequate.

2.6 State Relationships

For mixing-controlled, low-speed, adiabatic flames, having equal exchange coefficients of all species and heat, state relationships are easily found if the assumption of local thermodynamic equilibrium can be imposed. Then, scalar properties are determined from adiabatic flame computations for the mixture ratio of burner and ambient gases prescribed by the mixture fraction. This approach is effective for low mixture fractions, but often yields poor results at high mixture fractions—particularly for hydrocarbon fuels—where temperatures are too low to achieve equilibrium in times pertinent to flame processes [39,42,43].

Bilger [44] and Liew et al. [45] propose a laminar flamelet technique, to avoid the limitations of the thermodynamic equilibrium approximation. This is based on Bilger's [44] observation that measurements of scalar properties in laminar flames yield nearly universal correlations as a function of mixture fraction—relatively independent of local length scales and flame stretch—even when scalar properties do not satisfy thermodynamic equilibrium requirements in fuel-rich regions. State relationships are then assumed to be the same in a turbulent flame, e.g., observations in turbulent flames are assumed to be due to a succession of laminar flamelets (a wrinkled laminar flame) sweeping past a given location. This approach circumvents the current lack of information on the detailed chemistry of the reaction process, which will take many years to resolve for practical fuels. Instead, this information is replaced with data obtained from routine measurements in laminar flames.

Bilger suggested the laminar flamelet concept based on measurements reported by Tsuji and Yamaoka [46–48] for methane/air flames and Abdel-Khalik et al. [49] for n-heptane/air flames. Later measurements for laminar methane/air flames by Mitchell et al. [50] also support the concept. Use of state relationships developed in this manner has provided good predictions of scalar properties in turbulent methane/air, propane/air, ethylene/air, hydrogen/air and carbon monoxide/air diffusion flames as well [45,51–55].

Figures 1–3 are examples of state relationships developed from measurements in laminar diffusion flames. Figure 1 is for hydrogen/air flames having reactants at normal temperature and pressure (NTP) [53]. Data shown include sampling measurements in a laminar flame by Gore et al. [53]; spectroscopic measurements in laminar and turbulent flames by Drake and co-workers [56,57] and Aeschliman et al. [58]; predictions using a finite-rate chemistry model by Miller and Kee [59]; and predictions based on local thermodynamic equilibrium using CEC76 [60]. These results were obtained for a variety of flame configurations, original sources should be consulted for details. Scalar properties are plotted as a function of local fuel equivalence ratio, which is a single-valued function of mixture fraction. Measurements reported by Gore et al. [53] and Drake and co-workers [56,57], and the equilibrium predictions, give nearly the same correlation for the concentration of major gas species. This is reasonable in view of the fast rates of reaction associated with hydrogen/air chemistry. The earlier measurements of Aeschliman et al. [58] and the predictions of Miller and Kee [59] depart from the others. In our opinion, this is due to greater uncertainties in these studies. The temperature measurements of Drake and co-workers [56,57] are ca. 200 K below adiabatic equilibrium predictions. They attribute this behavior to super-equilibrium concentrations of the radical species OH. Radiative heat losses could also be a factor in this observation. Overall, the concentrations of major gas species needed for radiation computations satisfy universal correlations for a variety of laminar flame conditions, local length scales and flame residence times—supporting the laminar flamelet concept. Similar results, closely approximating thermodynamic equilibrium conditions. have been obtained for carbon monoxide/air diffusion flames [54].

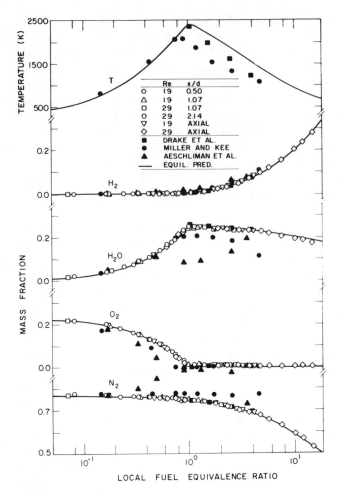

Figure 1 State relationships for hydrogen/air diffusion flames at NTP. From Gore et al. [53].

Figures 2 and 3 are illustrations of state relationships for a typical hydrocarbon system—ethylene/air [55]. The state relationships for major gas species are based on measurements in laminar flames. Equilibrium predictions, found from CEC76, also appear on the figures. Ethylene flames contain appreciable quantities of soot, nevertheless, nearly universal correlations of major gas species were still obtained—supporting the use of the laminar flamelet concept. It is also evident that this correlation is achieved in spite of appreciable departures from equilibrium for fuel-rich conditions.

Continuum radiation from soot dominates radiation properties of luminous flames like ethylene/air diffusion flames. Since soot particles are generally smaller than 0.1 μm, scattering from soot can be neglected during analysis of radiation. Furthermore, flame radiation is largely in the infrared ($\lambda > 1$ μm); therefore, the

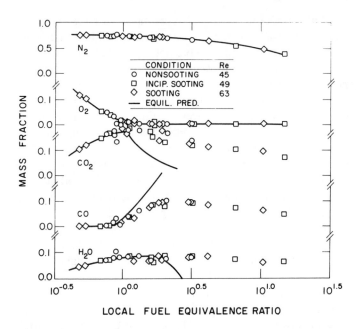

Figure 2 State relationships for ethylene/air diffusion flames at NTP. From Gore and Faeth [55].

Rayleigh limit of small particles can be used for the spectral absorption coefficient of soot, yielding the following expression [1]:

$$K_{\lambda s} = \frac{36\, n\kappa f_v/\lambda}{(n^2 - \kappa^2 + 2)^2 + 4n^2\kappa^2} \tag{11}$$

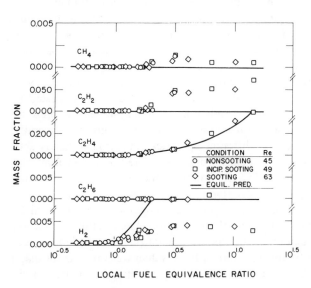

Figure 3 State relationships for ethylene/air diffusion flames at NTP (continued). From Gore and Faeth [55].

Where the complex index of refraction of soot has been taken to be $n-i\,\kappa$. From Eq. (11), it is clear that only the soot volume fraction and the real and imaginary parts of the refractive index of soot are needed to predict soot radiation properties. While there is controversy concerning the refractive index properties of soot, cf. [1,61–64] and references cited therein, estimates can be made; therefore, soot volume fraction is the main scalar property which must be found to compute radiation from luminous flames. Rational predictions of this property in diffusion flame environments are even less likely than predictions of major gas species concentrations for practical fuels; therefore, an alternative must be sought. Past measurements of soot properties in flames provide information which is helpful in this regard [61–64]. The measurements show that soot particle size and number density vary widely in laminar flames; however, soot volume fraction exhibits a greater tendency toward universality. In particular, strongly sooting materials, which yield the highest levels of continuum radiation, tend to have maximum soot volume fractions which are relatively independent of position in the flames. This suggests that an extension of the laminar flamelet concept to soot might be feasible.

The potential for a state relationship for soot volume fraction was examined by measurements in laminar ethylene/air diffusion flames [55]. The results are plotted in Fig. 4 for several positions above the burner exit and several burner flow rates (Reynolds numbers). These flow rates were adjusted to provide non-sooting, incipient sooting, and sooting flames, analogous to Santoro et al. [62]. The results exhibit a crude universality, which is approximated by the solid line denoted as the state relationship. Positions nearer the burner yielded much lower soot volume fractions, probably due to kinetic limitations. Similar behavior occurs for gas species near points of flame attachment, however, the region involved is smaller than for soot, since soot chemistry is slower. Spatial resolution was inadequate at higher positions. Clearly, soot volume fraction varies primarily with fuel equivalence ratio in the region considered in Fig. 4, reaching a maximum for a fuel-equivalence ratio near 1.5. Soot volume fraction declines as the flame is approached, due to oxidation of soot, and at high-fuel equivalence ratios, since temperatures are too low to decompose the fuel.

The notion that soot properties in turbulent flames can be associated with mixture fraction or a particular region of the flame has also been suggested by others. Tamanini [25] bases some of his radiation computations on the presence of a soot "layer" having a fixed soot concentration and temperature. Grosshandler and Vantelon [11] use a somewhat different approach, applying a steady-state assumption to a global kinetic mechanism in order to relate spectral soot absorption properties to a local Schvab-Zeldovich variable (which is proportional to mixture fraction) in a laminar opposed-flow flame.

While these findings are promising, the relatively slow rates of soot chemistry place residence-time limitations on soot volume fraction state relationships which have not been assessed adequately. Furthermore, laminar flame environments may be less representative of conditions in turbulent flames than was the case for major gas species concentrations. Unlike gas species, soot particles are too large to dif-

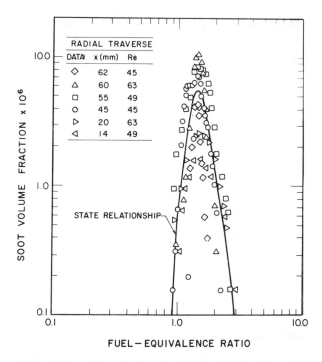

Figure 4 Soot volume fraction state relationship for ethylene/air diffusion flame at NTP. From Gore and Faeth [55].

fuse; therefore, they are convected with the local gas velocity, modified by effects of thermophoresis [63,64]. Such kinematic effects are probably very different in laminar and turbulent flames. Detailed assessment of the concept by measurements in turbulent flames has not yet been reported. Nevertheless, we will examine radiation predictions based on the state relationship illustrated in Fig. 4, as a general test of the idea.

3 EVALUATION OF FLAME STRUCTURE PREDICTIONS

3.1 Test Conditions

Flame structure predictions, using the methodology just outlined, have been evaluated for turbulent natural gas (largely methane) [39,40,43,51,66], propane [39,42,52], hydrogen [38,53], carbon monoxide [54] and ethylene [55] diffusion flames burning in air. These flames generally involved fully turbulent flow at the burner exit, so that reasonable performance should be expected from higher-order turbulence models. A few examples are considered in the following, emphasizing scalar properties, in order to demonstrate the performance of the structure analysis. Some of these flames are subsequently used to evaluate predictions of radiation properties; therefore, test conditions for both evaluations are summarized in Table 2.

Original sources should be consulted concerning the details of the experiments and experimental uncertainties, however, a few generalizations will provide perspective for the following comparisons between predictions and measurements. Time-averaged streamwise mean and fluctuating velocities were measured, while predictions are Favre averages. The differences between the two for mean velocities are comparable to experimental uncertainties in flames, ca. 10 percent [67,68]. However, Favre-averaged velocity fluctuations along the axis are roughly 20 percent lower than time averages and can be as much as 40 percent lower in high-temperature regions [67,68]. The analysis provides both time- and Favre-averaged scalar properties; therefore, both types of averaging are considered, since the degree of density weighting of the measurements used for evaluation is uncertain [51–55]. Experimental uncertainties of mean concentrations are generally less than 15 percent. Mean temperature measurements are not corrected for radiative heat losses, and are typically 100–200 K too low in the highest temperature regions of the flow [51–55].

3.2 Predictions and Measurements

Structure predictions and measurements along the axis of a turbulent carbon monoxide/air diffusion flame are illustrated in Fig. 5. Results are shown for mean and fluctuating streamwise velocities, mean temperature, and the mean concentrations of major gas species. The predicted values of streamwise velocity fluctuations were obtained by assuming isotropic turbulence ($\bar{u}''^2 = 2k/3$). If the usual level of anisotropy in jets is assumed ($\bar{u}''^2 = k$), predictions would be 20 percent higher. In view of the earlier discussion of effects of differences between time-

Table 2 Turbulent diffusion flame test conditions[a]

Reference	Burner diameter (mm)	Reynolds number	Heat release rate (kW)	Radiative heat loss (%)[b]
Hydrogen/Air				
Gore et al. [53]	5	3000, 5722	12.5, 23.8	10.2, 8.7
Carbon Monoxide/Air				
Razdan and Stevens [65]	5	11400	7.6	—
Gore et al. [54]	5	7475, 13140	5.3, 8.9	11.4, 8.4
Methane/Air				
Jeng and Faeth [51]	5	2920, 5850, 11700	6.8, 13.7, 27.4	14.0, 18.7, 18
Pfenning [27]	76.2, 101.6	4.6–5.0 × 10^6	100–200 × 10^3	—
Ethylene/Air				
Gore and Faeth [55]				
	5	6370, 12740	13.5, 27.0	36.0, 34.4

[a]Burner flow directed vertically upward in still air at normal temperature and pressure (except for Ra and Stevens [65] which had a 0.13 m/s coflow velocity).

[b]Percentage of heat release rate.

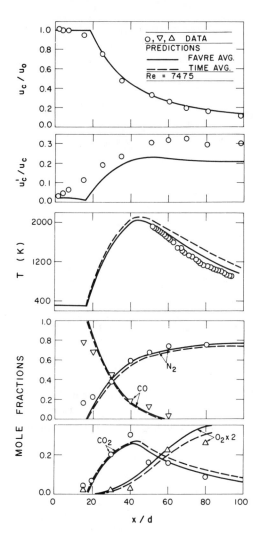

Figure 5 Properties along the axis of a turbulent carbon monoxide/air diffusion flame at NTP. From Gore et al. [54].

and Favre-averages, experimental uncertainties, and thermocouple radiation errors, the comparison between predictions and measurements is reasonably good. Comparable results were obtained for the other carbon monoxide flames listed in Table 2 [54].

Results for a hydrocarbon fueled flame are illustrated in Fig. 6. Predicted and measured scalar properties along the axis of a turbulent methane/air diffusion flame are shown. Predictions are based on the laminar flamelet method (assuming that the radiative heat loss is 20 percent of the energy release of reaction when finding temperatures) and a partial equilibrium approach computed for the limits of adiabatic flow and 20 percent radiative heat loss [51]. The comparison between predictions and measurements is also reasonably good.

The results illustrated in Figs. 5 and 6 are representative of findings for other

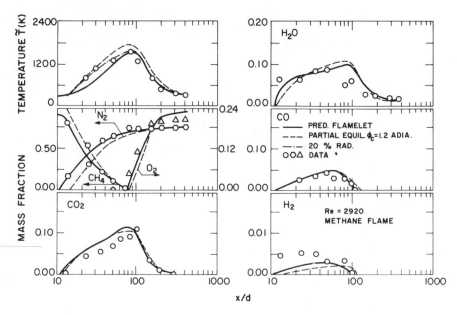

Figure 6 Scalar properties along the axis of a turbulent methane/air diffusion flame at NTP. From Jeng and Faeth [51].

turbulent diffusion flames. This includes hydrogen/air [53], large-scale methane/air [66], propane/air [52] and ethylene/air [55] flames. Thus, it is reasonable to use these scalar property predictions for radiation calculations. This extension is considered next.

4 ANALYSIS OF SPECTRAL RADIATION INTENSITIES

4.1 Overall Description

Measurements of the radiation properties of turbulent diffusion flames that we will consider involved both spectral radiation intensities and radiative heat fluxes. A narrow-band radiation analysis is used to provide predictions for comparison with these measurements.

The narrow-band radiation analysis treats band radiation from gases and continuum radiation from soot for species relevant to hydrogen, carbon monoxide and hydrocarbon combustion in air. The infrared region (1–6 μm wavelength range) is considered, allowing for the following gas bands: 1.14, 1.38, 1.87, 2.7 and 6.3 μm bands of H_2O; 2.7 and 4.3 μm bands of CO_2; 2.3 and 3.3 μm bands of CH_4; and the 4.7 μm band of CO. N_2, O_2 and H_2 are transparent in the infrared and are ignored, while concentrations of minor species are small and their contributions to radiation can be neglected. Scattering from gases is negligible and can also be ignored for soot—under the small-particle assumption [1].

Narrow-band analyses of flames [15,18,53–55,69,70] generally follow methods developed by Ludwig et al. [5]. The Goody statistical narrow-band model [6] is used in conjunction with the Curtis-Godson approximation for an inhomogeneous path. Both collisional and Doppler broadening are included. We will consider two methods to treat turbulence/radiation interactions: (1) the mean-property method, where radiative properties are computed using time-averaged scalar properties along the path and (2) the stochastic method, where interactions between turbulent fluctuations and radiation are treated using Monte Carlo methods.

4.2 Mean-Property Method

Given the scalar properties, the spectral radiation intensity along a line-of-sight is found by solving the equation of transfer

$$\frac{\partial I_\lambda}{\partial s} + \rho K_\lambda \left(I_\lambda - \frac{e_{\lambda,b}}{\pi} \right) = 0 \tag{12}$$

where $e_{\lambda,b}$ is the blackbody monochromatic emissive power. Paths through flames from a cold boundary are considered; therefore, assuming that the spectral radiance is small at $s = 0$, Eq. (12) can be rewritten as:

$$I_\lambda(s) = \int_0^1 \frac{e_{\lambda,b}(s')}{\pi} \frac{\partial}{\partial s'} [\tau_\lambda(s, s')] ds' \tag{13}$$

where

$$\tau_\lambda(s, s') = \exp\left(- \int_{s'}^s K_\lambda(s'') \rho(s'') ds'' \right) \tag{14}$$

We assume that turbulent fluctuations have no effect on radiation properties for the mean-property method. The properties needed for the calculations are *time-averaged* scalar quantities, supplied from predictions of the flow structure model, as described in Section 2.3.

4.3 Stochastic Method

The stochastic method seeks to provide insight concerning turbulence/radiation interactions using a model of turbulence. It rests on the assumption that the turbulent flow field consists of many eddies and that the properties of each eddy are uniform and statistically independent of one another. The size of each eddy is estimated from the dissipation length scale of the eddy, which is assumed to be invariant with time, cf. Gosman and Ioannides [71] and Shuen et al. [72] as follows:

$$L_e = \frac{C_\mu^{3/4} k^{3/2}}{\epsilon} \tag{15}$$

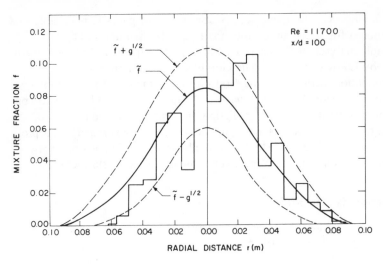

Figure 7 Typical realization of the optical path using the stochastic method. For a turbulent methane/air diffusion flame from Jeng et al. [69].

According to the present structure model, scalar properties needed for the radiation calculation are only functions of the mixture fraction, f. The instantaneous properties of each eddy are also only functions of the instantaneous value of f. The flow structure model provides $\tilde{f}$ and g at each point and these quantities are used to find the most probable value, μ, and the variance, σ, of the Favre-averaged clipped Gaussian distribution function of f. This Favre-averaged PDF is used to calculate the time-averaged PDF from Eq. (6). The cumulative distribution function of mixture fraction is then constructed from the time-averaged PDF. This distribution function is randomly sampled by selecting a random number in the range 0–1 and computing the instantaneous mixture fraction from the cumulative distribution function. This procedure is repeated for each eddy, to yield a realization of instantaneous mixture fractions along the radiation path.

A typical realization of a flame, using the stochastic method, is illustrated in Fig. 7. Results are shown for a radial path through a round methane/air diffusion flame, near the flame tip [69]. Although Favre-averaged quantities appear in the illustration for reference purposes, the selection of f for each eddy is based on the time-averaged PDF of mixture fraction, as noted earlier.

The instantaneous temperature, species concentrations, and soot volume fraction of an eddy are found from the instantaneous mixture fraction, in conjunction with the state relationships, e.g., Figs. 1–4. During the calculation of radiative heat transfer, the optical path length is divided according to eddy size. The properties of each eddy are assumed to be the same as those at the center of each eddy. The equation of transfer, described in connection with the mean property method, is then solved for this realization to yield the spectral radiation intensities. Random sampling in this manner continues until statistically significant mean spectral radiation intensities are found for the wavelength of interest.

5 EVALUATION OF SPECTRAL INTENSITY PREDICTIONS

5.1 Test Conditions

Test conditions for the turbulent diffusion flames are summarized in Table 2. Experimental uncertainties for spectral radiation intensities are generally less than 20 percent [53–55,69].

5.2 Narrow-Band Model

The narrow-band model has been widely tested for conditions where effects of turbulence are small [5,15,18,19,40,69]. In particular, Jeng [40] evaluates narrow-band predictions using the following measurements: the 2.7 μm band of H_2O and CO_2 [5,6,73,74], the 4.3 μm band of CO_2 [75–77], the 4.7 μm band of CO [78], and the 1.5–5.5 μm range of mixtures of CO_2, CO and H_2O [79]. In general, predictions agreed reasonably well with measurements, e.g., maximum differences were less than 25 percent.

Narrow-band predictions and the measurements of Grosshandler [79] appear in Fig. 8. The test conditions are for nonisothermal and nonhomogeneous mixtures of CO_2, H_2O, and CO; therefore, they are representative of flame environments. Predictions due to Grosshandler [79] and Jeng [40] are seen to be in reasonably good agreement with the measurements.

5.3 Nonluminous Flames

A number of assessments of narrow-band models, using a variety of structure models, have been reported for nonluminous turbulent diffusion flames, e.g., hy-

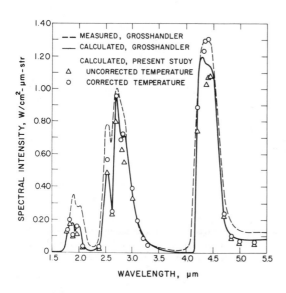

Figure 8 Spectral radiation intensities for a nonisothermal, nonhomogeneous gas mixture of carbon monoxide, carbon dioxide and water vapor. From Jeng et al. [69].

drogen/air [24,53], carbon monoxide/air [54] and a variety of hydrocarbon/air [15,18,19,22,40,55,66,69,70] flames. In the following, we will consider a sample of these results. Results shown only involve use of the structure and radiation models described earlier in this paper. Emphasis is placed on effects of turbulence/radiation interactions, which are evaluated by observing differences between the mean property and stochastic predictions. General operating conditions of the flames which are considered are summarized in Table 2.

Predicted and measured spectral radiation intensities (in the 1.5–6.0 μm wavelength range) are illustrated for a turbulent carbon monoxide/air diffusion flame in Fig. 9. These results are for radial paths through the flame at $x/d = 50$ and 90. The flame tip (the position where the fuel disappears along the axis) is roughly at $x/d = 50$ for this flame. For these conditions, the 2.7 and 4.3 μm radiation bands of CO_2 dominate the spectrum. Predictions using both the mean-property and stochastic methods are shown. The stochastic method predicts higher spectral intensities than the mean-property method, particularly for the 4.3 μm band. The differences are not large, however, since carbon monoxide/air flames have a rather slow variation of radiance properties with mixture ratio, which reduces effects of turbulent fluctuations.

Predicted and measured spectral radiation intensities (in the 1.0–4.0 μm wavelength range) are illustrated for a turbulent hydrogen air diffusion flame in Fig. 10. These results are for radial paths through the flame at $x/d = 50$, 90 and 130. The flame tip is roughly at $x/d = 110$ for these conditions. The 1.38, 1.87 and 2.7 μm radiation bands of water vapor dominate the spectrum. At longer wavelengths, the spectra begin to rise toward the rather broad 6.3 μm water vapor band. For hydrogen/air flames, effects of turbulent fluctuations are large, with

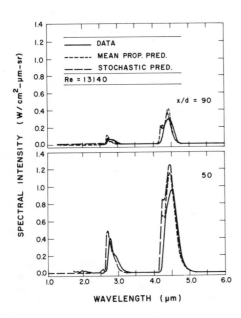

Figure 9 Spectral radiation intensities for a turbulent carbon monoxide/air diffusion flame at NTP. From Gore et al. [54].

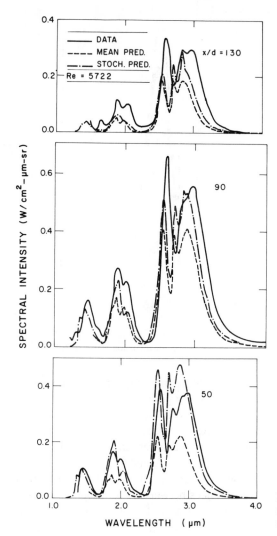

Figure 10 Spectral radiation intensities for a turbulent hydrogen/air diffusion flame at NTP. From Gore et al. [53].

as much as a 2:1 difference between stochastic and mean-property predictions. This occurs since radiation properties of hydrogen/air diffusion flames vary rapidly near the stoichiometric condition, which is the region that dominates the results illustrated in Fig. 10. In general, the measurements are bounded by the two prediction methods, with the stochastic method providing slightly better quantitative accuracy than the mean-property method.

The effect of combined hydrogen and carbon combustion is illustrated in Fig. 11. Predicted and measured spectral radiation intensities (in the 1.5–5.5 μm wavelength range) are illustrated for a turbulent natural gas (96 percent methane by volume)/air diffusion flame. These results are for a radial path through the flame, just before the flame tip at $x/d = 100$. Recent measurements in this lab-

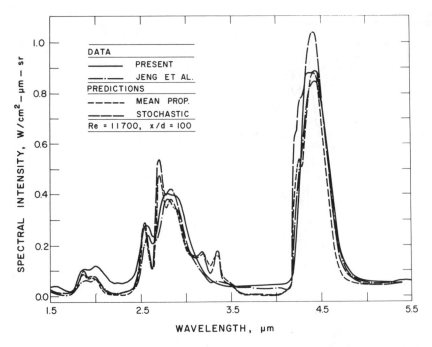

Figure 11 Spectral radiation intensities for a turbulent methane/air diffusion flame at NTP.

oratory, as well as earlier measurements by Jeng et al. [69] for the same flame, are illustrated. The two sets of measurements are in good agreement. At this condition, the spectrum is dominated by the 1.87 and 2.7 μm bands of water vapor and the 2.7 and 4.3 μm bands of carbon dioxide. Predictions also find the 3.3 μm band of carbon monoxide, but this is not very apparent in the measurements. The most intense bands are the 2.7 μm CO_2/H_2O combined band and the 4.3 μm CO_2 band. These bands behave very similar to those observed for pure carbon monoxide combustion in Fig. 9, e.g., differences between mean-property and stochastic predictions are moderate since radiance properties for carbon dioxide change relatively slowly with mixture fraction. Since methane has a higher percentage of hydrogen than other hydrocarbons, one would expect that nonluminous flames of most hydrocarbons would exhibit smaller effects of turbulence radiation interactions in the strongest gas bands than seen in Fig. 11, but experimental confirmation of such behavior is not yet available.

To summarize, it appears that effects of nonluminous turbulence/radiation interactions are highest in the flame zone of diffusion flames burning high concentrations of hydrogen, where differences between mean-property and stochastic predictions can reach 2:1—at least for the flames considered to date. The effect of nonluminous turbulence/radiation interactions is smaller in carbon monoxide and nonluminous hydrocarbon-fueled diffusion flames, where differences between mean-property and stochastic predictions are on the order of 20–30 percent (which is comparable to uncertainties in the narrow-band and flame structure models).

5.4 Luminous Flames

Predictions and measurements of spectral radiation intensities have been completed for turbulent ethylene/air diffusion flames [55]. Both spectral absorption and emission measurements were made for horizontal paths through the flames, at various distances above the burner exit.

The absorption measurements were used to test predictions of soot volume fractions based on the state relationship illustrated in Fig. 4. The radiation source was at HeNe laser beam at a wavelength of 632.8 nm. This wavelength is below the range of significant radiant emission from luminous flames, reducing uncertainties due to flame temperature predictions. This wavelength is also above the range where the Rayleigh limit for small particles is no longer acceptable.

Representative extinction results are illustrated in Fig. 12. Predicted and measured time-averaged transmittivities, $I_\lambda/I_{\lambda 0}$, are plotted as a function of radial position for chord-like paths through the flame at various heights above the burner

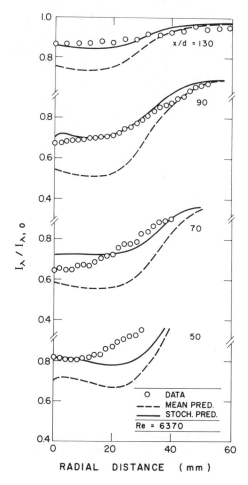

Figure 12 Absorption (632.8 nm wavelength) by soot in a turbulent ethylene/air diffusion flame. From Gore and Faeth [55].

exit. Predictions using both the mean-property and stochastic methods are shown. Absorption is only due to soot particles in the optical path. The flame contains appreciable quantities of soot, but most of it burns out in the reaction zone. Thus transmittivities tend toward unity along the periphery of the reaction zone. Both models broadly predict the trends of the measurements, however, the stochastic model provides the best absolute agreement. Mean property predictions of extinction are almost twice the stochastic predictions—suggesting substantial effects of turbulence/radiation interactions. The general behavior is supportive of universality of the soot volume fraction state relationship for the turbulent flames but with some differences from the results illustrated in Fig. 4 for laminar flames [55].

Predicted and measured spectral radiation intensities for turbulent ethylene/ air diffusion flames (for the 1–6 μm wavelength range) are illustrated in Fig. 13. The results are for radial paths through the flame at various heights above the burner exit. The mean position of the flame tip is in the range $x/d = 80$–100; therefore, the results appearing in Fig. 13 represent conditions before, near and after the tip of the flame. Spectral intensities are largest near the flame tip. The spectra are dominated by continuum radiation from soot. However, effects of the gas bands of water vapor (1.38, 1.87 and 2.7 μm) and carbon dioxide (2.7 and 4.3 μm) can still be seen, particularly the strong 4.3 μm band of carbon dioxide.

The comparison between predictions and measurements seen in Fig. 13 is encouraging, particularly since combined predictions of scalar structure, soot volume fractions and radiation properties are involved. While the mean-property method provides best quantitative agreement, this is probably fortuitous in view of poorer extinction predictions obtained using this approach (Fig. 12). Predictions of continuum radiation from soot are very sensitive to temperature estimates and the

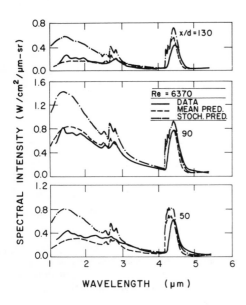

Figure 13 Spectral radiation intensities for a turbulent ethylene/air diffusion flame. From Gore and Faeth [55].

present assumption of optically-thin radiative heat losses from all parts of the flame is crude. For example, these flames lose roughly 35 percent of their energy of reaction by radiation and redistribution of energy within their structure by radiation is probably very significant, as suggested by Emmerich [29]. Therefore, more exact, coupled structure and radiation analysis could modify the relative performance of the mean-property and stochastic methods.

The stochastic predictions in Fig. 13 are 50–300 percent higher than the mean-property predictions in the continuum portions of the spectra—suggesting significant effects of turbulence/radiation interactions. Analysis showed that this effect was most strongly influenced by temperature fluctuations [55]. Radiant energy transfer from high to low temperature regions of the flame tends to reduce temperature fluctuations. The fact that this energy transport mechanism was ignored during present calculations probably explains why the stochastic analysis provided good results for extinction (where temperature is not a factor) but overestimated spectral radiation intensities in the infrared. We suspect that the present stochastic analysis generally causes overestimation of turbulence/radiation interactions due to this phenomena. Certainly, all evaluations thus far show that the stochastic predictions consistently overestimate measurements of spectral radiation intensities [53–55,60]. This deficiency can only be overcome by undertaking the added complications of coupled structure and radiation analysis. The extentions of the models which are required for coupled structure and radiation analysis will be discussed in Section 8.

6 ANALYSIS OF RADIATIVE HEAT FLUXES

6.1 Overall Description

Past methods used to compute radiative heat fluxes from turbulent flames include zone, Monte Carlo, and multiflux techniques. Lockwood and Shah [80] discuss the merits of these methods and propose a "discrete-transfer" method which gave good performance for several test problems involving gray gases. The method has some advantages for practical applications, e.g., geometrical adaptability and an ability to return any desired degree of precision. Lockwood and Shah [80] also suggest that it offers computational efficiencies in comparison to other methods. This approach will be considered here, following Ref. [70], since it involves summing predictions of intensities which is a convenient extension of the flame radiation results discussed thus far.

For the present, we continue consideration of uncoupled structure and radiation analysis. The problem posed is that we wish to determine the radiant flux incident on an increment of surface area near a turbulent flame burning in an infinite environment. Test results assessing the method were obtained for non-reflective flame surroundings; therefore, to simplify the discussion the following analysis will be for this limit. The extension of the method to other types of surfaces is straightforward but tedious.

6.2 Discrete Transfer Method

Under the limitations just described, the discrete-transfer method involves finding the spectral radiation intensity, $I_{\lambda i}$, along several paths, i, which pass through the point in question and then summing over both paths and wavelength to find the total radiative heat flux [80]. In the following, an incremental black surface is considered whose temperature is low so that re-emission of radiation from it can be ignored. Computations of spectral radiation intensities were described earlier. Previous findings showed that the mean-property method gives reasonably good results for the test flames; therefore, this approach will be used in the following. Application of the stochastic method is straightforward but more computationally intensive. The spectral radiance at the surface is found from the following equation:

$$J_\lambda = \sum_{i=1}^{n} [I_\lambda \cos(\theta)\Delta\omega]_i \tag{16}$$

where θ_i is the angle between path i and a normal to the surface and $\Delta\omega_i$ is the solid angle of the path when viewed from the surface. Naturally, the paths must be chosen to cover the full hemisphere viewed by the surface element, i.e.,

$$\sum_{i=1}^{n} \Delta\omega_i = 2\pi \tag{17}$$

The surface radiance is then found by summing the spectral radiance over all wavelengths of interest

$$J = \sum_{i=1}^{m} (J_\lambda \Delta\lambda)_i \tag{18}$$

The extension of this approach to consider coupled structure and radiation computations will be discussed subsequently.

A typical arrangement of paths, for a laboratory-sized flame, is illustrated in Fig. 14. The arrangement is for a heat flux sensor, having a relatively small absorption surface, at some distance from the flame. The figure shows the point where each path from the sensor intersects a plane through the flame axis and normal to the plane of the sensor and the axis. With this arrangement, the incremental solid angle needed in Eq. (16) for path i is given by

$$\Delta\omega_i = \left(\frac{\Delta A \cos \theta_A}{r^2}\right)_i \tag{19}$$

Where θ_{Ai} is the angle between the path and the normal to the incremental area ΔA_i in the axial plane, while r_i is the distance between the sensor and the point of intersection with the axial plane. For present conditions, the problem is symmetrical about the plane through the sensor and the flame axis, since the heat flux sensors used for measurements were directed at the flame axis. Sixty to eighty-four paths were used on the side computed for the results considered in the following.

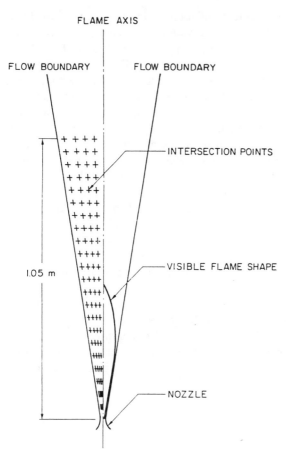

Figure 14 Intersection of radiation paths, used in discrete-transfer computations, with a normal plane through the flame axis. From Jeng and Faeth [70].

Flame structure computations were similar to the structure evaluation considered earlier, e.g., 30 radial nodes and about 1000 streamwise steps to cover $0 < x/d < 400$, which includes the region where significant contributions to the radiative heat flux were found. These results were then interpolated to find mean properties at 30 stations along each path. Richardson extrapolation suggests that the overall discretization error for these computations is less than 10 percent, which is well within the potential accuracy of the spectral intensity analysis [70].

7 EVALUATION OF RADIANCE PREDICTIONS

The radiative heat flux analysis has been evaluated for turbulent hydrogen/air [53], carbon monoxide/air [54], methane/air [66,70], and ethylene/air [55] diffusion flames. This involved mesurements of radiative heat fluxes at various positions around the flames. Uncertainties of these measurements were generally less than 10 percent.

Typical predictions and measurements of total radiative heat flux are illustrated in Fig. 15 for turbulent methane/air diffusion flames having various burner exit Reynolds numbers. These results are for a sensor facing the axis of the flame and being traversed in a vertical direction at a distance of 575 mm from the axis. The detector had a viewing angle of 150 degrees; therefore, the flame boundaries were entirely within the viewing angle of the sensor, except at points far from the burner exit where radiative heat fluxes are small in any event. The radiative heat flux reaches a maximum near the flame tip, roughly at $x/d = 100$. Although peak mean temperatures and species concentrations are the same for the three flames (calculations for all three use the same state relationships), radiative heat fluxes tend to decrease with decreasing flame Reynolds numbers due to smaller flame dimensions (an effect of buoyancy [51]). Predictions represent these trends reasonably well, with discrepancies on the order of 10–30 percent. Such levels of error are similar to the differences between predictions and measurements for the spectral intensities, discussed earlier, and are also representative of uncertainties in the narrow-band model and the structure model.

Recent measurements of Pfenning [27] provided an opportunity to evaluate this methodology for large-scale flames [66]. Several flames were tested having chemical energy release rates of 100–200 MW, with flame heights ca. 10 m. Mean temperatures were measured at several locations within the flames, using bare wire thermocouples. Radiative heat fluxes were also measured, using sensors having various positions and orientations with respect to the flames. Flame structure and radiation properties were predicted following Jeng and co-workers

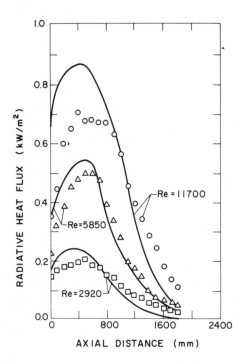

Figure 15 Total radiative heat flux distributions along the axis of turbulent methane/air diffusion flames at NTP. From Jeng and Faeth [70].

[51,69,70]. While these large flames appeared yellow, continuum radiation from soot was ignored, and radiation predictions only considered the nonluminous gas bands, similar to Figs. 11 and 15. This was based on the observation that laboratory-scale methane/air diffusion flames also appear to be yellow, but still exhibit negligible continuum radiation from soot, cf. Fig. 11. As noted earlier, effects of turbulence/radiation interactions are relatively small for methane/air diffusion flames, cf. Fig. 11; therefore, the mean-property method was used for radiative heat flux predictions.

Predictions tended to overestimate measured mean temperatures in the large methane/air diffusion flames by 200–300 K in the hottest portions of the flames [66]. However, this is typical of past experience for thermocouple measurements which are not corrected for radiation errors. Furthermore, local wind conditions were not entirely still, which tends to bias measured temperatures downward. The radiative heat flux predictions were also encouraging, tending to underestimate measurements by 10–20 percent, which is expected for mean property predictions with these reactants, cf. Fig. 15.

Results for the other flame systems evaluated thus far are similar to the findings shown here [53–55]. We conclude that the present methodology is promising for treating effects of flame scale and fuel type. However, luminous flames are sensitive to internal energy exchange by radiation, suggesting a need for coupled structure/radiation computations. This extension is considered next.

8 STRUCTURE AND RADIATION COUPLING

Earlier work has shown that radiation predictions are sensitive to assumed radiative heat loss fractions when the perturbation method, discussed thus far, is used [55,69]. Thus, some method of closing this technique, by providing a computed estimate of the radiative heat loss fraction, is needed. Furthermore, past measurements suggest that radiative energy exchange between various portions of flames is important [29,55]; therefore, radiative heat loss fractions vary with position in the flame and this additional complication must ultimately be considered—particularly for luminous flames. Strategies for dealing with these problems are briefly discussed in the following.

It is a relatively straightforward task to couple structure and radiation predictions using the perturbation approach described thus far, if the assumption of a constant radiative heat loss fraction is maintained. Computation of radiative heat fluxes to the surroundings, described in Section 6, then provides a means of estimating the total radiative heat loss from the flame and thus the radiative heat loss fraction. It has been found that the correlation of species concentrations as a function of mixture fraction is relatively independent of flame heat losses (and, thus, temperature levels); therefore, the species concentrations in conjunction with the radiative heat loss fraction can be used to estimate a new state relationship for temperature, using conservation of energy principles. The structure and radiation calculations are then repeated until closure is achieved by an iterative process.

The perturbation solution is satisfactory as long as the flame is optically thin. This assumption, however, is marginal even for relatively, small, nonluminous laboratory flames. For example, Jeng et al. [69] find that the transmittivity within the 4.3 μm band of CO_2 is only 0.075, for the relatively small (ca. 1 m high) natural gas flame whose spectral radiation properties are pictured in Fig. 11. Thus, there is appreciable radiant energy exchange between various portions of this flame, a circumstance that is magnified when larger-sized luminous flames are considered. As a result, any general approach for treating the radiative properties of turbulent flames must address fully-coupled structure and radiation predictions.

Thus far, fully-coupled structure and radiation analyses have neglected effects of turbulence/radiation interactions. Exact work at this limit can be based on Hottel's zone method [7], but this leads to lengthy computations. Multiflux methods offer significant simplifications and have been used frequently for coupled structure and radiation predictions, usually under the gray gas assumption [81,82]. However, the advantages of the discrete-transfer method [80] apply here and are consistent with the approach discussed thus far; therefore, this approach will be considered in the following.

As a first step, we ignore turbulence/radiation interactions. We also maintain all earlier assumptions of the structure analysis, except that radiant energy exchange is now considered. The total enthalpy of the flow then varies from point to point, and is governed by the following equation (within the other approximations of the turbulence analysis).

$$\bar{\rho}\tilde{u}\frac{\partial \bar{H}}{\partial x} + \bar{\rho}\tilde{v}\frac{\partial \bar{H}}{\partial r} = \frac{1}{r}\frac{\partial}{\partial r}(r\mu_{eff,H}\,\partial\bar{H}/\partial r) + \bar{Q}_r \tag{20}$$

where $\mu_{eff,H}$ is given by an expression analogous to $\mu_{eff,f}$ in Table 1 and $\sigma_H = \sigma_f$. $\bar{H}$ is the total enthalpy, including the enthalpies of formation of each species. $\bar{Q}_r$ is the source term due to absorption and emission of radiation. Evaluation of $\bar{Q}_r$ is discussed by Lockwood and Shah [80]. For computational cell i, this is given by

$$Q_{ri} = \sum_{k=1}^{n}\sum_{j=1}^{m}[\Delta I_\lambda(\omega)\delta A\delta\omega]_{ij}\,\delta\lambda_K \tag{21}$$

where $\Delta I_\lambda(\omega)_{ij}$ is the change in spectral intensity for path j crossing cell i and the $\delta\omega_{ij}$ are chosen to cover the full-spherical region observed from the cell. The expression is also summed over the spectral region where radiative energy transfer is significant. Solution of Eq. (20) provides the local value of $\bar{H}$, which then completes the prescription for state relationships at a point (assuming that this $\bar{H}$ is appropriate for all fluid appearing at a point). Since radiation involves contributions from the entire flow field, the parabolic nature of the computations discussed thus far is fundamentally lost. It is still easier to obtain numerically converged solutions using the parabolic formulation for the structure analysis, if the hydrodynamics still satisfy the boundary-layer approximations. The radiation computations can then be completed in an iterative fashion, updating the radiative source term for $\tilde{H}$ after each sweep. The perturbation solution discussed at the beginning of this section can provide an obvious first approximation to initiate the iteration process.

Consideration of turbulence/radiation interactions introduces several complexities. The foremost problem is that the total enthalpy varies with both mixture fraction (due to changes in absorption and emission properties with scalar properties) and position (due to radiative transport). Thus, scalar properties at any point will be governed by the joint PDF of H and f, raising fundamental questions concerning the degree of statistical independence of these properties. Given proper statistics at each point, clearly the stochastic method can be applied to find $\bar{Q}_r$ in Eq. (21) and the appropriate heat fluxes to the surroundings.

Finding $P(H, f)$, within the present moment formulation, requires a governing equation for the variance, h, of H. Using the conventional second-order turbulence modeling approximations, the governing equation for h can be written as follows:

$$\bar{\rho}\tilde{u}\frac{\partial h}{\partial x} + \bar{\rho}\tilde{v}\frac{\partial h}{\partial r} = \frac{1}{r}\frac{\partial}{\partial r}(r\mu_{\text{eff},h}\,\partial h/\partial r) + C_{h1}\mu_{\text{eff},h}(\partial\tilde{H}/\partial r)^2$$

$$-\frac{C_{h2}\bar{\rho}h\epsilon}{k} + 2\overline{H'Q'_r} \tag{22}$$

Within the stochastic method, $\overline{H'Q'_r}$ can be found, while estimates of the turbulence modeling constants can be obtained by analogy to those for other scalar properties; therefore, the formulation is closed if some approximation concerning statistical independence of f and H can be made. Beginning such analysis by assuming that these quantities are statistically independent is convenient, however, enthalpy variations are probably correlated with mixture fraction to a significant degree, e.g., the enthalpy is lower in high temperature regions due to radiative transport and higher elsewhere. Thus, a closure will have to be sought to quantify such effects. However, the need for this extension, in view of other approximations of the structure and radiation analysis, must still be assessed.

Moment formulations offer significant computational advantages for practical calculations; therefore, their continued development is needed for near-term applications. More exact approaches, however, offer opportunities for gaining insight concerning model approximations, even though these methods can only be applied to a very limited class of flows at present. This involves time-dependent solutions of the Navier-Stokes equations exemplified by the work of Baum and Rehm [83], Riley and Metcalfe [84], Boris et al. [85], Ghoniem et al. [86] and Johnson et al. [87]. These methods provide exciting prospects for studying turbulence/radiation interactions, among other flame properties.

9 CONCLUSIONS

Nonluminous and luminous radiation from turbulent diffusion flames have been discussed. Major conclusions are as follows:

1. Scalar property predictions for nonluminous flames have been very encouraging using the laminar flamelet method—at least for well-defined laboratory conditions. Estimates of flame radiation, based on these predictions, are typically within 20–30 percent of measurements. Effects of turbulence/radiation interactions are of the same order of magnitude for hydrocarbon fuels. However, fuels having high concentrations of hydrogen can yield spectral intensities on the order of twice the values based on mean properties, due to turbulent fluctuations.

2. Extension of the laminar flamelet concept to soot, and thus luminous flames, has also been encouraging. However, less universality of state relationships for soot and some effects of laminar flame hydrodynamics have been observed. Use of scalar properties found in this manner for predictions of the radiation properties of luminous flames has also been encouraging, but results for a wider range of conditions and additional direct assessment of estimates of soot properties in turbulent flames is needed. Turbulence/radiation interactions and coupled effects of radiation and structure are more important for luminous flames than for nonluminous flames, but have not been studied very extensively as yet.

3. The moment methods for analyzing turbulent flows, currently in widespread use, provide computational efficiencies needed to treat practical problems, although the physical basis of aspects of these models is questionable. Recent advances in numerical simulation of simplified turbulent flows, however, appear promising for studies of turbulence/radiation interactions in order to provide a better fundamental understanding of this process and its approximation using moment methods.

ACKNOWLEDGMENTS

The authors' research on radiation from turbulent flames is supported by the United States Department of Commerce, National Bureau of Standards, Grant No. 60NANB5D0576, with B. J. McCaffrey, of the Center of Fire Research, serving as scientific officer. The authors would also like to acknowledge the assistance of M.-C. Lai, L. D. Chen and D. Hully during portions of the work.

REFERENCES

1. C. L. Tien and S. C. Lee, Flame Radiation, *Prog. Energy Combust. Sci.*, vol. 8, pp. 41–59, 1982.
2. J. deRis, Fire Radiation—A Review, *Seventeenth Symposium (International) on Combustion*, p. 1003, The Combustion Institute, Pittsburgh, 1979.
3. H. Eickhoff, Turbulent Hydrocarbon Flames, *Prog. Energy Combust. Sci.*, vol. 8, pp. 159–168, 1982.
4. H. Siegel and J. R. Howell, *Thermal Radiation Heat Transfer*, McGraw-Hill, New York, 1972.
5. C. B. Ludwig, W. Malkmus, J. E. Reardon and J. A. Thomson, *Handbook of Infrared Radiation from Combustion Gases*, NASA, Rept. SP-3080, Washington, D.C., 1973.

6. R. M. Goody, *Atmospheric Radiation Theoretical Base,* vol. 1, Clarendon Press, Oxford, 1964.
7. H. C. Hottel and A. F. Sarofim, *Radiative Transfer,* McGraw-Hill, New York, 1967.
8. D. E. Negrelli, J. R. Lloyd and J. L. Novotny, A Theoretical and Experimental Study of Radiation Convection Interaction in a Diffusion Flame, *J. Heat Trans.,* vol. 99, pp. 212–220, 1977.
9. K. V. Liu, J. R. Lloyd and K. T. Yang, An Investigation of a Laminar Diffusion Flame Adjacent to a Vertical Flat Plate Burner, *Int. J. Heat Mass Trans.,* vol. 24, pp. 1959–1970, 1981.
10. D. K. Edwards and W. A. Menard, Comparison of Models for Correlation of Total Band Absorption, *Appl. Optics,* vol. 3, pp. 621–625, 1964.
11. W. L. Grosshandler and J. P. Vantelon, Predicting Soot Radiation in Laminar Diffusion Flames, *Comb. Sci. Tech.,* vol. 44, pp. 125–141, 1985.
12. G. H. Markstein, Relationship between Smoke Point and Radiant Emission from Buoyant Turbulent and Laminar Diffusion Flames, *Twentieth Symposium (International) on Combustion,* p. 1055, The Combustion Institute, Pittsburgh, 1985.
13. G. H. Markstein, Soot Formation, Radiant Emission and Absorption in Laminar Ethylene and Propylene Diffusion Flames, Factory Mutual Research, Rept. FMRC J.1.No. 0F0N7.BU, Norwood, MA, 1983.
14. G. H. Markstein and J. deRis, Radiant Emission and Absorption by Laminar Ethylene and Propylene Diffusion Flames, *Twentieth Symposium (International) on Combustion,* p. 1637, The Combustion Institute, Pittsburgh, 1985.
15. W. L. Grosshandler and R. F. Sawyer, Radiation from a Methanol Furnace, *J. Heat Trans.,* vol. 100, pp. 247–252, 1978.
16. W. L. Grosshandler, Radiative Heat Transfer in Nonhomogeneous Gases: A Simplified Approach, *Int. J. Heat Mass Trans.,* vol. 23, pp. 1447–1459, 1980.
17. W. L. Grosshandler and H. D. Nguyen, Application of the Total Transmittance Nonhomogeneous Radiation Model to Methane Combustion, *J. Heat Trans.,* vol. 107, pp. 445–450, 1985.
18. D. Karman and F. R. Steward, The Radiation Spectrum from a Test Furnace, *Int. J. Heat Mass Trans.,* vol. 27, pp. 1357–1364, 1984.
19. F. R. Steward and K. H. Guruz, The Effect of Solid Particles on Radiative Transfer in a Cylindrical Test Furnace, *Fifteenth Symposium (International) on Combustion,* p. 1271, The Combustion Institute, Pittsburgh, 1974.
20. J.-M. Souil, J.-M. Most and P. Joulain, Diffusion Flame Radiative Properties in Gas-Solid Combustion, *Comb. Sci. Tech.,* vol. 24, pp. 93–105, 1980.
21. J.-M. Souil, P. Joulain and E. Gengembre, Experimental and Theoretical Study of Thermal Radiation from Turbulent Diffusion Flames to Vertical Target Surfaces, *Comb. Sci. Tech.,* vol. 41, pp. 69–81, 1984.
22. D. C. Wilcox, Model for Fires with Low Initial Momentum and Nongray Thermal Radiation, *AIAA J.,* vol. 123, pp. 381–386, 1975.
23. W. Shackelford, Results of Spectroscopic Measurements on Six Foot Diameter LNG Pool Fires at Capistrano Test Site, TRW Systems Group, Rept. IOC4360.3.72-050, Redondo Beach, CA, 1972.
24. E. S. Fishburne and H. S. Pergament, The Dynamics and Radiant Intensity of Large Hydrogen Flames, *Seventeenth Symposium (International) on Combustion,* p. 1063, The Combustion Institute, Pittsburgh, 1979.
25. F. Tamanini, Reaction Rates, Air Entrainment and Radiation in Turbulent Fire Plumes, *Comb. Flame,* vol. 30, pp. 85–101, 1977.
26. G. H. Markstein, Scaling of Radiative Characteristics of Turbulent Diffusion Flames, *Sixteenth Symposium (International) on Combustion,* p. 1407, The Combustion Institute, Pittsburgh, 1977.
27. D. B. Pfenning, Blowout Fire Simulation Tests, Energy Analysts, Inc., Rept. 84-4-208, Norman, OK, 1984.
28. R. W. Claus, Spectral Flame Structure Radiance from a Tubular-Can Combustor, NASA, Technical Paper 1772, Washington, D.C., 1981.
29. V. Emmerich, Experimental Studies on the Radiant Heat of Soot-Containing Flames and Flame Gases, *Chem.-Ing.-Tech.,* vol. 55, pp. 490–495, 1983.

30. Y. S. H. Najjar and E. M. Goodger, Radiation and Smoke from a Gas Turbine Combustor using Heavy Fuels, *J. Heat Trans.*, vol. 105, pp. 82–88, 1983.

31. F. A. Williams and A. E. Fuhs, Apparent Emission Intensities from a Turbulent Flame Composed of Wrinkled Laminar Flames, *Jet. Prop.*, vol. 27, pp. 1099–1101, 1957.

32. G. Cox, On Radiant Heat Transfer from Turbulent Flames, *Comb. Sci. Tech.*, vol. 17, pp. 75–78, 1977.

33. V. P. Kabashnikov and G. I. Kmit, Influence of Turbulent Fluctuations on Thermal Radiation, *J. Appl. Spect.*, vol. 31, pp. 963–967, 1979.

34. W. L. Grosshandler and P. Joulain, Intermittency and Flame Radiation, *10th International Colloqium on Dynamics of Explosions and Reactive Systems*, Berkeley, CA, 1985.

35. R. Portscht, Studies on Characteristic Fluctuations of the Flame Radiation Emitted by Fires, *Comb. Sci. Tech.*, vol. 10, pp. 73–84, 1974.

36. F. A. Williams, *Combustion Theory*, 2nd ed., Benjamin Cummings Publishing, Menlo Park, CA, 1985.

37. F. C. Lockwood and A. S. Naguib, The Prediction of the Fluctuations in the Properties of Free, Round-Jet Turbulent Diffusion Flames, *Comb. Flame*, vol. 24, pp. 109–124, 1975.

38. R. W. Bilger, Turbulent Jet Diffusion Flames, *Prog. Energy Combust. Sci.*, vol. 1, pp. 87–109, 1976.

39. S.-M. Jeng, L.-D. Chen and G. M. Faeth, The Structure of Buoyant Methane and Propane Diffusion Flames, *Nineteenth Symposium (International) on Combustion*, p. 349, The Combustion Institute, Pittsburgh, 1982.

40. S.-M. Jeng, An Investigation of Structure and Radiation Properties of Turbulent Buoyant Diffusion Flames, Ph.D. thesis, The Pennsylvania State University, University Park, PA, 1984.

41. D. B. Spalding, *GENMIX: A General Computer Program for Two-Dimensional Parabolic Phenomena*, Pergamon Press, Oxford, 1977.

42. C.-P. Mao, G. A. Szekely, Jr., and G. M. Faeth, Evaluation of a Locally Homogeneous Flow Model of Spray Combustion, *J. Energy*, vol. 4, pp. 78–87, 1980.

43. H.-Z. You and G. M. Faeth, Buoyant Axisymmetric Turbulent Diffusion Flames in Still Air, *Comb. Flame*, vol. 44, pp. 251–275, 1982.

44. R. W. Bilger, Reaction Rates in Diffusion Flames, *Comb. Flame*, vol. 30, pp. 277–284, 1977.

45. S. K. Liew, K. N. C. Bray and J. B. Moss, A Flamelet Model of Turbulent Non-Premixed Combustion, *Comb. Sci. Tech.*, vol. 27, pp. 69–73, 1981.

46. H. Tsuji and I. Yamaoka, Structure Analysis of Counterflow Diffusion Flames in the Forward Stagnation Region of a Porous Cylinder, *Thirteenth Symposium (International) on Combustion*, p. 723, The Combustion Institute, Pittsburgh, 1971.

47. H. Tsuji and I. Yamaoka, The Counterflow Diffusion Flames in the Forward Stagnation Region of a Porous Cylinder, *Eleventh Symposium (International) on Combustion*, p. 970, The Combustion Institute, Pittsburgh, 1967.

48. H. Tsuji and I. Yamaoka, The Structure of Counterflow Diffusion Flames in the Forward Stagnation Region of a Porous Cylinder, *Twelfth Symposium (International) on Combustion*, p. 997, The Combustion Institute, Pittsburgh, 1969.

49. S. I. Abdel-Khalik, T. Tamura and M. M. El-Wakil, A Chromatographic and Interferometric Study of the Diffusion Flame Around a Simulated Drop, *Fifteenth Symposium (International) on Combustion*, p. 389, The Combustion Institute, Pittsburgh, 1975.

50. R. E. Mitchell, A. F. Sarofim and L. A. Clomberg, Experimental and Numerical Investigation of Confined Laminar Diffusion Flames, *Comb. Flame*, vol. 37, pp. 227–244.

51. S.-M. Jeng and G. M. Faeth, Species Concentrations and Turbulence Properties in Buoyant Methane Diffusion Flames, *J. Heat Trans.* vol. 106, pp. 721–727, 1984.

52. S.-M. Jeng and G. M. Faeth, Predictions of Mean Scalar Properties in Turbulent Propane Diffusion Flames, *J. Heat Trans.*, vol. 106, pp. 891–893, 1984.

53. J. P. Gore, S.-M. Jeng and G. M. Faeth, Spectral and Total Radiation Properties of Turbulent Hydrogen/Air Diffusion Flames, *J. Heat Trans.*, vol. 109, pp. 165–171, 1987.

54. J. P. Gore, S. M. Jeng and G. M. Faeth, Spectral and Total Radiation Properties of Turbulent Carbon Monoxide/Air Diffusion Flames, *AIAA J.*, vol. 25, pp. 339–345, 1987.

55. J. P. Gore and G. M. Faeth, Structure and Spectral Radiation Properties of Turbulent Ethylene/ Air Diffusion Flames, *Twenty-First Symposium (International) on Combustion,* p. 1521, The Combustion Institute, Pittsburgh, 1986.

56. M. C. Drake, M. Lapp, C. M. Penny and S. Warshaw, Measurements of Temperature and Concentration Fluctuations in Turbulent Diffusion Flames using Pulsed Raman Spectroscopy, *Eighteenth Symposium (International) on Combustion,* p. 1521, The Combustion Institute, Pittsburgh, 1981.

57. M. C. Drake, R. W. Pitz and M. Lapp, Laser Measurements of Nonpremixed Hydrogen-Air Flames for Assessment of Turbulent Combustion Models, AIAA Paper No. 84-0544, 1984.

58. D. P. Aeschliman, J. C. Cummings and R. A. Hill, Raman Spectroscopic Study of a Laminar Hydrogen Diffusion Flame in Air, *J. Quant. Spect. Rad. Trans.,* vol. 21, pp. 293–307, 1979.

59. J. A. Miller and R. J. Kee, Chemical Nonequilibrium Effects in Hydrogen-Air Laminar Jet Diffusion Flames, *J. Phys. Chem.,* vol. 81, pp. 2534–2542, 1977.

60. S. Gordon and B. J. McBride, Computer Program for Calculation of Complex Chemical Equilibrium Compositions, Rocket Performance, Incident and Reflected Shocks, and Chapman-Jouget Detonations, NASA, Rept. SP-273, Washington, D.C., 1971.

61. P. J. Pagni and S. Bard, Particulate Volume Fractions in Diffusion Flames, *Seventeenth Symposium (International) on Combustion,* p. 1017, The Combustion Institute, Pittsburgh, 1979.

62. R. J. Santoro, H. B. Semerjian and R. A. Dobbins, Soot Particle Measurements in Diffusion Flames, *Comb. Flame,* vol. 51, pp. 203–218, 1983.

63. J. H. Kent, H. Jander and H. G. Wagner, Soot Formation in a Laminar Diffusion Flame, *Eighteenth Symposium (International) on Combustion,* p. 1117, The Combustion Institute, Pittsburgh, 1981.

64. J. H. Kent and H. G. Wagner, Soot Measurements in Laminar Ethylene Diffusion Flames, *Comb. Flame,* vol. 47, pp. 53–65, 1982.

65. M. K. Razdan and J. G. Stevens, CO/Air Turbulent Diffusion Flames: Measurements and Modeling, *Comb. Flame,* vol. 59, pp. 289–301, 1985.

66. J. P. Gore, G. M. Faeth, D. Evans and D. B. Pfenning, Structure and Radiation Properties of Large-Scale Natural Gas/Air Diffusion Flames, *Fire and Materials,* vol. 10, pp. 161–169, 1986.

67. S. H. Starner and R. W. Bilger, Measurements of Scalar-Velocity Correlations in a Turbulent Diffusion Flame, *Eighteenth Symposium (International) on Combustion,* p. 921, The Combustion Institute, Pittsburgh, 1981.

68. G. M. Faeth and G. S. Samuelsen, Fast-Reaction Nonpremixed Combustion, *Prog. Energy Combust. Sci.,* vol. 12, pp. 305–372, 1986.

69. S.-M. Jeng, M.-C. Lai and G. M. Faeth, Nonluminous Radiation in Turbulent Buoyant Axisymmetric Flames, *Comb. Sci. Tech.,* vol. 40, pp. 41–53, 1984.

70. S.-M. Jeng and G. M. Faeth, Radiative Heat Fluxes Near Turbulent Buoyant Methane Diffusion Flames, *J. Heat Trans.,* vol. 106, pp. 886–888, 1984.

71. A. D. Gosman and E. Ioannides, Aspects of Computer Simulation of Liquid-Fueled Combustors, AIAA Paper No. 81-0323, 1981.

72. J.-S. Shuen, A. S. P. Solomon, Q.-F. Zhang and G. M. Faeth, Structure of Particle-Laden Jets: Measurements and Predictions, *AIAA J.,* vol. 23, pp. 396–404, 1985.

73. F. S. Simmons, H. Y. Yamada and C. B. Arnold, Measurement of Temperature Profile in Hot Gases by Emission-Absorption Spectroscopy, NASA Rept. CR-72491, 1969.

74. W. H. Giedt and K. Saido, Investigation of Absorption in the 2.7 μm Bands of Nonisothermal High Temperature H_2O, CO_2 and H_2O-CO_2 Mixtures, Dept. of Mechanical Engineering, Rept. 70-10, University of California, Davis, CA, 1970.

75. D. E. Burch and D. A. Gryvnak, Infrared Radiation Emitted by Hot Gases and its Transmission through Stagnant Atmospheres, Aeronutronic, Rept. U-1929, 1962.

76. C. C. Ferriso, High Temperature Infrared Emission Absorption Studies, *J. Chem. Phys.,* vol. 37, pp. 1955–1961, 1962.

77. D. K. Edwards, L. K. Glassen, W. C. Hauser and J. S. Tuchscher, Radiation Heat Transfer in Nonisothermal Nongray Gases, *J. Heat Trans.,* vol. 89, pp. 219–229, 1967.

78. M. M. Abu-Romia and C. L. Tien, Measurements and Infrared Radiation of Carbon Monoxide at Elevated Temperature, *J. Quant. Spect. and Rad. Trans.*, vol. 6, pp. 143–167, 1966.

79. W. L. Grosshandler, A Study of a Model Furnace Burning Methanol and a Methanol/Coal Slurry, Ph.D. thesis, The University of California, Berkeley, CA, 1976.

80. F. C. Lockwood and N. B. Shah, A New Radiation Solution Method for Incorporation in General Combustion Prediction Procedures, *Eighteenth Symposium (International) on Combustion,* p. 1405, The Combustion Institute, Pittsburgh, 1981.

81. A. D. Gosman and F. C. Lockwood, Incorporation of a Flux Model for Radiation into a Finite Difference Procedure for Furnace Calculations, *Fourteenth Symposium (International) on Combustion,* p. 661, The Combustion Institute, Pittsburgh, 1973.

82. R. G. Siddal and N. Selcuk, Two-Flux Modeling of Two-Dimensional Radiative Transfer in Axi-Symmetrical Furnaces, *J. Inst. Fuel,* vol. 49, pp. 10–20, 1976.

83. H. R. Baum and R. G. Rehm, Calculations of Three Dimensional Buoyant Plumes in Enclosures, *Comb. Sci. Tech.,* vol. 40, pp. 55–77, 1984.

84. J. J. Riley and R. W. Metcalfe, Direct Numerical Simulations of Chemically Reacting Turbulent Mixing Layers, AIAA Paper No. 85-0321, 1985.

85. J. P. Boris, E. S. Oran, M. J. Fritts and C. Oswald, Time Dependent Compressible Simulations of Shear Flows: Tests of Outflow Boundary Conditions, Naval Research Laboratory, Memorandum Rept. 5249, Washington, D.C., 1983.

86. A. F. Ghoniem, A. J. Chorin and A. K. Oppenheim, Numerical Modeling of Turbulent Flow in a Combustion Tunnel, *Phil. Trans. R. Soc. Lond.,* vol. A304, pp. 303–325, 1982.

87. S. C. Johnson, R. W. Dibble, R. W. Schefer, W. T. Ashurst and W. Kollmann, Laser Measurements and Stochastic Simulations of Turbulent Reacting Flows, AIAA Paper No. 84-0543, 1984.

SOME RECENT DEVELOPMENTS AND TRENDS IN FINITE ELEMENT COMPUTATIONAL NATURAL CONVECTION

Dominique H. Pelletier, J. N. Reddy, and Joseph A. Schetz

ABSTRACT

This chapter contains a review of the literature and results of penalty-finite-element analysis of two-dimensional natural convection (i.e., coupled conductive and convective heat transfer and fluid flow). The penalty function formulation of the equations governing steady, laminar flow of incompressible viscous fluids is presented and the associated finite-element model is described. Numerical results for a number of sample problems are presented and discussed.

1 INTRODUCTION

Buoyancy driven flows play an important role in many engineering problems of practical interest. As a few examples, these include thermal insulation of buildings [1,2]; heat transfer through double glazed windows [3,4]; cooling of electronic equipment [5,14,15]; circulation of planetary atmosphere [6]; crystal growth from melt [7]; cooling of nuclear reactor cores [8]; sterilization of canned food [9,16]; storage of spent nuclear fuel [10,11]; convective cooling of underground electric cable systems [12]; and analysis of solar collector systems [13].

Although most of these flows are fully three-dimensional and time dependent, the limitations imposed by both experimental and theoretical techniques have forced researchers to analyze only those fluid motions that are believed to lend themselves to approximation by two-dimensional models. The Navier-Stokes and energy equations describing these phenomena are highly nonlinear, and the strong coupling between the equations makes it difficult to obtain analytical solutions.

Recent advances in computer hardware have made it possible to investigate several finite difference and finite-element formulations of the Navier-Stokes equations (see, for example, [17–43]). One of the major difficulties associated with the solution of the Navier-Stokes equations in non-rectangular geometries is the application of the boundary conditions. Although attempts have been made to rectify this problem by constructing body fitted curvilinear meshes in finite difference methods [32], the finite element method has a definite advantage in that any complicated geometry can be suitably represented by using non-uniform and non-rectangular meshes, and appropriate boundary conditions of the model can be imposed in a natural way.

In spite of recent developments many complex internal and external flows remained unsolved. For practical engineering problems, only case by case studies can be attempted. There does not exist, at present, a general-purpose, foolproof method for solving buoyancy driven flows. In that sense, computational natural convection is still very much of an art. Much further work is needed, especially for three-dimensional time-dependent flows, and variable-property flows, even to solve some of the basic problems of computational fluid dynamics.

In the present paper the penalty finite element model for natural convection is described, and its application to some problems is discussed. The scaling and nondimensionalization of free convection problems are also presented. The weak form leading to the finite element discretization is described along with some of the computational details. Finally, numerical results for a number of representative problems are presented.

2 GOVERNING EQUATIONS

2.1 The Boussinesq Equations

The Navier-Stokes and energy equations describing the buoyancy driven flow of an incompressible fluid in a domain Ω can be written as:

$$u_{i,i} = 0 \tag{1a}$$

$$\rho u_j u_{i,j} + p_{,i} - \rho f_i - \rho g_i[1 - \beta(T - T_\beta)] - [\mu(u_{i,j} + u_{j,i})]_{,j} = 0 \tag{1b}$$

$$\rho c_p(u_j T_{,j}) - (kT_{,j})_{,j} - \mu\Phi - \rho q_s = 0 \tag{1c}$$

where u_i = velocity components
$\quad f_i$ = body force components
$\quad g_i$ = components of the gravitational force

p = pressure
T = temperature
ρ = density
μ = viscosity
k = thermal conductivity
c_p = specific heat at constant pressure
β = volume expansion coefficient
T_β = reference temperature for which buoyancy forces are zero
q_s = heat source per unit mass
Φ = viscous dissipation function

and, $u_{i,j} = \partial u_i / \partial x_j$, etc., and summation on repeated subscripts is implied.

We have assumed that the Boussinesq approximation holds: the effects of density variation are negligible except for the buoyancy force. Also, the flow is assumed to be laminar. The fluid properties c_p, μ, β, and k can be temperature dependent. A complete and detailed derivation of the Boussinesq equation can be found in [44].

The boundary conditions needed to complete the definition of the problem are a combination of prescribed velocities, tractions, temperatures and heat fluxes. Let Γ be the boundary of the fluid, then:

$$\Gamma = \Gamma_u \cup \Gamma_t = \Gamma_T \cup \Gamma_q$$

$$\Gamma_u \cap \Gamma_t = \phi$$

$$\Gamma_T \cap \Gamma_q = \phi$$

Here, Γ_u, Γ_t, Γ_T, and Γ_q represent the portions of the boundary on which the velocities, stresses, temperature and heat flux, respectively, are specified. The boundary conditions are:

$$u_i = u_i^* \text{ on } \Gamma_{u_i}$$

$$t_i \equiv \sigma_{ij} n_j = t_i^* \text{ on } \Gamma_{t_i}$$

$$T = T^* \text{ on } \Gamma_T$$

$$q = (kT_{,j})n_j = q^* \text{ on } \Gamma_q$$

where σ_{ij} are the components of stress tensor, t_i are the boundary stresses and n_j is the outward unit normal. The quantities with an asterisk denote specified quantities. The total stress components σ_{ij} can be related to the pressure P and velocity gradients (for the Newtonian case) by

$$\sigma_{ij} = -p\delta_{ij} + \mu(u_{i,j} + u_{j,i})$$

For two dimensional problems, Eq. (1) is written as

$$\frac{\partial u}{\partial x} + \frac{\partial v}{\partial y} = 0$$

$$\rho\left(u\frac{\partial u}{\partial x} + v\frac{\partial u}{\partial y}\right) - \frac{\partial p}{\partial x} + \rho f_x - \rho g_x[1 - \beta(T - T_0)]$$

$$- \frac{\partial}{\partial x}\left[2\mu\frac{\partial u}{\partial x}\right] - \frac{\partial}{\partial y}\left[\mu\left(\frac{\partial u}{\partial y} + \frac{\partial v}{\partial x}\right)\right] = 0$$

$$\rho\left(u\frac{\partial v}{\partial x} + v\frac{\partial v}{\partial y}\right) - \frac{\partial p}{\partial y} + \rho f_y - \rho g_y[1 - \beta(T - T_\beta)]$$

$$- \frac{\partial}{\partial x}\left[\mu\left(\frac{\partial u}{\partial y} + \frac{\partial v}{\partial x}\right)\right] - \frac{\partial}{\partial y}\left[2\mu\frac{\partial v}{\partial y}\right] = 0$$

$$\rho c_p\left(u\frac{\partial T}{\partial x} + v\frac{\partial T}{\partial y}\right) - \frac{\partial}{\partial x}\left(k\frac{\partial T}{\partial x}\right) - \frac{\partial}{\partial y}\left(k\frac{\partial T}{\partial y}\right) - \mu\Phi - \rho q_x = 0$$

2.2 Scale Analysis

Although the Navier-Stokes equations are required to properly describe the flow and heat transfer for the general case, useful physical insight is gained from a careful study of the simpler case where the boundary layer approximation holds. Figure 1 contains the coordinate system for the flow near a heated vertical flat plate at temperature T_0, the fluid at infinity is at temperature T_∞. The flow near the plate is upward. For this problem under a restricted range of conditions, the boundary layer approximation holds, and we follow closely the analysis of [45].

Consider the conservation of mass, momentum, and energy in the thermal boundary layer region $[x \approx 0(\delta_T), y \approx 0(H)]$

$$\frac{\partial u}{\partial x} + \frac{\partial v}{\partial y} = 0$$

$$\rho\left(u\frac{\partial v}{\partial x} + v\frac{\partial v}{\partial y}\right) = \mu\frac{\partial^2 v}{\partial x^2} + g\beta(T - T_\infty)$$

$$\rho c_p\left(u\frac{\partial T}{\partial x} + v\frac{\partial T}{\partial y}\right) = k\frac{\partial^2 T}{\partial x^2} \qquad (2)$$

A simple balance between convection and conduction yields

$$\frac{u\Delta T}{\delta_T}, \frac{v\Delta T}{H} \sim \frac{\alpha\Delta T}{\delta_T^2} \qquad (3)$$

where $\Delta T = T_0 - T_\infty$. The mass conservation produces

$$\frac{u}{\delta_T} \sim \frac{v}{H} \qquad (4)$$

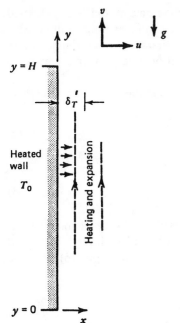

Figure 1 Coordinate system for free convection near a vertical flat plate (from [45]).

Hence from Eqs. (3) and (4) we obtain

$$v \sim \frac{\alpha H}{\delta_T^2}$$

where the thermal thickness δ_T is still an unknown.

The momentum equation states an interplay between three forces: inertia, friction, and buoyancy. Hence we have

$$\underset{\text{Inertia}}{u \frac{v}{\delta_T}, u \frac{v}{H}} \quad \sim \quad \underset{\text{Friction}}{\frac{\mu v}{\rho \delta_T^2}} \quad \sim \quad \underset{\text{Buoyancy}}{g \beta \Delta T} \tag{5}$$

The fluid layer can be ruled by either an inertia-buoyancy balance, or by a friction-buoyancy balance. Equation (5) can be rewritten, after some manipulation, as

$$\underset{\text{Inertia}}{\left(\frac{H}{\delta_T}\right)^4 \mathrm{Ra}^{-1}\mathrm{Pr}^{-1}} \quad \sim \quad \underset{\text{Friction}}{\left(\frac{H}{\delta_T}\right)^4 \mathrm{Ra}^{-1}} \quad \sim \quad \underset{\text{Buoyancy}}{1} \tag{6}$$

where the Rayleigh number is

$$\mathrm{Ra} = \frac{g \beta \Delta T H^3 \rho}{\alpha \mu}$$

and $\mathrm{Pr} = \mu c_p / k$ is the Prandtl number.

One sees from Eq. (6) that the balance between inertia and friction is governed by a fluid property, the Prandtl number. High Prandtl number fluids will be governed by a friction-buoyancy balance, whereas low Prandtl number fluids will be governed by an inertia-buoyancy balance. The two boundary layer regimes are illustrated in Fig. 2.

For high Prandtl number fluids, one easily finds that the momentum and thermal boundary layer thicknesses are given as

$$\delta \sim H \, \mathrm{Ra}^{-1/4} \mathrm{Pr}^{1/2} \qquad (7a)$$

$$\delta_T \sim H \, \mathrm{Ra}^{-1/4} \qquad (7b)$$

$$\frac{\delta}{\delta_T} \sim \mathrm{Pr}^{1/2} > 1 \qquad (8)$$

Most of the literature on convection refers to δ as the velocity boundary layer thickness. In fact, from Fig. 2, the free convection boundary layer is governed by two length scales (δ and δ_T) (see [45]) and not by a single one (δ) as is the

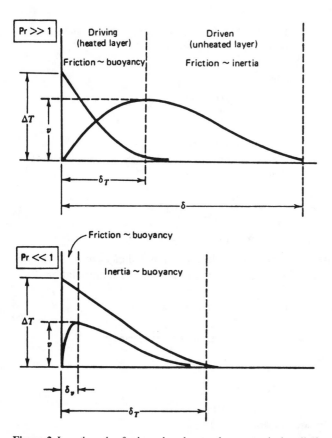

Figure 2 Length scales for boundary layers along a vertical wall (from [45]).

case in forced convection. The velocity scale is reached with a thin layer δ_T, while the velocity decays to zero within a thick layer δ.

For low Prandtl number fluids, one can show that

$$\delta_T \sim H(\text{Ra Pr})^{-1/4} \tag{9}$$

$$v \sim \frac{\alpha}{H}(\text{Ra Pr})^{1/2} \tag{10}$$

where $\alpha = k/(\rho c_p)$. Notice the emergence of a new dimensionless group. The product Ra Pr plays the same role as Ra for high Prandtl number fluids. For this case (see Fig. 2) the velocity profile is as wide as the temperature profile. Because of the nonslip boundary condition at the wall, the location of the velocity peak δ_v is an important length scale. A balance of buoyancy and friction forces yields

$$\delta_v \sim H \, \text{Gr}^{-1/4} \tag{11}$$

where the Grashof number Gr is defined by

$$\text{Gr} = \frac{g\beta\Delta T H^3}{v^2} = \frac{\text{Ra}}{\text{Pr}} \tag{12}$$

and one finally finds

$$\frac{\delta_v}{\delta_T} \sim \text{Pr}^{1/2} < 1$$

The Grashof number is a relevant dimensionless group only in the δ_v scale for $\text{Pr} < 1$. From a heat transfer point of view the important groups in external natural convection are the Rayleigh number for $\text{Pr} > 1$ and Ra Pr for $\text{Pr} < 1$ (see [45]). It should also be noted that only the 1/4 power of Ra, Ra Pr, Gr have a physical meaning:

$$\text{Ra}^{1/4} \sim \frac{\text{Wall height}}{\text{Thermal boundary layer thickness}} \quad \text{if Pr} > 1$$

$$(\text{RaPr})^{1/4} \sim \frac{\text{Wall height}}{\text{Thermal boundary layer thickness}} \quad \text{if Pr} < 1$$

$$\text{Gr}^{1/4} \sim \frac{\text{Wall height}}{\text{Wall shear layer thickness}}$$

These numbers have a simple geometric meaning: the slenderness of the boundary layer.

2.3 Nondimensionalization

The use of dimensionless parameters in the numerical solution process is advantageous because

1. They provide an estimate of the order of difficulty of the problem.
2. They provide a measure of the relative importance of the terms in the equations (they indicate which physical phenomena are dominant).
3. They reduce the potentially large differences in orders of magnitude that might occur between terms in the field equations.

It is usually difficult to compare numerical solutions from more than one author for a given problem. The reason is that the nondimensionalizations often seem to be made indiscriminately. This is especially true of the velocity scaling. Following [46], we introduce reference quantities as

Length: L_0
Velocity: U_0
Pressure: p_0
Temperature: ΔT_0

The logical choice for L_0 should be that seen in the previous section: the location of the velocity scale (see Fig. 2). Unfortunately the value of this length scale is unknown a priori. The alternative is to select a physical dimension for L_0.

The following choices for U_0 are possible:

$$U_0 = \frac{\nu}{L_0}, \; U_0 = \frac{\alpha}{L_0}$$

$$U_0 = (\alpha u)^{1/2} \frac{1}{L_0}, \; U_0 = \frac{\beta g \Delta T L_0}{\nu}, \; U_0 = \frac{\alpha}{L_0 \sqrt{Ra\,Pr}}$$

where $\alpha = k/(\rho c_p)$ is the thermal diffusivity of the fluid. It is important to note that each velocity scale usually has a specific physical meaning. Furthermore, for each scaling the dimensionless parameters appear as coefficients of different terms as described below:

1. $U_0 = \nu/L_0$: This is a viscous diffusion velocity and the parameters appear as:

$$Inertia = Gr \; (buoyancy) + viscous$$
$$Convection = (Pr)^{-1} \; conduction$$

This velocity scale implies that the inertia and viscous terms are of the same order of magnitude.
2. $U_0 = \alpha/L_0$: This is a thermal diffusion velocity for which

$$(Pr)^{-1} \; Inertia = Ra \; (buoyancy) + viscous$$
$$Convection = conduction$$

This scale indicates a balance between convection and conduction.

3. $U_0 = (\alpha\mu/\rho)^{1/2}/L_0$: This hybrid scale has no clear explicit implication; we have

$$\text{Inertia} = \text{Ra (buoyancy)} + (\text{Pr})^{1/2} \text{ (viscous)}$$
$$\text{Convection} = (\text{Pr})^{-1/2} \text{ conduction}$$

4. $U_0 = \beta g \Delta T L_0^3/(\rho/\mu)$: This scale implies that the buoyancy and viscous forces are of equal magnitude and is generally applicable for Gr < 1 and Ra < 1. We have

$$\text{Gr (Inertia)} = \text{buoyancy} + \text{viscous}$$
$$\text{Convection} = (\text{Pr})^{-1} \text{ conduction}$$

5. $U_0 = \alpha/[L_0(\text{Ra Pr})^{1/2}]$: This scale is similar to scale 2 but implies a balance between inertia and buoyancy and offer some numerical advantages for solutions at high Ra; see [35]. This is the form adopted here. We have

$$\overline{\text{Inertia} = \text{buoyancy}} + \sqrt{\text{Ra}/\text{Pr}} \text{ viscous}$$
$$\sqrt{\text{Ra Pr}} \text{ Convection} = \text{conduction}$$

Since the parametric values are usually large, it is apparent that different scalings for a given problem cause different terms in the equations to be of disparate magnitude. This can be the cause of considerable numerical problems and errors if the wrong scaling is used for a given problem (see [46]). The considerations of this and the previous section should make it possible to choose the appropriate velocity and length scales for a given problem.

The following simple technique makes it possible to use any of the above dimensionless forms with a single computer program written for the dimensional form of the equations of motion Eq. (1). The equations of motion are nondimensionalized and then compared to Eq. (1). By setting the fluid properties to the values given in Table 1 the chosen dimensionless form is obtained.

When using a dimensionless set of equations, the geometry of the domain must be scaled by L_0 and the Dirichlet boundary conditions scaled by U_0 and ΔT_0. The natural boundary conditions must also be scaled according to the dimensionless form selected.

Table 1 Dimensionless forms obtained through values of fluid properties

Fluid property	Form				
	1	2	3	4	5
ρ	1	1	1	1	$(\text{Ra}/\text{Pr})^{1/2}$
g	1	1	1	1	1
β	Ra/Pr	RaPr	Ra	Gr	1
μ	1	Pr	$\text{Pr}^{1/2}$	Gr	1
c_p	1	1	1	1	Pr
k	Pr^{-1}	1	$\text{Pr}^{-1/2}$	Pr^{-1}	1

3 VARIATIONAL AND FINITE ELEMENT FORMULATIONS OF THE FULL NAVIER–STOKES EQUATIONS

Many problems of practical interest involve either recirculation or upstream influence of the dependent variables, and these require that the fully elliptic, Navier-Stokes equations must be considered. This section presents the variational and finite element formulation of the Boussinesq form of the Navier-Stokes equations for solving free convection problems.

3.1 Weak Form and Boundary Conditions

It is now well accepted that no functional in the classical sense can be constructed for the full Navier-Stokes equations. In other words, there is no functional whose Euler-Lagrange equations are the Navier-Stokes equations. Nevertheless, it is possible to obtain a weak form of the Navier-Stokes equations [42,43,47]. A weak form of a problem is defined to be an integral statement obtained by multiplying the given equation with a weight function, integrating over the domain, and trading the differentiation from the problem dependent variable to the weight function such that the natural boundary conditions of the problem are accommodated into the integral statement (see [47] for details). The corresponding finite element formulation is straightforward. Details of the procedure are readily available in [34–43,47]; hence, only an outline of the procedure is presented here.

When an approximation solution (u, T, p) is substituted in Eq. (1), a set of residual equations is obtained in the form

$$\text{Continuity: } F_1(u) \qquad \equiv R_1 \tag{13}$$

$$\text{Momentum: } F_2(u, T, p) \equiv R_2 \tag{14a}$$

$$\text{Energy: } \qquad F_3(u, T) \qquad \equiv R_3 \tag{14b}$$

where R_1, R_2, R_3 denote the residuals in Eqs. (1a), (1b), and (1c), respectively. The variational (weak) forms of these equations are obtained as follows: (1) multiply the residuals with appropriate weighting functions (say δp, δu_i, and δT, respectively); (2) integrate the products over the domain of a typical element Ω^e and trade the differentiation equally between the weighting functions and the dependent variables; and (3) include the natural boundary conditions in the integrals (see [47] for details). The procedure is outlined below:

Step (1):

$$\int_{\Omega^e} R_1 \delta p \, dA = 0$$

$$\int_{\Omega^e} R_2 \delta u_i \, dA = 0 \tag{15}$$

$$\int_{\Omega^e} R_3 \delta T \, dA = 0$$

or

$$0 = \int_{\Omega^e} u_{i,i} \delta p \, dA \tag{16}$$

$$0 = \int_{\Omega^e} \delta u_i \{ \rho u_j u_{i,j} - \rho f_i - \rho g_i [1 - \beta(T - T_\beta)] + p_{,i} [\mu(u_{i,j} + u_{j,i})]_{,j} \} dA \tag{17}$$

$$0 = \int_{\Omega^e} \delta T \{ \rho c_p u_j T_{,j} - (kT_{,j})_{,j} - \mu \phi - \rho q_s \} dA \tag{18}$$

Step (2):

Integration by parts of the pressure and viscous terms in Eq. (17), and the conduction terms in Eq. (18) yields

$$0 = \int_{\Omega^e} \{ \delta u_i \rho u_j u_{i,j} - \delta u_i \rho f_i - \delta u_i \rho g_i [1 - \beta(T - T_\beta)]$$

$$+ \delta u_{i,j} \mu(u_{i,j} + u_{j,i}) \} dA - \int_{\Gamma^e} \delta u_i [-P + \mu(u_{i,j} + u_{j,i})] n_j ds \tag{19}$$

$$0 = \int_{\Omega^e} (\delta T \rho c_p u_j T_{,j} - \delta T_{,j} T_{,j} - \mu \phi \delta T + \rho q_s \delta T) dA - \int_{\Gamma^e} \delta T (kT_{,j}) n_j ds \tag{20}$$

The integration by parts has two consequences. First, it reduces the differentiability requirements on the velocity and temperature fields [because the weak forms in Eqs. (19) and (20) now contain only the first derivatives of u_i and T]. Second, it allows the incorporation of the natural boundary conditions into the weak form as described below.

Step (3):

The coefficients of δu_i and δT in the boundary integrals of Eqs. (19) and (20) represent the natural boundary conditions of the problem. We denote them by

$$t_i \equiv [-P + \mu(u_{i,j} + u_{j,i})] n_j \qquad \text{(boundary stresses)} \tag{21}$$

$$q_n \equiv (kT_{,j}) n_j \qquad \text{(heat flux)} \tag{22}$$

and substitute into Eqs. (19) and (20) to obtain the final weak forms. Note that the continuity equation remains unmodified. This completes the variational formulation of the equations.

A comment is in order on the terminology that is quite common among finite element developers. Equations (16)–(18) are called the Galerkin integrals, because the Galerkin method makes use of these expressions to determine approximate solutions to p, u_i, and T in the form

$$p = \sum_{j=1}^{m} p_j \phi_j, \ u_i = \sum_{j=1}^{n} u_i^j \psi_j, \ T = \sum_{j=1}^{n} T_j \psi_j \tag{23}$$

(hence $\delta p = \sum_{j=1}^{m} \delta p_j \psi_j$, etc.). The Petrov-Galerkin method (or, the general weighted residual method) is one which seeks solutions of the form (23) by solving Eqs. (16)–(18) for a choice of weighting functions other than the approximation functions. Although most researchers refer to the finite element model based on Eqs. (16), (19), and (20) as the Galerkin finite element model, it is strictly an incorrect name because we are no longer using the Galerkin intergrals in Eq. (15); instead we used the weak form to develop the finite element model. Since for a weak formulation, both the Ritz and Galerkin methods give the same equations, we shall refer to the model as the Ritz model, because the Ritz method is always applied to only weak formulations.

Note that the pressure does not appear explicitly in Eq. (1a); it appears only in Eq. (1b) which describes the relations between the components of the velocity vector. Hence, Eq. (1a) can be viewed as the constraint on the velocity field. Two techniques exist to enforce the continuity (constraint) equation: the Lagrange multiplier method [which leads to Eqs. (16), (19), and (20)] and the penalty function method. Both methods have been described elsewhere (see [40,47,48]), and we focus on the penalty function method here.

3.2 The Penalty Function Method

The penalty function method is an approximate technique of including constraints into variational formulations. The penalty function method transforms a given constrained variational problem into an (actually, a sequence of) unconstrained variational problem(s) by the introduction of a penalty on the infringement of constraints [38]. Equations (1b), (1c), or alternatively Eqs. (19) and (20) can be viewed as those relating the velocity components u_i and the temperature T, and Eq. (1a) as a constraint condition on the velocity field. That is, we wish to solve Eqs. (19) and (20) subject to the constraint in Eq. (1a). This point of view enables us to employ the penalty function method to formulate the problem variationally and, subsequently, by the finite element method.

To fix ideas, suppose that we seek a generalized solution $\Lambda = (u_1, u_2)$ to the weak form (nonlinear, in general)

$$\delta I(\Lambda, \delta\Lambda) = 0 \qquad \text{in } \Omega^e \tag{24}$$

subject to the constraint

$$G(\Lambda) = 0 \qquad \text{in } \Omega^e \tag{25}$$

Here, δ denotes the variational symbol and Ω^e is a typical finite element. In the penalty method, we seek the solution to Eqs. (24)–(25) by solving the modified weak problem:

$$\delta I(\Lambda, \delta\Lambda) + \delta\left\{\frac{\gamma}{2}\int_{\Omega^e} [G(\lambda)]^2 \, dA\right\} = 0 \tag{26}$$

where γ is the penalty parameter, which is a preselected function or constant whose value, in general, has an effect on the accuracy of the solution [36–48].

For sufficiently large values of γ the variational problem in Eq. (26) is equivalent to that of Eqs. (24)–(25) [38,47]. It can also be shown that an approximation to the Lagrange multiplier associated with the constraint in Eq. (24) (which usually has an important physical meaning in engineering problems) is given by the formula

$$\lambda_\gamma = \gamma G(\Lambda_\gamma) \tag{27}$$

where Λ_γ is the solution of the modified problem (26).

Returning to the problem at hand, we apply the penalty function concept described above to the problem associated with Eqs. (1a), (1b), (1c):

$$\delta I(\Lambda, \delta\Lambda) + \delta I_p(\Lambda, \delta\Lambda) = 0 \tag{28}$$

where $\Lambda = (u_i, T)$, $\delta\Lambda = (\delta u_i, \delta T)$, and

$$
\begin{aligned}
\delta I(\Lambda, \delta\Lambda) = &\int_{\Omega^e} \{\rho \delta u_i u_j u_{i,j} + \rho \delta u_i f_i + \rho \delta u_i g_i[1 - \beta(T - T_\beta)] \\
&+ \delta u_{i,j}\mu(u_{i,j} + u_{j,i}) + \rho \delta T c_p u_j T_{,j} \\
&+ \rho \delta T c_p u_j T_{,j} - \delta T_{,j} T_{,j} - \mu \Phi \delta T + \rho q_s \delta T\} dA \\
&- \int_{\Gamma^e} \{t_i \delta u_i + q \delta T\} ds
\end{aligned}
\tag{29}
$$

$$I_P(\Lambda) = \frac{\gamma}{2} \int_{\Omega^e} (\nabla \cdot u)^2 dA \tag{30}$$

An approximation to the Lagrange multiplier, which is equal to the negative of the pressure p, can be shown to be given by

$$p_\gamma = -\gamma u_{i,i} \tag{31}$$

Equation (31) does not form the basis of the penalty method, as implied in a number of papers [36,37]. It is a consequence of applying the penalty function method. In describing the penalty function method, others often replace the divergence free condition with Eq. (31). This is confusing, because there are no physical grounds to *assume* that the velocity field is related to the pressure through Eq. (31).

3.3 Finite Element Formulation

3.3.1 Reduced integration and consistent penalty model. The finite element model of the variational Eq. (28) is simple (see [38,47]). Over a typical element Ω^e of the finite element mesh, a generic dependent variable u is interpolated by

$$u = \sum_{j=1}^{n} u_j \phi_j \tag{32}$$

where ϕ_j are the element interpolation functions. All integrals in Eqs. (28)–(30) are computed using Gauss quadrature. To ensure that the penalty finite element solution is accurate, one must use reduced integration to evaluate the penalty functional. The need for reduced integration can be explained (see [38–40,47–49]) in terms of the condition for existence and uniqueness of solutions to the discretized penalty problem. It has been found that numerical integration of the penalty functional with one less number of Gaussian points will ensure proper behavior of the penalty solution.

Alternatively, the consistent (or mixed) penalty method can be used to construct the penalty variational formulation that will be stable and convergent [50]. This method uses the following weak form of Eqs. (1b), (1c), and (31):

$$\int_{\Omega^e} \{\rho\delta u_i u_j u_{i,j} - \rho\delta u_i f_i + \rho\delta u_i g_i[1 - \beta(T - T_B)] + p\delta u_{i,i}$$

$$+ \delta u_{i,j}\mu(u_{i,j} + u_{j,i}) + \rho\delta T c_p u_j T_{,j} + \delta T_{,j}T_{,j} - \mu\Phi\delta T + \rho q_i\delta T\}dA$$

$$= \int_{\Gamma^e} \{t_i\,\delta u_i + q\delta T\}ds \tag{33}$$

$$\int_{\Omega^e} p\delta p\,dA = \int_{\Omega^e} -\lambda\delta p u_{i,i}\,dA \tag{34}$$

The explicit introduction of interpolation functions for the pressure provides great freedom in the construction of the penalty functional and directly identifies the discrete approximation of the constraint. It can be shown (see [51]) that to every penalty finite element approximation there corresponds an equivalent mixed approximation (velocity-pressure). Consequently, the consistent formulation provides for the construction of very accurate and stable penalty finite element approximations by using the same interpolation used in the best mixed methods [51].

3.3.2 Finite element equations.

The velocity, temperature and pressure are interpolated over a typical element Ω^e according to the expressions

$$u_i = \phi^T \mathbf{U}_i \tag{35a}$$

$$T = \theta^T \mathbf{T} \tag{35b}$$

$$p = \psi^T \mathbf{P} \tag{35c}$$

where $\mathbf{U}_i$, $\mathbf{T}$, and $\mathbf{P}$ are vectors of element nodal unknowns and ϕ, θ, ψ are vectors of element interpolation functions.

Equations (35) are substituted into Eqs. (33) and (34), the pressure unknowns are solved for in Eq. (34) in terms of $\mathbf{U}_i$ and substituted into Eq. (33), thus eliminating the pressure. The resulting equations can be written as a single matrix equation for 2-D problems (extension to 3-D proceeds similarly)

$$
\begin{bmatrix}
2K_{11} + K_{22} + \lambda^{-1}C_1M_p^{-1}C_1^T & K_{12} + \lambda^{-1}C_1M_p^{-1}C_2^T & B_1 \\
K_{21} + \lambda^{-1}C_2M_p^{-1}C_1^T & K_{11} + 2K_{22} + \lambda^{-1}C_2M_p^{-1}C_2^T & B_2 \\
0 & 0 & L_{11} + L_{22}
\end{bmatrix}
\begin{bmatrix} U_1 \\ U_2 \\ T \end{bmatrix}
$$

$$
+ \begin{bmatrix}
A_1(U_1) + A_2(U_2) & 0 & 0 \\
0 & A_1(U_1) + A_2(U_2) & 0 \\
0 & 0 & D_1(U_1) + D_2(U_2)
\end{bmatrix}
\begin{bmatrix} U_1 \\ U_2 \\ T \end{bmatrix}
$$

$$
= \begin{bmatrix} F_1 \\ F_2 \\ G \end{bmatrix} \tag{36}
$$

The coefficient matrices shown in Eq. (36) are defined by:

$$
C_i = \int_{\Omega^e} \frac{\partial \phi}{\partial x_i} \psi^T dA
$$

$$
M_P = \int_{\Omega^e} \psi \psi^T dA
$$

$$
K_{ij} = \int_{\Omega^e} \mu \frac{\partial \phi}{\partial x_j} \frac{\partial \phi^T}{\partial x_i} dA
$$

$$
L_{ij} = \int_{\Omega^e} k \frac{\partial \theta}{\partial x_i} \frac{\partial \theta^T}{\partial x_j} dA
$$

$$
A_i(U_j) = \int_{\Omega^e} \rho \phi u_j \frac{\partial \phi^T}{\partial x_i} dA
$$

$$
D_i(U_j) = \int_{\Omega^e} \rho c_P \theta U_j \frac{\partial \theta^T}{\partial x_i} dA
$$

$$
B_i = \int_{\Omega^e} \rho g_i \beta \phi \theta^T dA
$$

$$
F_i = \int_{\Gamma^e} t_i^* \phi ds + \int_{\Omega^e} \rho f_i \phi dA + \int_{\Omega^e} \rho g_i (1 + \beta T_\beta) \phi dA
$$

$$
G = - \int_{\Gamma^e} q^* \theta ds + \int_{\Omega^e} q_s \theta dA + \int_{\Omega^e} \mu \Phi \theta dA
$$

The model just described allows great freedom in the choice of the velocity and pressure approximations. Unfortunately, if a stable and accurate formulation is sought, one cannot mix any approximation of the velocity with any approximation of the pressure. There are mathematical restrictions for proper approximation of a divergence-free velocity field expressed through the so-called "Ladyzhensakaya Babuska-Brezzi or LBB" condition [47–51].

For planar and axisymmetric problems, the 9-noded biquadratic velocity and linear pressure element $(Q_2 - P_1)$ [48–50] provides the most accurate and most effective solutions to the Navier-Stokes equations. This element is free of any parasitic pressures and provides accurate pressure solutions. This element is used in the numerical solution of problems presented in this paper.

3.3.3 Computational savings of the penalty method.

To illustrate the computational savings afforded by the penalty method, we present the characteristics of the global stiffness matrix for three problems using penalty and velocity-pressure formulations: flow in an annulus (Problem 1), flow in a square cavity (Problem 2), and flow in a complex geometry (Problem 3). The grids used are shown in Fig. 3. The number of nodes, elements, bandwidth, etc. for the three problems are summarized in Table 2. As can be seen from Table 2, the penalty formulation results in a reduction of 20% in the number of unknowns, 35% in the size of the stiffness matrix, and 45% in the solution time. The savings are significant.

3.4 Solution Procedure

The element equations are assembled into the global system matrix in the standard fashion [48,52]. Because of the presence of the nonlinear convective terms, the resulting system of algrebraic equations is nonlinear, and an iterative solution scheme must be used to solve them. The most frequent methods are: successive substitution (Picard iteration), Newton-Raphson, and Quasi-Newton [53]. In the Picard iteration method the nonlinear terms for the current iteration are evaluated using the solution from the previous iteration. This scheme has a fairly large radius of convergence, but the rate of convergence can be very low. The Newton-Raphson method has a superior rate of convergence. Its convergence rate is quadratic as long as the initial solution vector is within the radius of convergence. Unfortunately, the radius of convergence of the Newton-Raphson method is much smaller than that of successive substitution. For both methods, the *L-U* factorization must

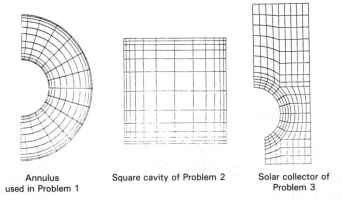

Annulus
used in Problem 1   Square cavity of Problem 2   Solar collector of
                                                                        Problem 3

Figure 3 Grids for the three sample problems.

Table 2 Comparison of the storage and computational times required by the mixed and penalty finite element models

Variable	Problem 1		Problem 2		Problem 3	
	Penalty	UVP	Penalty	UVP	Penalty	UVP
NUMNP	667	667	625	625	779	779
NELEM	154	153	114	144	178	178
NEQ	1753	2205	1633	2065	2000	2534
NEM	335,203	495,541	340,003	503,095	434,532	672,600
MINB	96	112	104	122	109	130
MAXB	135	168	147	183	772	976
TLU	690	1190	750	1312	1015	1830

NUMNP = Number of nodes; NELEM = number of elements; NEQ = number of degrees of freedom (excluding Dirichlet boundary conditions); NEM = size of the stiffness matrix in words; MINB = average half-bandwidth of the matrix; MAXB = maximum half-bandwidth; TLU = factorization time of the stiffness matrix on IBM-3081 in seconds.

be performed at each iteration, which is a very costly procedure for large problems. In the Quasi-Newton update procedure the L-U decomposition of the Jacobian of the system of nonlinear equations is updated in a simple manner at each iteration. Details are given in [53]. This algorithm has superlinear convergence, and in practice its rate of convergence approaches that of Newton-Raphson, while the cost of one Quasi-Newton iteration is typically 10 to 20% of that of a Newton-Raphson iteration. The reader is referred to [53] and [71–74] for complete details on the mathematical foundation and the implementation of the methods.

The following strategy was found suitable throughout the present study:

1. Perform one to three steps of successive substitution with a reasonable initial guess of the velocity field: either a Stokes solution or the solution of a problem at lower Rayleigh number to bring the solution within the radius of convergence of Quasi-Newton methods.
2. Switch to Quasi-Newton and iterate until convergence.

When this algorithm is applied, the following linear matrix equation is obtained:

$$Ax = b$$

where the coefficient matrix A is large (usually several thousand equations), sparse, banded and unsymmetric. A compacted skyline storage mode is adopted to store the matrix. Only coefficients between the leftmost and rightmost nonzero entries in each row are stored. All equations corresponding to known degrees of freedom are eliminated. For most problems, the matrix is too large to fit in core even with the use of such a storage scheme. The matrix is segmented into blocks and stored on low speed disk drives [54], and the system of equations is solved by direct Gaussian elimination.

4 APPLICATIONS

We now present a few sample calculations illustrating the two dimensional capability of the method described previously for solving the Navier-Stokes equations. A modified version of the code FIDAP [55] was used to perform most of these simulations. FIDAP is a multi-purpose code that allows the user to formulate and solve his problem with a wide range of methods and options to suit his individual needs and preferences.

4.1 The Square Cavity

The problem of interest here is the unit square cavity containing a viscous fluid. The two vertical walls are kept at a uniform temperature differing by a constant amount ΔT, the left wall being hotter. The horizontal faces are insulated. The Prandtl number if kept constant at 0.72 and the Rayleigh number assumes the following range of values: 10^4, 10^5, 10^6, and 10^7.

The cavity is covered by a rectangular mesh of graded quadratic elements (see Fig. 3). For Ra $= 10^4$, the velocity field and streamlines are nearly symmetric as shown in Fig. 4. No sharp boundary layers can be seen. The flow at Ra $= 10^4$ shows the onset of a stratified core in the cavity, a behavior characteristic of higher Rayleigh number flows. At Ra $= 10^5$ (see Fig. 5) three major changes can be seen:

1. Secondary circulation exists in the core.
2. There is a substantial temperature gradient near the vertical walls.

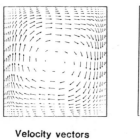

Velocity vectors

Streamline contours

Isotherms

Vorticity contours

Figure 4 Square cavity at Ra $= 10^4$, Pr $= 0.72$.

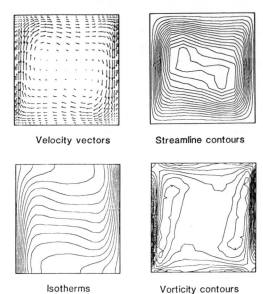

Velocity vectors   Streamline contours

Isotherms   Vorticity contours

Figure 5 Square cavity at Ra = 10^5, Pr = 0.72.

3. The lower left and upper right corners have become much more active than the other two.

The basic flow pattern is established, and no new features appear as the Rayleigh number is increased to 10^6 and 10^7 (see Figs. 6 and 7). Noticeable in these flows is the intensification of the vortices and the thinning of the boundary layers.

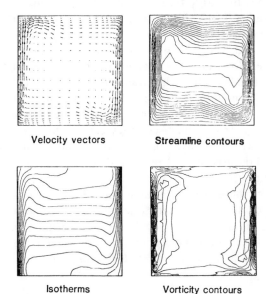

Velocity vectors   Streamline contours

Isotherms   Vorticity contours

Figure 6 Square cavity at Ra = 10^6, Pr = 0.72.

Velocity vectors

Streamline contours

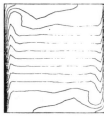

Isotherms

Vorticity contours

Figure 7 Square cavity at Ra = 10^7, Pr = 0.72.

The boundary layer thickness is estimated as (see [56])

$$\frac{\delta}{D} \sim Ra^{-1/4}$$

where D is the length of the side of the cavity. Thus, for Ra = 10^6 a boundary layer thickness of 0.0316 is predicted with 0.0178 for Ra = 10^7. The node closest to the boundary is located at a distance of 0.012 off the wall, and hence, the mesh chosen is not fine enough to resolve properly the fluid motion and thermal characteristics of the more difficult flows. Also, the mesh is not dense enough in the corners to capture the complex corner flow. Figure 8 contains the dimensionless heat transfer coefficient, the Nusselt number, on the hot wall. For Ra = 10^3, 10^4 and 10^5, the predictions of the Nusselt number are in excellent agreement with the accurate predictions of [35,57]. At higher Rayleigh numbers the Nusselt number prediction is higher than that of Ref. [35]. This discrepancy can be attributed to the following reasons: (1) the coarse mesh used in the corners and (2) the standard Gauss point evaluation method used here (Upson et al. [50] argue that their consistent flux technique is more accurate).

The results are also compared with those of [35,75,76] in Table 3. The results of [75] are the standard against which all other numerical techniques have been compared. Table 3 is a summary of the maximum velocities and the location at which they occur. The average Nusselt number was computed to be 9.1 for the case where Ra = 10^6. Reference [35] obtained a value of 9.1699, and [75] quotes 9.027. The values are seen to be in good agreement.

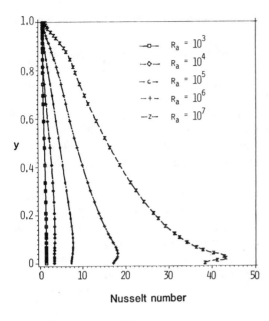

Figure legend:
- $R_a = 10^3$
- $R_a = 10^4$
- $R_a = 10^5$
- $R_a = 10^6$
- $R_a = 10^7$

Nusselt number

Figure 8 Hot wall Nusselt number for the square cavity.

4.2 Convection in an Inclined Cavity

Following [35], a series of computations were performed with the cavity tilted at 30, 45, 60, and 90 degrees. The cavity in the previous simulation corresponds to 0 degrees, while Benard Flow [57] corresponds to 90 degrees. Each tilted cavity simulation used a solution at a smaller tilt angle as an initial guess. The Benard

Table 3 Locations of velocity maxima

Ra	Source	V_{max} on $y = 0.5$	U_{max} on $x = 0.5$
10^3	Present	3.7 at 0.180	3.65 at 0.811
	[35]	3.704 at 0.166	3.656 at 0.812
	[75]	3.697 at 0.178	3.649 at 0.81
	[76]	3.69 at 0.18	3.64 at 0.813
10^4	Present	19.6 at 0.121	16.16 at 0.822
	[35]	19.675 at 0.1187	16.193 at 0.822
	[75]	19.167 at 0.119	16.718 at 0.823
	[76]	19.7 at 0.12	16.2 at 0.82
10^5	Present	68.9 at 0.0672	34.97 at 0.857
	[35]	68.896 at 0.0663	34.62 at 0.856
	[75]	68.59 at 0.066	34.73 at 0.855
	[76]	68.6 at 0.066	34.8 at 0.86
10^6	Present	221 at 0.0376	64.92 at 0.855
	[35]	220 at 0.0237	64.593 at 0.888
	[75]	216.36 at 0.0379	64.63 at 0.85
	[76]	222 at 0.039	63.9 at 0.85

results for Ra $= 10^6$ were obtained by working up from a lower Rayleigh number solution.

The solution at 30 degrees (see Fig. 9) clearly shows major changes from its 0 degree counterpart. The central core is becoming isothermal; it is no longer stably stratified. The thermal boundary layer has thickened, and this results in a lower Nusselt number (see Fig. 10).

Further increases of the tilt angle to 45 and 60 degrees (see Figs. 11–12) result in a continuation of the flow pattern established at 30 degrees, except for the unicellular-characteristic and an essentially isothermal core. The Nusselt number further decreases (see Fig. 10). The velocity field and streamlines are approaching overall symmetry.

The velocity field of the Benard solution at low Ra (see Fig. 13) qualitatively resembles that obtained at 0 degrees inclination. Closer investigation, however, reveals the possibility of recirculation eddies in the corners. At Ra $= 10^5$ (see Fig. 14), the presence of the eddies is confirmed in the velocity vector plot. At Ra $= 10^6$, the strength and extent of the recirculation bubbles have increased. The core of the cavity is almost isothermal and irrotational as illustrated in Fig. 15. Although quite weak, these recirculation bubbles significantly affect the heat transfer.

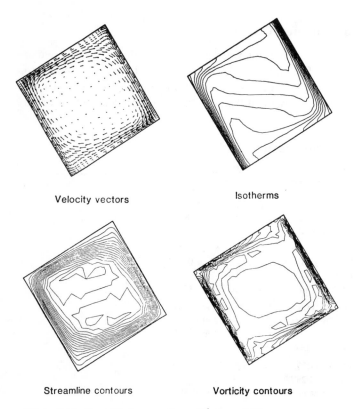

Velocity vectors   Isotherms

Streamline contours   Vorticity contours

Figure 9 Cavity at 30 degrees, Ra $= 10^6$, Pr $= 0.72$.

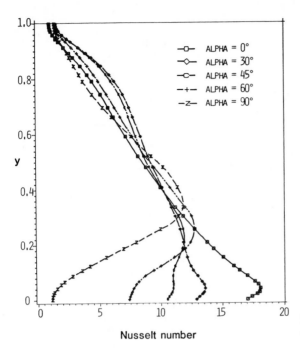

Nusselt number

Figure 10 Hot wall Nusselt number for cavities at an angle.

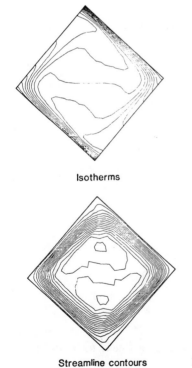

Isotherms

Streamline contours

Figure 11 Cavity at 45 degrees, Ra $= 10^6$, Pr $= 0.72$.

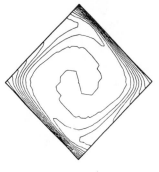

Isotherms

Streamline contours

Figure 12 Cavity at 60 degrees, Ra = 10^6, Pr = 0.72.

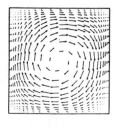

Velocity vectors

Streamline contours

Isotherms

Vorticity contours

Figure 13 Benard flow, Ra = 10^4, Pr = 0.72.

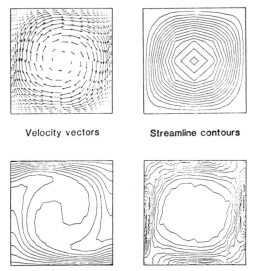

Velocity vectors   Streamline contours

Isotherms   Vorticity contours   **Figure 14** Benard flow, Ra = 10^5, Pr = 0.72.

The maximum heat-transfer coefficient has markedly decreased and moved to the center of the wall (Fig. 16).

The increase in Ra from 10^4 to 10^6 produces smooth increases in the Nusselt number for the three Benard flows. Thus, if maximization of the heat-transfer rate is of primary concern, the designer should avoid any inclined configurations.

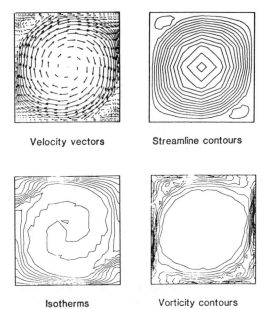

Velocity vectors   Streamline contours

Isotherms   Vorticity contours   **Figure 15** Benard flow, Ra = 10^6, Pr = 0.72.

4.3 Convection in an Annulus

This example is typical of a problem arising in the design of an annular receiver tube for a solar collector. A number of simplifying assumptions were made: the flow is laminar, the fluid has constant properties, and the boundaries are maintained at constant temperature. In spite of these simplifications, the simulation is still representative of the difficulties encountered. The Rayleigh number is based on the inner cylinder diameter.

Results of the analysis of natural convection in an annulus between two concentric cylinders are presented in Figs. 17–18. Due to symmetry with respect to the vertical centerline, only half of the annulus was considered. The ratio of outer to inner radius is 2.6, the same as that used in [57]. The inner cylinder is hot and the outer one is cold. The annulus is covered with a mesh of quadratic elements (see Fig. 3). Figures 17–18 contain results for $Ra = 10^4$ and 5×10^4 with $Pr = 0.706$. For these cases, experimental results are available in [13]. The streamlines and isotherms obtained in this work compare very well with those of [13]. As can be seen even at moderate Ra, the boundary layers are very thin. Mesh refinement will be required for solutions at higher Rayleigh numbers. The present results show the increasing importance of convective heat transfer with increasing Rayleigh numbers.

Figures 19 and 20 contain results on the effect of eccentricity on the previous problem. The Rayleigh number is based on inner cylinder diameter. Convergence difficulties were encountered at Rayleigh numbers greater than 10^4. The center of the flow has moved upward; for a given Rayleigh number, the effect on convec-

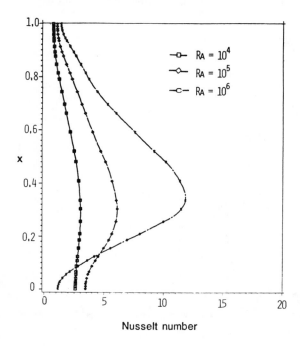

Nusselt number

Figure 16 Hot wall Nusselt number for Benard flow.

Velocity vectors

Streamline contours

Isotherms

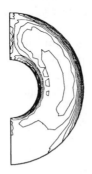

Vorticity contours

Figure 17 Annulus, Ra = 10^4, Pr = 0.706.

tion is stronger for the eccentric case, as can be seen from the curvature of the isotherms.

4.4 A More Complicated Geometry

This example was treated by Gartling and Nickel [11] and illustrates the versatility of the FEM in handling relatively complex geometries. The mesh of quadratic elements used is shown in Fig. 3. This geometry is typical of the end part of a flat solar collector. For simplicity, symmetry was invoked. The Rayleigh number is based on the cylinder radius (Rayleigh numbers based on the diameter would be 8 times higher). Results are presented for Ra = 10^3 and 10^4 in Figs. 21–22. Convergence difficulties were experienced at higher Rayleigh numbers, indicating the need for mesh refinement near the solid walls. As can be seen, the effects of convection increase with increasing Rayleigh number.

4.5 Free Convection near a Vertical Flat Plate with a Discontinuous Wall Temperature Distribution

The problem is that of free convection near a vertical flat plate. The bottom half of the plate is held at a temperature T_1, while the top half is maintained at a temperature T_2. The equations of motion are nondimensionalized with respect to H_0, the half-length of the plate, and $\Delta T = T_1 - T_0$ using $U_0 = \alpha/[H_0(\mathrm{RaPr})^{1/2}]$. The mesh and boundary conditions are shown in Fig. 23. The aim of the calculations was to simulate the experimental work of [59]. The experiments were run at a Grashof number of 3×10^7, which is a very high value making this simulation difficult.

The first of several simulations was that of the case for which $T_1 = T_2 = 1$, the flat plate at a uniform temperature. In order to obtain solutions at high Rayleigh numbers, a sequence of simulations at increasing values of Rayleigh numbers were run with $\mathrm{Ra} = 10^3$, 10^4, 5×10^4, and 10^5. Convergence difficulties were encountered at Rayleigh numbers greater than 10^5. We were unable to obtain convergence at $\mathrm{Ra} = 2.5 \times 10^5$ or 5×10^5. A few attempts with a more refined mesh were also unsuccessful. It is not clear yet whether the difficulties are due to an inappropriate mesh or to the absence of a steady state solution. Solutions obtained for $\mathrm{Ra} = 10^5$ are now presented for various temperature distributions. Figures 24–25 contain the solution found for $T_1 = T_2 = 1$. The thin momentum boundary layer is seen on the velocity vector plot, while the streamline contour plot clearly shows the entrainment due to buoyancy forces. The temperature con-

Velocity vectors

Streamline contours

Isotherms

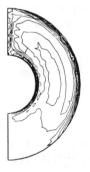

Vorticity contours

Figure 18 Annulus, $\mathrm{Ra} = 5 \times 10^4$, $\mathrm{Pr} = 0.706$.

Velocity vectors   Streamline contours

Figure 19 Offset cylinders, Ra = 10^3, Pr = 0.706.

Isotherms   Vorticity contours

tour plot shows the nice plume characteristic of high Rayleigh number flows. This solution compares very well with the analytical boundary layer solution of [44] when comparisons are made away from the leading or trailing edges. The pressure contour plots indicate the presence of strong leading and trailing edge singularities.

Simulations were also carried out with various combinations of temperatures. No difficulties were encountered in obtaining a solution with $T_1 = 1$ and $T_2 = 1.25$. The only difference caused by the increased buoyancy forces on the top half of the plate is a slight thinning of the velocity and thermal boundary layers (see Fig. 26).

Next, we tried decreasing the value of T_2. This results in lower buoyancy forces on the top half of the plate, so the flow is retarded on the top part of the plate. As long as T_2 is greater than 0 (the ambient temperature), the flow will maintain its boundary layer character. When T_2 is decreased below 0 (the top half of the plate is cooled) the buoyancy forces will oppose the vertical flow, and when T_2 is low enough, separation should occur. Obtaining solutions for values of T_2 lower than T_1 proved to be a difficult task. The nature of the flow changes rapidly and strongly with decreasing values of T_2. It was necessary to carry out the sim-

Velocity vectors

Streamline contours

Isotherms

Vorticity contours

Figure 20 Offset cylinders, Ra = 10^4, Pr = 0.706.

Streamline contours

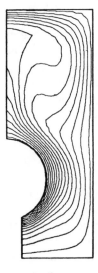

Isotherms

Figure 21 Complex geometry, Ra = 10^3, Pr = 0.72.

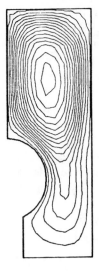

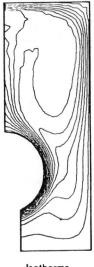

Streamline contours   Isotherms

Figure 22 Complex geometry, $Ra = 10^4$, $Pr = 0.72$.

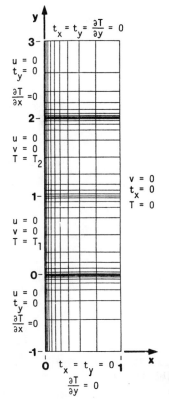

Figure 23 Grid and boundary conditions for flat plate with discontinuous temperature.

ulation with the following sequence of values of T_2 obtained from trial and error: 0.9, 0.8, 0.5, 0.2, 0.1, 0.05, 0.0, -0.05, -0.1, -0.15, -0.2, -0.35, -0.45, -0.5, -0.52, -0.55, and -0.57. We were unable to obtain solutions for T_2 less than -0.57 with the current mesh.

As T_2 is reduced to 0.5, a stagnation point appears at the point of discontinuity in temperature, as evidenced in the pressure contour plot shown in Fig. 27. The isotherms indicate the appearance of a very weak plume. With T_2 further reduced to 0, the boundary layer thickens on the top half of the plate (see Fig. 28). It can be seen that the streamlines are moving away from trailing edge of the plate (compare with Fig. 24). Figures 29 and 30 show that the plume strengthens and that the midplate stagnation point influence extends to most of the flat plate. Clearly, this regime could not be properly described within the boundary layer approximation.

Further reduction of T_2 enhances the trends described previously until T_2 reaches a value of -0.35, the point of incipient separation. The boundary layer is very

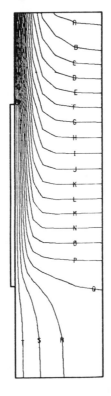

Velocity vectors   Streamline contours

Figure 24 Flat plate $T_1 = T_2 = 1$, Ra $= 10^5$, Pr $= 0.7$.

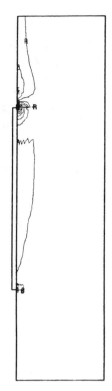

Isotherms   Pressure contours

Figure 25 Flat plate $T_1 = T_2 = 1$, Ra = 10^5, Pr = 0.7.

thick and shows points of inflection (see Fig. 31). The thermal plume shows signs of changing into a jet, and strong pressure gradients can be seen (see Fig. 32).

The flow separates for T_2 lower than -0.35. For $T_2 = -0.5$, a strong jet and recirculation bubble appear, and the thermal plume shows some very complex features (see Fig. 33). The surface pressure assumes a very complicated form; pressure peaks increase in magnitude, and the trailing edge region feels some very high pressure gradients due to the presence of the recirculation bubble (see Fig. 34). Due to the size of the recirculation bubble and the extent of the perturbation in the flow field, the current mesh is very likely inappropriate. The mesh should extend further above the plate, and it should be refined to better capture the features of the flow.

As mentioned earlier, tracking the solution as the temperature T_2 was changed was a very difficult task. Parametric studies of this type are current practice in a design environment where assessment of the impact of various design changes must be performed routinely. It is therefore desirable to investigate means of automating the solution tracking procedure. Some answers may be found in the use of parameter-stepping and bifurcation search techniques [60].

4.6 Flat Plate in a Channel

This problem illustrates the effect of changing the free stream boundary conditions to those of a solid wall (no slip). This configuration is very similar to that of fin-cooled electronic equipment. The value of Ra is 10^5 and $T_2 = T_1 = 1$. Figure 35 shows that the velocity profile is much fuller in the plate region. The vertical pumping action is enhanced by the presence of the right hand sidewall. The thermal boundary layer is thinner than its infinite medium counterpart.

Finally, we show the solution obtained for the above configuration with Ra $= 10^3$, $T_1 = 1$ and $T_2 = -1$. Because of the thermal inversion, the flow is downward at the trailing edge and upward at the bottom leading edge. The two boundary layers collide at the midsection of the plate and stream outward from the plate in a jet-like fashion. This behavior was observed in [59]. The isotherms show perfect symmetry (see Fig. 36).

5 SUGGESTIONS FOR FUTURE RESEARCH

A number of the points presented here were previously identified in a recent workshop on free convection [61]. Even though that report emphasized the need for

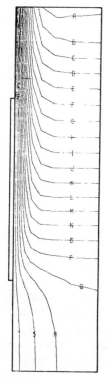

Streamline contours   Isotherms   **Figure 26** Flat plate $T_1 = T_2 = 1.25$.

 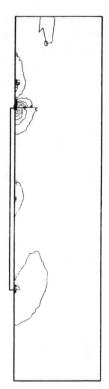

Isotherms   Pressure contours   **Figure 27** Flat plate $T_1 = 1$, $T_2 = 0.5$.

gathering experimental data for numerous problems, we feel that parallel numerical investigation will be very useful. Such calculations may uncover some unexpected physical phenomena and help focus the effort of experimentalists. Due to the wide variety of free convection problems encountered in practice, there is no doubt that new numerical difficulties will emerge, and they will require solution to help further our understanding of free convection phenomena. We outline here but a few of the interesting areas of research that require attention.

5.1 Boundary Layer Flows

The limitations of first order boundary layer theory need to be explored with reference to boundary layers with separation, stagnation points and flow reversals. This is also true for flows with discontinuous boundary conditions and other than quiescent ambient conditions. Predictive tools and experimental results are needed for these flows.

Many buoyancy-induced boundary layers that occur in nature and in applications are unsteady, three-dimensional, or conjugate problems. Much remains to be done with regard to these complex boundary layer flows. Again experimental results accompanied by predictive tools are required.

For turbulent boundary layers, experimental data is needed which will support modelling efforts. In addition, the database should be extended to include high Grashof numbers and variable fluid properties. Turbulent boundary layers with chemical reaction are of considerable interest both from the practical and from the academic standpoint.

Since natural convection plays an important role in many engineering problems as indicated in the Introduction, there exists a long list of topics which deserve attention. These include enhancement of buoyancy-induced flows, high Ra cases, electric and magnetic effects, variable properties and the adequacy of the Boussinesq approximation, double diffusion systems, low Prandtl number effects, interfacial effects, and natural convection with change in phase.

5.2 Convection in Enclosures

From a recent workshop [61], the state of the art of natural convection relevant to the present discussion can be summarized as follows:

- The effect of tilt is roughly understood: one can use certain scaling laws to obtain approximate heat transfer results.

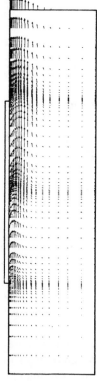

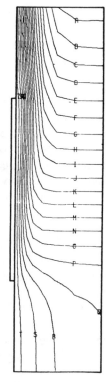

Velocity vectors   Streamline contours   **Figure 28** Flat plate $T_1 = 1$, $T_2 = 0$.

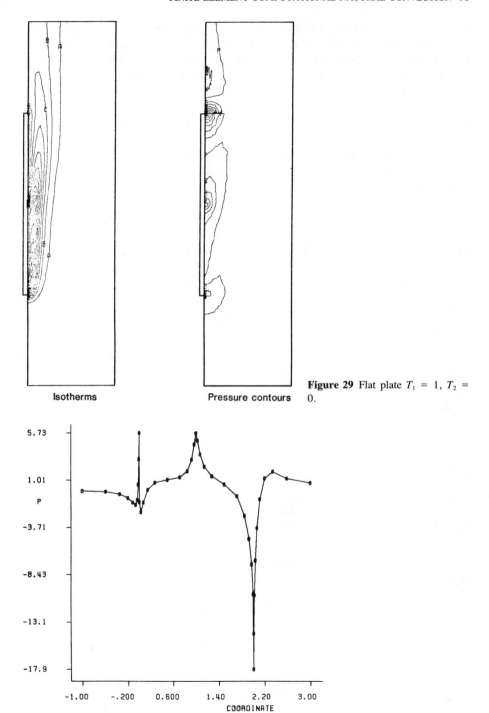

Isotherms   Pressure contours

Figure 29 Flat plate $T_1 = 1$, $T_2 = 0$.

Figure 30 Flat plate $T_1 = 1$, $T_2 = 0$, surface pressure.

- Very little work has been done either experimentally or analytically for cavities containing baffles or partial divisions particularly for large systems such as rooms.
- There are too many incompletely defined two-dimensional studies.
- Data for complicated engineered geometries are almost nonexistent.
- Doubts exist about what equations to use in analysis of natural convection in porous media.
- Very little heat transfer data are available for low Prandtl number fluids.

Many researchers operate in an assumed two-dimensional world. As a result, a great deal of data and many theoretical predictions exist without knowing exactly under what circumstances they are valid. The whole question of 2-D or 3-D approximations (i.e., representations) needs attention.

Thermal boundary conditions (wall conductance) on all the enclosure boundaries need further study. The thermal boundary conditions can lead to stabilizing or destabilizing the flow and corresponding large changes in the heat transfer coefficient. Both experimental and theoretical work is needed.

At all angles of inclination, stability bounds are needed for the onset of trans-

Velocity vectors

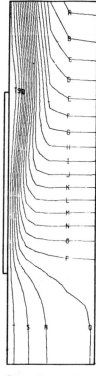

Streamline contours

Figure 31 Flat plate $T_1 = 1$, $T_2 = -0.35$.

Isotherms

Pressure contours

Figure 32 Flat plate $T_1 = 1$, $T_2 = -0.35$.

verse or longitudinal rolls. The effects of Prandtl number and aspect ratio need to be delineated. In the heated from above case, heat transfer measurements are needed at Rayleigh numbers above the stability boundary.

At low W/L or H/L, wall conductance has a very strong effect on heat transfer. Limited measurements have been reported only for almost insulating or almost perfectly conducting walls. No heat transfer data exists for low W/L and H/L with $W < H$ and angles of tilt.

There are very few high Rayleigh number data for any enclosures. Such data are needed for design. Almost no low Prandtl number data are available.

Enclosures in communication with one another are common engineering systems that have received almost no attention. Very little work has been done either experimentally or analytically for enclosures containing baffles or partial divisions. This work would address convection of air in and between rooms in passive solar design applications, for example.

Future research needs are of two types. First, one must address those areas where the research results have direct application to design of a thermal device or system. Second, on a different time scale, one can look into the physical processes governing the phenomena of interest.

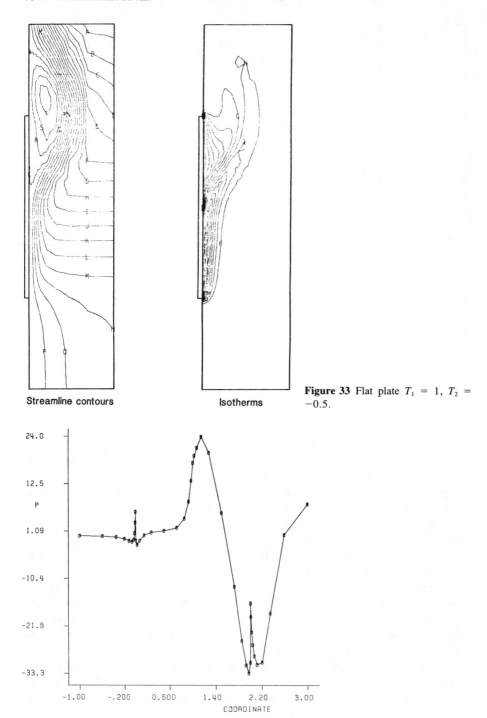

Streamline contours   Isotherms

Figure 33 Flat plate $T_1 = 1$, $T_2 = -0.5$.

Figure 34 Flat plate $T_1 = 1$, $T_2 = -0.5$, surface pressure.

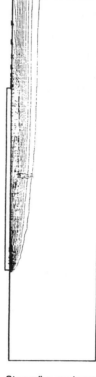

Velocity vectors   Streamline contours

Figure 35 Flat plate in a channel T_1 $= T_2 = 1$, Ra $= 10^5$.

A large number of geometries require thermal analysis. A study should be initiated whose goal is to unify in some manner heat transfer correlations for non-simple geometries.

All the experimental efforts should be accompanied by analysis. Under most circumstances this means solution by numerical techniques. There is little doubt that computational studies will provide much needed information on these topics.

5.3 Mixed Convection

Mixed convection processes may be considered in terms of external flow over surfaces and in free boundary flows, such as plumes, wakes and buoyant jets, and in terms of internal flows in tubes and channels and in enclosed fluid regions. The external flow problems have received much greater attention because of greater simplicity in analysis and in experimentation as compared to internal flows. However, internal mixed convection flows are also of particular interest in heat exchangers, in reactors, in electronic systems and in heat rejection and energy storage systems. Consequently, recent years have seen a growing interest in internal mixed convection. A very important consideration in the study of mixed convec-

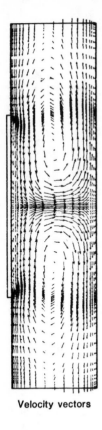

Velocity vectors

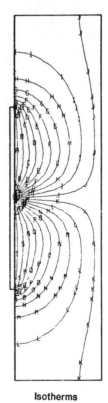

Isotherms

Figure 36 Flat plate in a channel T_1 = 1, T_2 = 1, Ra = 10^3.

tion is that of the limits of the flow conditions and regions for which neither the forced nor the natural convection mechanism are dominant and both need to be considered in conjunction.

Some of the pressing needs in the mixed convection area can be stated as:

1. Flow separation in mixed convection flows due to the opposing flow situation needs a detailed investigation, both analytically and experimentally. The non-boundary layer effects and reverse flow near separation need to be considered. Some results were presented in this chapter.
2. There is considerable need for work on three-dimensional flow problems, such as cross flow over spheres, vertical cylinders and plates. These problems are generally of greater practical interest. New and more efficient numerical methods need to be studied for three-dimensional flows, and new approximate analytical approaches should also be considered. The stability of three-dimensional flows is particularly relevant to the actual situation and needs to be considered.
3. Analytical and experimental work is needed on turbulent mixed convection flows, even for flat surfaces. Turbulence models, such as the $k - \epsilon$ model, employed for natural and forced convection flows, need to be extended to

these flows. Here, good experimental data is sorely needed to determine the effect of buoyancy on turbulence and to provide inputs for the turbulence modeling efforts.

4. Mixed convection due to moving surfaces has received little attention, though this flow is of interest in many industrial applications such as rotating machinery and extrusion processes. The transient effects at the onset of the motion as well as the steady state circumstance are of interest.

5. Mixed convection in enclosures, particularly for turbulent flows, is in need of a detailed investigation. These flows are of interest in heat rejection to water bodies, energy storage and extraction systems such as solar ponds, nuclear power systems and in fires in rooms with forced ventilation. The transient effects in these flows also need to be considered.

5.4 Turbulence Modeling

Direct simulation of practical turbulent flows will not be tractable for the foreseeable future, so modeling is necessary. Since the ultimate turbulence model is not expected, it is important to be able to identify the general features of a given problem, so that the simplest, and yet sufficiently accurate, model can be selected. Systematic research to enable us to do so by way of developing a catalog of tools should prove to be very fruitful.

Very little work on turbulence modeling is known for general three-dimensional flows. Formulating a $k - \epsilon - g$ model for three-dimensional cases would lead to so many additional field equations so as to make it impractical. Rational simplifications must be introduced, and their validity must be demonstrated by comparison with experimental data. This is one very challenging area of needed research in natural convection.

In the current development of new turbulence approaches mentioned previously, there are none ready to be applied directly to natural convection phenomena. Such an eventuality, however, should be kept in mind from a development point of view, so that natural convection studies can simultaneously benefit from any significant breakthrough in these developments. More specifically, there is a need to encourage communications with researchers in atmospherical sciences and geophysics to determine whether recent advances in these fields have a relevance to the development of turbulence models in buoyant flows.

5.5 Miscellaneous Problems

Other areas of potentially fruitful research include investigation of the validity of the Boussinesq approximation and effects of variable properties [62,63], the study of conjugate convection problems, and double diffusive systems [64]. The reader will also find in [64,65,66] a wealth of problems that have received little or no attention from an analytical or computational point of view. Other formulations and numerical approaches that prove to be accurate and/or efficient should be carefully evaluated [68–70]).

6 CONCLUSIONS

A finite element model based on the penalty function method was presented for the solution of the Boussinesq approximation of the Navier-Stokes equations. A variety of nondimensional forms of the equations of motion were presented and discussed. A number of applications outlining the robustness and flexibility of the penalty finite element method were outlined. The complex flow of natural convection near a flat plate with discontinuous temperature was solved at low and moderate Rayleigh numbers for a number of boundary conditions. This flow exemplifies the difficulties that the analyst is faced with in computational free convection. Finally a few directions for future research have been pointed out.

REFERENCES

1. S. Ostrach, Natural Convection in Enclosures, *Advances in Heat Transfer,* Vol. 8, pp. 161–227, 1972.
2. G. K. Batchelor, Heat Transfer by Free Convection Across a Closed Cavity Between Boundaries at Different Temperatures, *Quarterly Journal of Applied Mathematics,* Vol. 12, pp. 209–233, 1954.
3. J. W. Elder, Numerical Experiments with Free Convection in a Vertical Slot, *Journal Fluid Mechanics,* Vol. 24, pp. 823–843, 1965.
4. A. E. Gill, The Boundary Layer Regime for Convection in a Rectangular Cavity, *Journal Fluid Mechanics,* Vol. 26, pp. 515–536, 1966.
5. B. D. Pedersen, et al., Development of a Compressed Gas-Insulation Transmission Line, *IEEE Winter Power Meeting,* Paper 71 TP 193 PWR, 1971.
6. J. E. Hart, Stability of Thin Non-Rotating Hadley Circulation, *Journ. Atmos. Sci.,* Vol. 29, pp. 687–697, 1972.
7. J. P. Carruthers, Crystal Growth From Melt, *Treatise on Solid State Chemistry,* Vol. 5, Plenum Press, 1975.
8. B. S. Petuklov, Actual Problems of Heat Transfer in Nuclear Power Engineering, *Int. Seminar on Future Energy Production,* Hemisphere, pp. 151–163, Washington, D.C., 1976.
9. J. Hiddink, et al., Natural Convection Heating of Liquids in Closed Containers, *App. Sci. Res.,* Vol. 32, pp. 217–237, 1976.
10. D. K. Gartling, Convective Heat Transfer Analysis by the Finite Element Method, *Comp. Meth. Appl. Mech. Eng.,* Vol. 12, pp. 365–382, 1977.
11. D. K. Gartling, NACHOS—A Finite Element Computer Program for Incompressible Flow Problems, Sandia National Labs Report, SAND77-1333, Albuquerque, NM, 1978.
12. J. C. Chato, and R. S. Abdulhadi, Flow and Heat Transfer in Convectively Cooled Underground Electric Cable Systems; part 1—Velocity Distributions and Pressure Drop Correlations; part 2—Temperature Distributions and Correlations, *ASME Journ. Heat Transfer,* Vol. 100, pp. 30–40, 1978.
13. T. H. Kuehn, and R. J. Goldstein, An Experimental and Theoretical Study of Natural Convection in the Annulus Between Horizontal Cylinders, *Journ. Fluid Mech.,* Vol. 74, pp. 695–719, 1976.
14. D. Torok, Augmenting Experimental Studies for Flow and Thermal Field Prediction by the Finite Element Method, *Computers in Engineering,* 1984, Las Vegas, NV.
15. D. Torok, Augmenting Experimental Methods for Flow Visualization and Thermal Performance Prediction in Electronic Packaging using Finite Elements, Fundamentals of Natural Convection/Electronic Equipment Cooling, *2nd National Heat Transfer Conference,* Niagara Falls, NY, August 5–8, 1984.
16. M. S. Engelman, and R. L. Sani, Finite Element Simulation of an In-Package Pasteurization Process, *Numerical Heat Transfer,* Vol. 6, pp. 41–54, 1983.

17. A. J. Chorin, Numerical Solutions of the Navier-Stokes Equations, *Mathematics of Computation,* Vol. 22, pp. 745–762, 1968.
18. G. P. Williams, Numerical Integration of the Three-Dimensional Navier-Stokes Equations for Incompressible Flows, *J. Fluid Mech.* Vol. 37, pp. 727–750, 1969.
19. K. Aziz, and J. D. Hellums, Numerical Solution of the Three-Dimensional Equations of Motion for Laminar Natural Convection, *The Physics of Fluids,* Vol. 10(2), pp. 314–324, 1967.
20. G. D. Mallinson and G. de Vahl Davis, The Methods of the False Transient for the Solution of Coupled Elliptic Equations, *J. Comp. Phys.,* Vol. 12, pp. 435–461, 1973.
21. G. de Vahl Davis and G. D. Mallinson, An Evaluation of Upwind and Central Difference Approximations by a Study of Recirculating Flow, *Computers and Fluids,* Vol. 4, pp. 29–43, 1976.
22. G. D. Mallinson and G. de Vahl Davis, Three-Dimensional Natural Convection in a Box: A Numerical Study, *J. Fluid Mech.,* Vol. 83(1), pp. 1–31, 1977.
23. D. F. Roscoe, The Solution of the Three-Dimensional Navier-Stokes Equations Using a New Finite Difference Approach, *Int. J. Num. Meth. Eng.,* Vol. 10, pp. 1299–1308, 1976.
24. D. N. de G. Allen and R. V. Southwell, Relaxation Methods Applied to Determine the Motion, in Two Dimensions, of Viscous Fluid: Part A—Fixed Cylinder, *Quart. J. Mech. and Appl. Math.,* Vol. 8, pp. 129–145, 1955.
25. S. M. Richardson and A. R. H. Cornish, Solution of Three-Dimensional Incompressible Flow Problems, *J. Fluid Mech.,* Vol. 82, pp. 309–319, 1977.
26. S. C. R. Dennis, D. B. Ingham and R. N. Cook, Finite Difference Methods for Calculating Steady Incompressible Flows in Three Dimensions, *J. Comp. Phys.,* Vol. 33, pp. 325–339, 1979.
27. S. C. R. Dennis and J. D. Hudson, A Difference Method for Solving the Navier-Stokes Equations, *Proc. Int. Conf. Num. Meth. Laminar and Turbulent Flows,* Swansea, U.K., Pentech Press, London, 1978.
28. M. Nallasamy and K. K. Prasad, Numerical Studies on Quasilinear and Linear Elliptic Equations, *J. Comp. Phys.,* Vol. 15, pp. 429–448, 1974.
29. D. B. Spalding, A Novel Finite-Difference Formulation for Differential Expressions Involving Both First and Second Derivatives, *Int. J. Num. Meth. Eng.,* Vol. 4, pp. 551–559, 1972.
30. R. K. Agarwal, A Third-Order-Accurate Upwind Scheme for Navier-Stokes Solutions at High Reynolds Numbers, AIAA Paper No. 81-0112, 1981.
31. R. K. Agarwal, A Third-Order-Accurate Upwind Scheme for Navier-Stokes Solutions in Three Dimensions, McDonnell Douglas Research Laboratories, Report No. MDRL 81-20; ASME Winter Annual Meeting, Washington, D.C. 1981.
32. A. Ghosh, C. W. Mastin, and J. F. Thompson, Locally One-Dimensional Scheme for the Solution of Three-Dimensional Navier-Stokes Equations, in *Computers in Flow Predictions and Fluid Dynamics Experiments* (eds. K. N. Ghia et al.), American Society of Mechanical Engineers, New York, pp. 3–10, 1981.
33. B. Tabarrok and R. C. Lin, Finite Element Analysis of Free Convection Flows, *Int. J. Heat Mass Transfer,* Vol. 20, pp. 945–952, 1977.
34. D. K. Gartling, Convective Heat Transfer Analysis by the Finite Element Method, *Computer Meth. Appl. Mech. Eng.,* Vol. 12, pp. 365–382, 1977.
35. C. D. Upson, P. M. Gresho and R. L. Lee, Finite Element Simulations of Thermally Induced Convection in a Closed Cavity, Informal Report, UCID-18602, Lawrence Livermore National Laboratory, Livermore, CA, 1979.
36. J. C. Heinrich, R. S. Marshall, and O. C. Zienkiewicz, Penalty Function Solution of Coupled Convective Heat Transfer, *Int. Conf. Numer. Meth. in Laminar and Turbulent Flow,* Swansea, 1978.
37. R. S. Marshall, J. C. Heinrich and O. C. Zienkiewicz, Natural Convection in a Square Enclosure by a Finite-Element Penalty Function Method Using Primitive Fluid Variables, *Num. Heat Transfer,* Vol. 1, pp. 315–330, 1978.
38. J. N. Reddy, On Penalty Function Methods in the Finite Element Analysis of Fluid Flow, *Int. J. Numer. Meth. in Fluids,* Vol. 2, pp. 151–171, 1982.
39. J. N. Reddy, Penalty Finite Element Methods for the Solution of Advection and Free Convection

Flows, Finite Element Methods in Engineering, (ed. A. P. Kabaila and V. A. Pulmano). The University of New South Wales, Sydney, Australia, pp. 583–598, 1979.

40. J. N. Reddy and A. Satake, A Comparison of Various Finite Element Models of Natural Convection in Enclosures, *J. Heat Transfer,* Vol. 102, pp. 659–666, 1980.

41. P. M. Gresho, S. T. K. Chan, R. L. Lee and C. D. Upson, Solution of the Time-Dependent, Three-Dimensional Incompressible Navier-Stokes Equations Via FEM, *Proc. Int. Conf. Num. Meth. for Laminar and Turbulent Flow,* (ed. C. Taylor and B. A. Shreffler) Pineridge Press, Swansea, UK, pp. 27–39, 1981.

42. J. N. Reddy, Finite-Element Simulation of Natural Convection in Three-Dimensional Enclosures, *AIAA/ASME 3rd Joint Thermophysics, Fluids, Plasma and Heat Transfer Conference,* St. Louis, Mo., 1982.

43. J. N. Reddy, Penalty-Finite-Element Analysis of 3-D Navier-Stokes Equations, *Computer Meth. App. Mech. and Eng.,* Vol. 35, pp. 87–106, 1982.

44. S. Ostrach, An Analysis of Laminar Free-Convection Flow and Heat Transfer About a Flat Plate Parallel to the Direction of the Generating Body Force, *NACA Rept. 1111,* 1953.

45. A. Bejan, *Convection Heat Transfer,* John Wiley, New York, 1984.

46. S. Ostrach, Natural Convective Heat Transfer in Cavities and Cells, *J. Heat Transfer,* Vol. 1, pp. 365–378, 1982.

47. J. N. Reddy, *Applied Functional Analysis and Variational Methods in Engineering,* McGraw-Hill, New York, 1986.

48. J. N. Reddy, *An Introduction to the Finite Element Method,* McGraw-Hill, New York, 1984.

49. J. T. Oden, Penalty Methods and Reduced Selective Integration for Stokesian Flows, *Third Int. Conf. on Finite Elements in Flow Problems,* Banff, Canada, 1980.

50. M. S. Engelman, Sani, R. L., Gresho, P. M., Bercovier, M., Consistent vs. Reduced Integrative Penalty Methods for Incompressible Media Using Several Old and New Elements, *Int. J. Num. Meth. Fluids,* Vol. 2, pp. 25–42, 1982.

51. M. Bercovier, Perturbation of Mixed Variational Problem. Application to Mixed Finite Element Methods, *RAIRO Analyse Numerique,* Vol. 12, pp. 211–236, 1978.

52. O. C. Zienkiewicz, *The Finite Element Method,* 3rd Edition, McGraw-Hill, New York, 1977.

53. M. Engelman, G. Strang and K. J. Bathe, The Application of Quasi-Newton Methods in Fluid Mechanics, *Int. Journ. Num. Meth. Eng.,* Vol. 17, pp. 707–718, 1981.

54. Y. Hasbani and M. S. Engelman, Out of Core Solution of Linear Equations with Non-Symmetric Coefficient Matrix, *Comp. Fluids,* Vol. 7, pp. 13–31, 1979.

55. M. S. Engelman, FIDAP: A Fluid Dynamics Analysis Program, *Advances in Engineering Software,* October 1982.

56. A. E. Gill, The Boundary Layer Response for Convection in a Rectangular Cavity, *Journal of Fluid Mechanics,* Vol. 24, pp. 823–843, 1954.

57. J. N. Reddy, Penalty-Finite Element Methods in Conduction and Convection Heat Transfer, R. W. Lewis, K. Morgan, and B. A. Schrefler (eds.), in *Numerical Heat Transfer,* Vol. II, John Wiley, London, pp. 145–178, 1983.

58. D. J. Tritton, *Physical Fluid Dynamics,* Van Nostrand Reinhold, New York, 1977.

59. J. A. Schetz and R. Eichorn, Natural Convection with Discontinuous Wall-Temperature Variations, *J. Fluid Mech.* Vol. 18, Part 2, pp. 167–176, 1964.

60. C. P. Jackson and H. Wintersk, A Finite-Element Study of the Benard Problem using Parameter-Stepping and Bifurcation Search, *Int. J. Num. Meth. Fluids,* Vol. 4, pp. 127–145, 1984.

61. K. T. Yang and J. R. Lloyd, Proceedings of a Workshop on Natural Convection, July 18–21, 1982, Beaver Run, Brenchenridge, Colorado. Sponsored by N.S.F. and University of Notre Dame.

62. H. Shankatullahl and B. Gebhort, The Effect of Variable Properties on Laminar Natural Convection Boundary-Layer Flow Over a Vertical Isothermal Surface in Water, *Numerical Heat Transfer,* Vol. 2, pp. 215–232, 1979.

63. P. Vasseur, L. Robillard and B. Chandra Shekar, Nautral Convection Heat Transfer of Water Within a Horizontal Cylindrical Annulus with Density Inversion Effects, *Journal of Heat Transfer,* Vol. 105, pp. 117–123, 1983.

64. S. Ostrach, Natural Convection with Combined Driving Forces, *PCH Physics Chemical Hydrodynamics,* Vol. 1, pp. 233–247, 1980.
65. K. E. Torrance, (ed.), Natural Convection in Enclosures, 19th National Heat Transfer Conference, Orlando, Florida, July 27–30, 1980.
66. I. Catton and R. N. Smith, (ed.), Natural Convection, 20th ASME/AICHE National Heat Transfer Conference, Milwaukee, Wisconsin, August 2–5, 1981.
67. I. Catton and K. E. Torrance, Natural Convection in Enclosures—1983, 21st National Heat Transfer Conference, Seattle, Washington, July 24–28, 1983.
68. T. C. Chawla, G. Leaf and W. J. Minkowycz, A Collocation Method for Convection Dominated Flows, *Int. J. Numer. Meth. Fluids,* Vol. 4, pp. 271–281, 1984.
69. C. T. Yang and S. N. Atluri, An "Assumed Deviatoric Stress-Pressure-Velocity" Mixed Finite Element Method for Unsteady, Convective, Incompressible Viscous Flow, Part I: Theoretical Development, *Int. J. Numer. Meth. Fluids,* Vol. 3, pp. 377–398, 1983.
70. C. T. Yang and S. N. Atluri, An "Assumed Deviatoric Stress-Pressure-Velocity" Mixed Finite Element Method for Unsteady, Convective, Incompressible Viscous Flow, Part II: Computational Studies, *Int. J. Numer. Meth. Fluids,* Vol. 4, pp. 43–69, 1984.
71. J. Dennis and J. More, Quasi-Newton Method, Motivation and Theory, SIAM Review, Vol. 19, p. 46, 1965.
72. H. Matthies, and G. Strong, The Solution of Nonlinear Finite Element Equations, *Int. J. Num. Meth. Eng.,* Vol. 11, p. 1155, 1977.
73. K. J. Bathe, and A. Cimento, Some Practical Procedures for the Solution of Nonlinear Finite Element Equations, *Comp. Meth. Appl. Mech. Eng.,* Vol. 22, p. 59, 1980.
74. C. G. Broyden, A Class of Methods for Solving Nonlinear Simultaneous Equations, *Math. Comp.,* Vol. 19, p. 577, 1965.
75. G. de Vaht Davis, Natural Convection in a Square Cavity—A Benchmark Solution, *Int. J. Num. Meth. Fluids,* Vol. 3, pp. 249–264, 1983.
76. K. H. Winters, A Numerical Study of Natural Convection in a Square Cavity, United Kingdom Atomic Energy Authority, AERE-R9747, August 1980.

THREE

THERMAL STRATIFICATION MODELING FOR OCEANS AND LAKES

B. Henderson-Sellers and A. M. Davies

ABSTRACT

A review of thermal stratification models as applied to lakes and oceans suggests that there has, to date, been little interaction between applications to the two systems, despite the fact that the modeling approximations used in each sphere have resulted in almost identical model formulations. These can be further subdivided, in each case, into integral or mixed layer models and differential models employing a parametrization of eddy diffusion coefficients. Attention has been focused on contrasts between model formulations and applicability without introducing the several inter-model comparison case studies recently made in the literature (e.g., Martin, 1985). The computational requirements for stratification models intended for incorporation into coupled atmosphere-ocean general circulation climate models are also considered, since this is likely to be one of the imminent key areas of application for such thermodynamic models.

NOMENCLATURE

a_1, b_1	coefficients in Eq. (20)
a_2, b_2	coefficients in Eq. (21)
a_3, b_3	constants in Eqs. (105) and (106)
$A(z)$	cross-sectional area as a function of depth (lake model)
A_s	surface area (in lake model)
b	buoyancy
b'	fluctuations in buoyancy

B'	nondimensionalized buoyancy flux [Eq. (53)]
$B*$	buoyancy flux
B_0	surface buoyancy flux
c_D	aerodynamic drag coefficient
c_e	empirical coefficient in relating rate of change of turbulent kinetic energy (TKE) to friction velocity
c_i	velocity of long internal waves at thermocline interface
c_p	specific heat (at constant pressure)
c_1	constant in Eq. (26)
c_2	constant in Eq. (100) (= 15.0)
C	generalized drag coefficient [Eq. (36)]
C_m	heat capacity of mixed layer (ML)
C_K, C_S, C_T, Λ_L	empirical coefficients in Eq. (85)
d	compensation depth
D	dissipation
D'	nondimensionalized dissipation [Eq. (53)]
$D*$	dissipation [Eq. (52)]
D_T	rate of turbulent decay [Eq. (104)]
E	TKE produced at base of thermocline by interfacial shear
E_k	rate of mechanical energy/input
E_p	potential (buoyant) energy of the isothermal ML
E_w	total input kinetic energy
$\mathbf{f}$	Coriolis vector
f	Coriolis parameter $(= |\mathbf{f}|)$
f_T	fraction of thermals (see discussion of Fig. 7)
$f(R_i), g(R_i)$	stability functions used in Eqs. (98) and (99)
$f'(R_i)$	function of Richardson number given by Eq. (83) or (84)
$\mathbf{F}$	damping term in Eq. (1)
F_h	TKE lost by internal gravity waves
F_s	TKE produced by wind-wave induced current shear
g	acceleration due to gravity
G'	nondimensionalized input KE
$G*$	input KE [Eq. (52)]
h	depth of mixed layer
h_m	depth of mixed layer calculated by convective adjustment
$H*$	stability parameter
J_0	normalized radiation penetration term [Eq. (39)]
k	von Karman's constant
$k*$	parameter in Ekman profile
K_1, K_2, K_3	coefficients in arctangent model of radiation penetration [Eq. (10)]
K_H	turbulent diffusion coefficient for heat
K_{H0}	neutral value of the turbulent diffusion coefficient for heat
K_q	turbulent diffusion coefficient for TKE
K_M	turbulent diffusion coefficient for momentum

K_{M0}	neutral value for the turbulent diffusion coefficient for momentum
l	mixing length scale
l_0	reference length scale given by Eq. (102)
L	Monin-Obukhov length scale
$L*$	length of lake in wind direction
L_c	depth of convective mixing
L_1	Obukhov length scale
L_2	length scale $= \overset{*}{w}/f$
m	fraction of input KE available to TKE budget (after taking dissipation into account)
m_1	fraction of total input KE available to TKE budget
n	fraction of buoyant (convective) energy available to TKE budget
N	Brunt-Väisälä frequency
p	pressure
P_0	neutral value of the turbulent Prandtl number ($\sim$1)
$P*$	nondimensionalized mixed layer entrainment
q^2	turbulent kinetic energy (TKE)
q_m^2	turbulent kinetic energy per unit mass
Q	heat flux source term
Q_e	interfacial heat flux
Q_n	net surface heat flux
$\dot{Q}_0$	$= \partial\theta/\partial t$
R	empirical constant in Eq. (6)
R_b	bulk Richardson number
R_g	gradient Richardson number
R_i	Richardson number
s	empirical coefficient in ECM
$S*$	penetrative shortwave radiation
S'	nondimensionalized radiation penetration [Eq. (53)]
S_H, S_M	stability functions used in Eqs. (96) and (97)
S_k	contribution of convective mixing to TKE budget
S_p	rate of consumption of potential energy
t	time
T	temperature
T_h	temperature at depths below the mixed layer (viz. $z < -h$)
T_m	temperature of mixed layer calculated by convective adjustment
T_s	temperature of the mixed layer
u, v	horizontal velocity components
u_f	convective velocity
u_s	surface current
$\overline{U}$	$= h\,\overline{v}$
U	wind speed

v	vector velocity $= (u, v, w)$; or (u, v) in two dimensions
V	volume (of lake)
w_e	interfacial entrainment velocity $(= dh/dt)$
$\overset{*}{w}$	friction velocity in water
$\overset{*}{w}_s$	surface value of friction velocity, $\overset{*}{w}$
W	Wedderburn number
$W*$	convectively generated TKE [Eq. (52)]
x, y	horizontal coordinates
z	depth (measured vertically upwards from sea surface)
z_A	depth of surface absorption layer [Eq. (9)]
$Z*$	neutral depth parameter
α	coefficient of (volumetric) expansion of water
α_H	coefficient of molecular diffusion for heat
β	proportion of shortwave energy absorbed in upper few centimeters
Γ	temperature gradient below the mixed layer
ΔF	energy leakage out of ML to deeper water
ΔQ	perturbation in atmospheric forcing
Δt	time step
ΔT	temperature increment of ML due to perturbation
ΔT_0	temperature increment of deeper water due to perturbation
Δz	incremental depth
$\Delta \rho$	density difference between two layers of two-layer representations
ϵ	dissipation rate
η	extinction coefficient
η_1, η_2	extinction coefficients in Eq. (6)
$\eta*$	empirical coefficient in Eq. (86)
θ	total heat content
γ	constant in Eq. (102) $(= 0.2)$
$\gamma*$	constant in Eq. (120)
λ	climate feedback parameter
Λ	Heaviside function [Eq. (55)]
ρ	water density
ρ_a	air density
ρ_0	reference density
ρ_w	water density
σ	$= E_k/E_p$ [Eq. (90)] or $1/4m$ [Eq. (93)]
τ	shear stress
τ_a	surface wind drag
τ_w	drag at the water surface
$\phi(z)$	penetrative shortwave component
ϕ_c	sensible energy flux
ϕ_e	evaporative energy flux
ϕ_{li}	incoming longwave energy flux (after reflective losses)

ϕ_{lo}	outgoing longwave energy flux
ϕ_N	net available energy flux
ϕ_r	energy flux of precipitation
ϕ_{si}	incident shortwave energy flux (after reflective losses)
ϕ_{sp}	$[=(1 - \beta)\phi_{si}]$ penetrative shortwave radiation

Superscripts

$'$ (prime)	perturbation quantities
$^-$ (overbar)	indicates mean value

1 INTRODUCTION

In water bodies of all sizes, the temperature patterns and the way they change over both seasonal and diurnal (strictly diel) (i.e., over a 24 hour period) cycles are determined by a wide range of influences. Under different circumstances, these forcing processes may be an imbalance in the surface energy budget, penetration of shortwave radiation, convection, turbulent mixing, advection, currents and physicochemical characteristics of the water (e.g., turbidity, salinity). Knowledge of the temperature characteristics of water bodies may be of importance directly (e.g., controlling the temperature—and hence acceptability—of potable water in reservoirs; e.g., Gray, 1986), in the influence on chemical transports (e.g., in eutrophication studies, e.g., Gulliver and Stefan, 1982) and as a component part of a larger scale model (e.g., global climate modeling, Semtner, 1984).

Mathematical models are available on different time and space scales as an aid to understanding the complex interactions occurring between these various processes. In this paper we concentrate upon the downward transfer of heat and momentum; the various mathematical models used to understand this phenomenon; and various means of parameterizing the sub-grid scale processes involved. The assumption is made that at depth the fluid is quiescent, and therefore shallow sea regions in which tidally induced turbulence at the sea bed is important in determining the thermocline depth (Simpson and Hunter, 1974; James, 1977) will be excluded. Consequently the topic of the review will be restricted to lakes and oceans. Also here we will concentrate on one-dimensional (vertical) models rather than the more complex three-dimensional ocean circulation models (e.g., Bryan, 1975; Cox, 1975). The emphasis here will be on time scales of the order of a month or longer, rather than the shorter time scale (of the order of a few days) associated with a major wind event. It is considered worthwhile to review here one-dimensional (vertical) thermal stratification models highlighting the differences, deficiencies and ranges of applicabilities of existing models and contrasting their use in both limnological and oceanographic studies—two branches of the hydrological sciences which have to date failed to benefit from intensive interaction.

It is important to stress that the modeling approaches discussed in this review, as applied to *both* lakes and oceans (Section 5), neglect or parametrize many of the dynamical effects within the water body. In the case of lakes, where the influence of the boundaries may be evidenced by seiches, wind set-up and the need to take into account variations in cross-sectional area, there is unlikely to be any large-scale lateral advection (except in the case of lakes and reservoirs with a relatively large throughflow). In the open ocean and coastal seas, lateral advection could provide large influxes of heat at depth to disturb the vertical temperature profile; while short period seiches do not exist, although basin modes can be present.

A second major contrast exists in terms of the wave field as engendered by a wind affected by fetch and, in the case of lakes, the wind direction relative to obstacles/topography along the shoreline. At present wave effects are seldom modeled directly (but see, e.g., Tucker and Green, 1977) in stratification models, although sheltering effects (e.g., Henderson-Sellers, 1977; Ford and Stefan, 1980; Imberger and Parker, 1985) and fetch effects (e.g., Smith, 1979) have been considered for lakes.

Despite these various phenomena in lakes, cf. oceans, point models for the two systems, often designed independently, have, perhaps fortuitously, developed along parallel lines by virtue of assumptions neglecting all the dynamic processes discussed above. It is, however, likely that this close parallelism will fade as models become more elaborate and are able to incorporate these once-neglected dynamic effects (e.g., Imberger and Hamblin, 1982; Davies, 1986). The present discussion concentrates on areas of mutual interest. In this context the most significant difference becomes the need to include cross-sectional area variations with depth, $A(z)$, in lake models (see Fig. 12), but not in ocean models, which can be assumed to describe a water column. (At what spatial scale a lake is big enough to demand one or more column models is an unresolved question, although for short time scales, a full three-dimensional representation would be necessary to model major wind events accurately.)

Figure 1a illustrates the major pathways of energy transformation within a water body. However it does not illustrate the comparative magnitudes of these various processes, which will differ as a function of time and space (vertically downwards in the one-dimensional structure imposed here for modeling simplification). It should be noted initially that the forcing of largest magnitude on the long time scale in this system is the energetics associated with the surface energy budget and shortwave penetrative component (Section 2), a point noted by Kraus and Rooth (1961); and it is often only the smaller time and space scales that are influenced by the transformations shown on the left hand side of Fig. 1a. Nonetheless these energy transformations (and their source terms at the surface and at the thermocline interface) can be of paramount importance in the deepening of the thermocline and cannot be neglected if the phenomena are to be included in a two- or three-dimensional model or if such effects (e.g., large-scale lateral advection of water masses and hence of energy) are to be incorporated in the one-dimensional model by some type of appropriate parameterization (e.g., Environmental Laboratory, 1982; Henderson-Sellers, 1984, for discussion of the inclusion of the effects of lake inflows into a one-dimensional schematization).

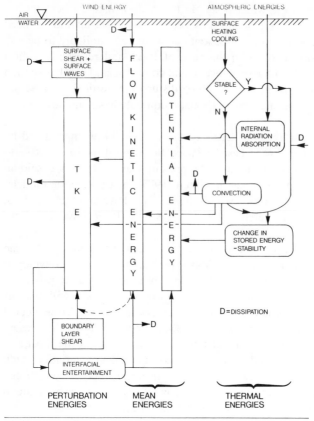

ENERGY BALANCE OF THE UPPER MIXED LAYER OF A LAKE

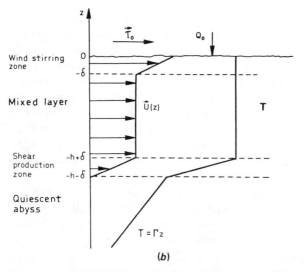

Figure 1 (*a*) Energy balance of the upper mixed layer of a lake. (*b*) Definition sketch showing the vertical structure of the mixed-layer model (after Niiler, 1975).

The equations describing the vertical profiles of temperature and velocity in a water body have been stated or derived by many authors. In order to elucidate the discussion to follow they are reiterated here with some brief comment. [These may be expressed in vector form (e.g., Niiler, 1975) or component form (e.g., Mellor and Durbin, 1975).] Neglecting the horizontal advection terms, these are given by

$$\frac{\partial \mathbf{v}}{\partial t} + \mathbf{f} \times \mathbf{v} = -\frac{\partial}{\partial z}(\overline{\mathbf{v}'w'}) - \frac{\mathbf{F}}{\rho_0} \tag{1}$$

$$\frac{\partial T}{\partial t} + \frac{\partial}{\partial z}(\overline{w'T'}) = \frac{Q(z, t)}{\rho_w c_p} \tag{2}$$

where $\mathbf{v}$ is the current vector; $\mathbf{f}$, the Coriolis vector; T, the temperature; t, the time; Q, a heat flux (source) term; ρ, the water density; ρ_0, a reference density; c_p, the specific heat of water; and primes (') indicate perturbed quantities (due to turbulent fluctuations). The term $\mathbf{F}$ is described by Niiler (1975) as a "damping term" which he notes must be related to the dissipation. This agrees with the equations presented by Mellor and Durbin (1975) in which this is given explicitly as the molecular viscosity term. The different methods of solution (discussed below) refer to the different methods of closure of these equations (essentially via the correlation terms $\overline{w'\mathbf{v}'}$, $\overline{w'T'}$), the neglect of differing processes (shown in Figure 1a) as well as differing parametrizations of the terms retained.

The mode of closure selected can be used to divide the models into those using an eddy diffusion closure scheme (e.g., Mellor and Durbin, 1975; Henderson-Sellers, 1985b) and those in which closure is expressed in terms of the rate of descent of the thermocline w_e given simply as the rate of change of (ML) depth h by

$$w_e = \frac{dh}{dt} \tag{3}$$

(see also Fig. 2). In this case the closure equation is that for the turbulent kinetic energy (TKE) budget (Section 3). However, a more useful division is between *differential* models which solve the primitive equations in a multi-level finite difference scheme and *bulk* or *integral* models which make an a priori assumption about the existence of a homogeneous ML (Fig. 1b). Consequently differential models are able to predict the existence and depth of a mixed layer; and in contrast the assumed existence of this homogeneous mixed layer by the bulk model permits the equations to be simplified by integration over the depth h of the ML, although the concept has recently been extended (Monismith, 1985) in a three-layer parameterization. In these bulk models it is temporal variations of the ML depth, h, which are calculated [Eq. (3)]. The characteristics of the ML are thus represented by a single equation, which decreases the computational time required for solution at the expense of resolution within the ML. These integral models have been widely used in both limnological and oceanographic studies, although it is being realized that they have some limitations in terms largely of their instability over longer simulation periods (Niiler, 1977; see discussion in Section 5) and their

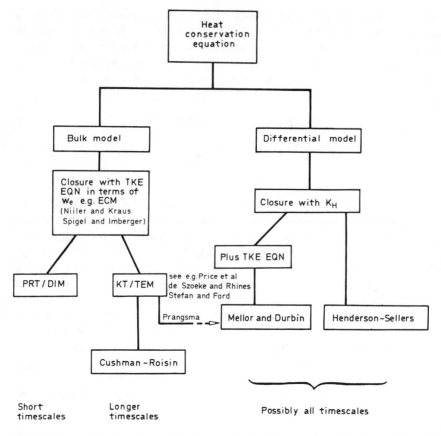

Figure 2 Flow chart describing various forms of mixed layer and stratification models, differentiating between ECM, DIM, TEM, and differential models.

inability to simulate successfully storm events (Price et al., 1978). Secondly, simplifications may be made within the framework of the bulk ML models, leading to the identification of the DIM (dynamic instability model; e.g., Pollard et al., 1973) and the TEM (turbulent erosion model; e.g., Kraus and Turner, 1967). These different approximations are in fact valid over different time scales (e.g., de Szoeke and Rhines, 1976; Niller, 1977). These problems will be discussed in more detail later.

2 SYSTEM SCALE ENERGY BUDGET

It is important to differentiate between the processes determining the energy budget on the system scale (such as surface energy balance; see later in this section) and those responsible for smaller scale turbulent sources and transformations (see Section 4) (although it should be noted that the associated temporal scales may often be similar). In this and the following section each of the processes shown in Fig. 1a will be described and the source terms discussed briefly. In Sections 4 and 5 the way in which different models include, neglect or ignore each of these processes will be evaluated.

On the system space scale, the water body heats up/cools down (increase/decrease of stored thermal energy) largely by atmospheric forcing (including solar radiation). Typical total daily heat input to a water body may be of the order of 2.5 MJ m^{-2} (i.e., an average energy flux of 25 W m^{-2}) (at latitudes of approximately 50°N). This is composed of a range of radiative and nonradiative heat fluxes (e.g., longwave atmospheric radiation, shortwave solar radiation, evaporation), all of which have magnitudes of tens or hundreds of watts per square meter. A proportion of the received (shortwave) solar radiation can penetrate to depth. The remainder of the energy fluxes interact with the upper few millimeters of the water body (McAlister and McLeish, 1969; Wu, 1985a). Of this total absorbed energy, the majority (in the warming period) is used to heat up the water directly, as well as altering the potential energy of the system (right hand side of Fig. 1a).

Consequently it is necessary to evaluate the surface energy budget (SEB) in order to assess the available energy both at the surface and at depth (see e.g., Haney, 1971; Henderson-Sellers, 1986). This can be expressed as

$$\phi_N = \phi_{si} - \phi_{lo} + \phi_{li} - \phi_e - \phi_c \pm \phi_r \tag{4}$$

where ϕ_{si} is the incident shortwave (after reflective losses), ϕ_{lo} is the longwave energy loss, ϕ_{li} is the incoming longwave radiation (after reflective losses) due to atmospheric emissions, ϕ_e is the evaporative (latent energy) flux, ϕ_c the sensible energy, and ϕ_r the energy associated with precipitation (usually considered to be negligible). The incoming shortwave energy (of which approximately 60% penetrates the surface layer) is determined by astronomical and meteorological factors and can best be included in the SEB equation from observations. Of the other terms it is the evaporative flux which is hardest to measure and hence which often needs calculation.

2.1 Solar Radiation Penetration

Although longwave radiation exchanges together with the nonradiative energy fluxes are concentrated in the uppermost 1–2 mm of the water body (McAlister and McLeish; 1969), the shortwave radiative flux is partly absorbed by this surface layer and partly penetrates to greater depth. Simpson and Dickey (1981a) comment that in some oceanic studies this differential absorption (with wavelength) is ignored (e.g., Denman, 1973; Alexander and Kim, 1976) such that the radiation term is given by a decaying exponential (where z is measured vertically upward and $z = 0$ is the surface):

$$\phi(z) = \phi_{si} \exp{(\eta z)} \tag{5}$$

where η is the extinction coefficient, ϕ_{si} is the total penetrative shortwave radiation, and $\phi(z)$ the portion of ϕ_{si} reaching depth z. Later observations (Paulson and Simpson, 1977) showed that irradiance decreased more rapidly with depth than predicted by Eq. (5) for the upper few meters such that a two term representation would be more appropriate:

$$\phi(z) = \phi_{si}[R \exp (\eta_1 z) + (1 - R) \exp (\eta_2 z)] \tag{6}$$

where R is an empirical constant and η_1 and η_2 are two extinction coefficients. Some modelers (Spigel et al., 1986) even use a three term exponential representation for $\phi(z)$.

In lake studies an alternative approach has been adopted (e.g., Elder and Wunderlich, 1968; Huber and Harleman, 1968) by which a portion, β, of the incoming solar radiation (after reflective losses have been taken into account, ϕ_{si}, is assumed to be absorbed in the upper layers (~0.6 m) to account for the energy associated with wavelengths at the red end of the solar spectrum. The value of β has been taken as 0.4 (Dake and Harleman, 1969) or 0.5 (e.g., Hurley Octavio et al., 1977), although Williams et al. (1981) have pointed out that this fraction should, more realistically, be related to the water turbidity (Fig. 3) as

$$\beta = 0.265 \ln \eta + 0.614 \tag{7}$$

The turbidity itself may be quantified as an extinction coefficient, η, which is inversely proportional to the Secchi disc measurement (see, e.g., Brezonik, 1978), although in hypertrophic lakes, especially, the concentration of phytoplankton, suspended inorganic particulates and dissolved yellow substances (*gelbstoff*) are important in determining absorption and scattering (Effler et al., 1984). The distribution of shortwave radiation with depth, $\phi(z)$, is then given by

$$\phi(z) = (1 - \beta)\, \phi_{si} \exp(\eta z) \tag{8}$$

However, as pointed out by Zaneveld and Spinrad (1980), the exponential decay begins at the base of the surface absorption layer (at $z = -z_A$), so that the correct form for Eq. (8) is

$$\phi(z) = (1 - \beta)\phi_{si} \exp[\eta(z - z_A)] \tag{9}$$

where $z_A \approx 0.6$ m. For clear waters the discrepancy is minimal; but for turbid waters ($\eta \approx 1.0$), Eq. (8) may underestimate by a factor of approximately 2.

An alternative approach, proposed by Zaneveld and Spinrad (1980) and Zaneveld et al. (1981), utilizes a single functional form for $\phi(z)$ to encompass both the surface absorbed flux and the penetrative radiation:

$$\phi(z) = \phi_{si} \exp (K_1 z)[1 + K_2 \tan^{-1}(K_3 z)] \tag{10}$$

where K_1, K_2, and K_3 are empirical constants (Jerlov, 1976). It can be shown that K_1 corresponds to η and K_2 can be related to η and β as

$$K_2 \equiv \frac{2[1 - (1 - \beta) \exp(-\eta z_A)]}{\pi} \tag{11}$$

The third parameter, K_3, of the arctangent model has no direct equivalence in the

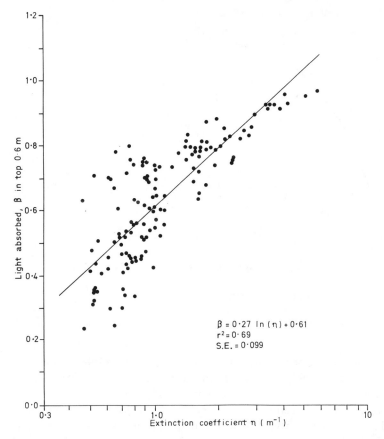

Figure 3 Light absorbed in the upper 0.6 m as a function of extinction coefficient (after Williams et al., 1981).

(η, β) model, being a measure of the rapidity of falloff with depth within the upper layers of the water body. Figure 4 compares the radiation penetration described by the Zaneveld and Spinrad (1980) model (with $K_1 = 0.1$, $K_2 = 0.231$, $K_3 = 4$) compared with the simple extinction model [Eq. (8)] in which $\eta = 0.1$ and $\beta = 0.40$. Although simulations of the stratification cycle which use the arctangent model show no significant difference from the (η, β) model, it is probably preferable since it is easier to employ in a prognostic model using the values of K_1, K_2, K_3 of Jerlov (1976) and/or observed values of K_1 ($\equiv \eta$).

Sensitivity experiments to these different parameterizations have been reported for the ocean by, e.g., Simpson and Dickey (1981a,b) and for lakes by Henderson-Sellers (1985a) and will be discussed in more detail in Section 5.

2.2 Convection

In the warming season both the SEB and the penetrative solar radiation are positive quantities and a stable thermal profile persists. However under certain cir-

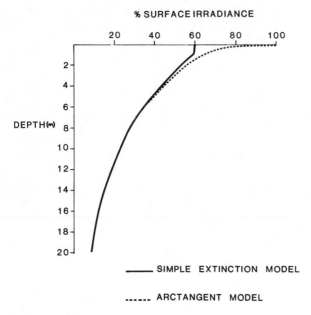

Figure 4 Comparison of light penetration using simple extinction model and arctangent model.

cumstances (e.g., autumnal or overnight cooling, e.g., Woods (1980), Shay and Gregg (1986); enhanced energy removal from the surface under extreme atmospheric situations—see discussion below) it is possible that the SEB can become negative. In such a case, convection results such that cold surface water descends and is replaced by warmer underlying water—a major mechanism in polar regions for the formation of deep bottom water. Mixing takes place until a stable thermal profile is re-attained. The convective cell may be totally contained within the ML (nonpenetrative convection) or may be responsible for assisting in deepening of the thermocline (penetrative convection). In addition to altering the overall thermal content of the ML, such convective processes, on a smaller (energy) scale, may provide an additional source term ($\sim 2.5 \times 10^{-5}$ W m^{-2}) in the TKE balance which itself is responsible for small scale moderation of ML depth, current structure etc. Again such convective contributions to the TKE are made over the depth of the convective sublayer, which may be less than or equal to the depth of the ML (Woods, 1980).

The depth to which this buoyant* convection penetrates can be considered in two distinct ways: either in terms of an understanding of the physical processes (e.g., Tennekes, 1973; Cushman-Roisin, 1981) or, allied but possibly highly simplified, in terms of an appropriate modeling parameterization which will maintain

---

*Since the vertical coordinate is positive vertically upward yet gravity acts vertically downward, this is actually a negatively buoyant situation. However, here we will follow convention and use the term buoyant convection.

the features of interest in thermocline modeling without necessarily representing all the physical processes on all time and space scales accurately (e.g., Dake and Harleman, 1969; see Fig. 5). Indeed many authors have expressed the opinion that the *processes* of penetrative convection, as discussed here, are poorly understood (e.g., Dillon et al., 1981; Patterson et al., 1984; Shay and Gregg, 1984). Despite this, some reasonable approximations can be made which will permit adequate description within the limits of thermal stratification modeling without digressing into a full consideration of the structure of the boundary layer following the extensive and excellent atmospheric boundary layer literature (for review see Nieuwstadt and van Dop, 1982). If no regard is taken of the (small) contribution that is actually made during convective overturning to the TKE budget, then a simple heat balance approach may be used (e.g., Dake and Harleman, 1969). This instantaneous convective adjustment gives an isothermal mixed layer, depth h_m and temperature T_m, where

$$T_m(t) = \frac{\int_{-h_m}^{0} A(z)T(z, t)dz}{\int_{-h_m}^{0} A(z)dz} \tag{12}$$

For an oceanic water column, the cross-sectional area is constant with depth so that $A(z) = A_s$ for all depths and hence the new mixed layer temperature is given by

$$T_m(t) = \frac{\int_{-h_m}^{0} T(z, t)dz}{h_m} \tag{13}$$

This convectively adjusted ML depth (Kraus and Rooth, 1961) is also discussed by Kraus and Turner (1967) and Woods (1980). They define the compensation depth as that level above which there is a net upward heat flux and below which

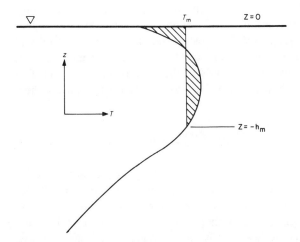

Figure 5 Convective mixing at surface associated with unstable temperature profiles (after Dake and Harleman, 1969).

the net flux is downward. Kraus and Turner (1967) calculate that for exponential absorption of radiation (see earlier discussion), the compensation depth d is given by

$$d = \frac{1}{\eta} \ln\left(1 + \frac{\phi_N - \phi_{sp}}{\phi_{si}}\right) \tag{14}$$

where ϕ_{sp} is the penetrative shortwave radiation. Woods (1980), noting that Kraus and Turner's (1967) definition is based upon the 24 hour balance between solar radiation absorption and surface heat loss, extends the concept to instantaneous values. This depth, being independent of the TKE budget, is used as a diagnostic tool for stratification on several time scales. This convective sublayer, Woods (1980) notes, is normally less than the depth of the mixed layer, being less than 1 m during daylight hours. If the compensation depth lies below the ML depth, then convection occurs throughout the whole depth of the ML, subsequently possibly eroding the thermocline (penetrative convection).

The use of the concept of compensation depth also permits Woods (1980) to evaluate the contribution, S_k, made by convective mixing to the TKE budget (see also Section 3). This is estimated from the rate of consumption of potential energy, S_p, such that

$$S_k \leq S_p = \int_{-L_c}^{0} Q(z)\left(\frac{\alpha g z}{c_p}\right) dz \tag{15}$$

where g is the acceleration due to gravity, $Q = -\partial\phi/\partial z$ and L_c is the maximum depth of convection given by

$$L_c = \begin{cases} d & \text{if } d < h \\ h & \text{if } d \geq h \end{cases} \tag{16}$$

Woods (1980) stresses that diurnal variability be taken into account [cf. Kraus and Turner (1967) who used the compensation depth/convective mixing calculation for 24-hour averages] since $d < h$ for most of the day but $d \geq h$ mainly at night, thus invoking the diurnal variability of Eq. (16). In the latter case $(d \geq h)$ the TKE from wind-driven convection extends to the base of the ML such that the rate of convective energy production, S_k, depends only on the surface heat loss, Q_n, and the ML depth

$$S_k = S_p = \frac{0.5 \, Q_n h \alpha g}{c_p} \qquad (d > h) \tag{17}$$

Consequently it is evident that the existence and depth of penetration of convective mixing is determined by the relationship between the penetration of solar radiation and the net surface energy balance; the latter of which may become negative whenever there are significantly larger radiative losses than gains (e.g., at night) or when the nonradiative losses become large. In general the nonradiative terms are dominated by evaporation, such that, for example, in storm conditions with high winds, such evaporative energy losses can be significantly increased (e.g., Shay and Gregg, 1984). When surface heating is intense and wind speeds low, the compensation depth d is small so that downward transfer of heat begins relatively close to the surface. Overnight surface cooling abstracts energy from a greater depth: a depth determined at least in part by the strength of the stratification of the waters. Woods (1984) equates this depth to that determined by a convective adjustment calculation (Gill and Turner, 1976) and hence is analogous to the convectively mixed layer depth described above which has been created by thermal instabilities near the surface of the water column and penetrating to a depth dependent upon ambient stability and wind speed.

Furthermore, in their study of the sensitivity of models to parameterization of the shortwave penetration (discussed above), Simpson and Dickey (1981a) deduce that whereas surface temperatures increase monotonically with time for an ocean with a positive surface radiation balance, there is a transition over the wind speed range of $10–20$ m s^{-1} such that at this higher wind speed, surface temperatures decrease with time. At such high wind speeds they therefore deduce that the temperature regime is wind dominated, since the chosen parameterization of solar radiation penetration has little effect. These conclusions together suggest that for low wind speeds it is highly important to parameterize successfully/accurately the radiation penetration term; while at higher wind speeds this becomes unimportant and instead correct parameterization of the wind-induced turbulence becomes the major priority.

By consideration of the system energy levels involved in lake dynamics, it is thus evident that the specification of the wind stress, which is responsible for adding kinetic energy to the water body, is less important on the long time scale (of order a month) than accurate calculation of the surface energy budget. When high wind events occur, deepening of the upper mixed layer may result, *not* directly, as a consequence of increases to the total system energy (viz. added TKE) (as proposed by, e.g., Pollard et al., 1973); Fischer et al., 1979, p169; Davis et al., 1981; Caldwell, 1983), but as a result of changes in the SEB (especially increases in the evaporation rate) which leads to a rapid energy loss at the water surface (and to the overall water column) and consequently to enhanced convective mixing. This interacts with a more vigorous dissipation of TKE throughout the ML caused directly by the increased wind speed. The enhanced turbulence levels also cause the ML to deepen. (Further simulation results supporting these conclusions are discussed in Section 5.3.)

These conclusions are corroborated by the result of Mellor and Durbin (1975, depicted in their figures 11 and 12 and reproduced here as Fig. 6), about which they made no interpretative comments. These authors considered the imposition of a *constant* surface flux of "$\pm 10^{-2}$ cm K s^{-1}" ($\approx \pm 420$ W m^{-2}) with wind stresses of 2×10^{-4} m^2 s^{-2} and 1×10^{-4} m^2 s^{-2}, respectively (corresponding to wind speeds of approximately 9.1 and 6.5 m s^{-1}, respectively, using the authors' value of 0.002 for the aerodynamic drag coefficient). For the cooling case, the temperature profiles show little contrast, except for slight differences in overall mixed layer depth; thus demonstrating that in the cooling situation the temperature profile is largely determined by the negative surface heat flux and more or less

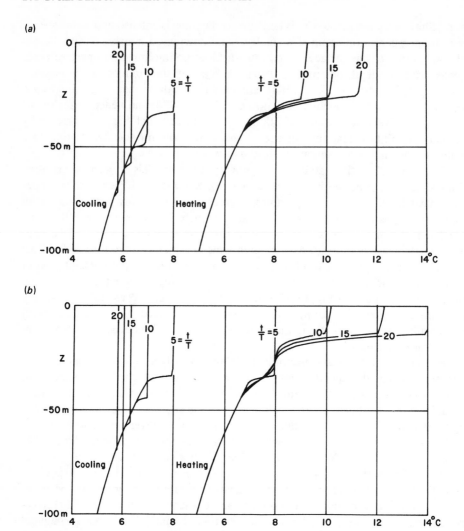

Figure 6 Temperature profiles as a function of nondimensional time, t/T (where t is time and $T = 2\pi/f$). The effects of sudden heating or cooling for two wind different stresses (upper: stress of 2 $\times$ 10^{-4} m² s⁻²; lower: stress of 1 $\times$ 10^{-4} m s⁻²) are shown (after Mellor and Durbin, 1975, *Journal of Physical Oceanography,* American Meteorological Society, Boston).

independent of wind speed (viz. input of KE). In contrast, for their heating case, less wind-induced turbulence restricts the depth of the mixing and permits the surface temperatures to rise more rapidly when the surface stress is only 1×10^{-4} m² s⁻².

Nevertheless, it should be stressed that the time scale of interest may well primarily determine the type of model (and its implicit modeling assumptions) to be used. Filyushkin and Miropol'skiy (1981) point out that over the time scale of years, there is a repetitive seasonal cycle of temperature in the ocean, closely correlated with the "march of the seasons"; whereas the impact of the stochastically distributed storm events do nothing to disturb the year-to-year cycling, only disturbing on a short time scale the temperature characteristics of the water body.

2.3 Kinetic Energy

The total input kinetic energy E_w is equal to the shear stress, τ_w, multiplied by the water velocity at the interface u_s, such that

$$E_w = \tau_w u_s = \rho_w \overset{*}{w}{}^2 u_s \tag{18}$$

where ρ is the water density, $\overset{*}{w}$ the (water) friction velocity, and u_s the surface velocity (although it should be noted that in an oceanic environment with swell and waves, the surface level itself is ill-defined). The friction velocity is related to the wind speed by

$$\overset{*}{w}{}^2 = \frac{\rho_a}{\rho_w} c_D U^2 \tag{19}$$

where ρ_a is the air density, and both the aerodynamic drag coefficient c_D and the wind velocity U must relate to the same height of measurement in the atmosphere (usually 2 or 10 m).

Although a constant value of $c_D \approx 1.2 \times 10^{-3}$ is often assumed, it should be noted that many authors (e.g., Wieringa, 1974; Garratt, 1977; Smith, 1980; Wu, 1980) have deduced formulas for c_D either as

$$c_D = a_1 U^{b_1} \tag{20}$$

or

$$c_D = (a_2 + b_2 U) \tag{21}$$

where $a_1 \approx 5 \times 10^{-4}$, $a_2 \approx 6 - 8 \times 10^{-4}$, $b_1 \approx 0.3 - 0.5$ and $b_2 \approx 6.5 \times 10^{-5}$ (for U in m s^{-1}). These formulas are reviewed by Smith (1981) from which it may be concluded that c_D can be given by the value for smooth surfaces at low wind speeds, as a constant at intermediate values and as an increasing function of U at higher wind speeds. A similar relationship is proposed by Large and Pond (1981), based on an extensive analysis of a large number of observations. They suggest that for neutral conditions

$$c_D = \begin{cases} 1.2 \times 10^{-3} & 4 \leq U < 11 \\ (0.49 + 0.065\,U)10^{-3} & 11 \leq U \leq 25 \end{cases} \tag{22}$$

Additionally, the effect of atmospheric stability may be included (e.g., Deardorff, 1968; Schwab, 1978), as also can the influence of fetch/wave age (Wu, 1985b; Geernaert et al., 1986; Huang et al., 1986). (However, these effects will not be included for the present discussion.) Inclusion of such additional independent variables may increase the precision of these results but not significantly (see, e.g., Weber, 1983).

The problem in evaluating Eq. (18) numerically is, in addition to the difficulty of defining the surface level, that the functional form of surface velocity (as a function of either friction velocity or of wind speed) is not well determined. It is generally assumed that there exists a linear relationship between the surface velocity and friction velocity (e.g., Kondo et al., 1979) or wind speed (see, e.g., summary in Smith, 1979). Bloss and Harleman (1979) note values of Madsen (1977) and Kondo et al. (1979) of the order of 30 for $u_s/\overset{*}{w}_s$ (a value also substantiated by Csanady, 1979) but suggest that in lakes the ratio must be nearer unity. There seems little justification for this latter assumption since from Eq. (19), with $c_D \approx 1.2 \times 10^{-3}$ and $U \approx 5$ m s^{-1}, $\overset{*}{w}_s = O(0.005$ m s$^{-1})$; whereas observations of u_s at this wind speed (e.g., Smith, 1979; Churchill and Pade, 1981) give $u_s \approx 0.1$ m s^{-1} suggesting a ratio $u_s/\overset{*}{w}_s = O(20)$. This compares well with the values of 15.9 and 22 given by Wu (1973) and Wu (1983), respectively.

Alternatively, the ratio $u_s/\overset{*}{w}_s$ can be related to the "wind factor," u_s/U, by assuming that the shear stress is continuous across the interface:

$$\tau_a = \rho_a c_D U^2 = \tau_w = \rho_w \overset{*}{w}_s^2 \tag{23}$$

where τ_a is the surface wind drag and τ_w is the drag at the water surface. Hence

$$\overset{*}{w}_s = \left(\frac{\rho_a c_D U^2}{\rho_w}\right)^{1/2} \tag{24}$$

The ratio of $u_s/\overset{*}{w}_s$ is then given straightforwardly as

$$\frac{u_s}{\overset{*}{w}_s} = u_s\left(\frac{\rho_w}{\rho_a c_D U^2}\right)^{1/2}$$

$$= \frac{u_s}{U}\left(\frac{\rho_w}{\rho_a c_D}\right)^{1/2} \tag{25}$$

The value of the wind factor, u_s/U, is generally assumed to have a value of about 3% for large water bodies, possibly decreasing to as little as 1.5% for enclosed stretches of water (lakes/reservoirs). The difference can be explained by a simple interpretation of figure 1 of the study of Weber (1983). His calculated curve for $u_s(U)$ gives a wind factor of about 3.1–3.4% for a fully developed sea; but he notes that the value of u_s is composed of approximately equal contributions from the Ekman flow and from the wave-induced flow. Hence in lakes where wave-induced flow is severely limited or indeed negligibly small, the total surface current is likely to be about half that of a comparative situation at sea and hence the ratio $u_s/U \approx 0.015$. In enclosed sea regions the situation is more complex due to the presence of land boundaries which can set up gradient currents, which will influence this ratio. The level of turbulence at depth, due to tidal motion is also important (Davies, 1985). The presence of new surface stratification in the ocean or in a lake can also be important (Davies, 1986), as can diurnal variations of heating and cooling (Woods and Strass, 1986).

For a range of values of u_s/U between 0.015 and 0.030 and taking $c_D = 1.2 \times 10^{-3}$, the corresponding value of the ratio $u_s/\overset{*}{w}_s$ is thus in the range

12.5 (lakes) to 25.0 (oceans). Representing this ratio as c_1 permits the input KE [Eq. (18)] to be rewritten as

$$E_w = \rho_w \overset{*}{w}{}^2 u_s = c_1 \rho_w \overset{*}{w}{}^3 \qquad (26)$$

This is the form quoted by, e.g., Niiler and Kraus (1977) and Garwood (1977). Hence Eq. (26) is appropriate for the calculation of input KE where the value of c_1 would appear to be of the order of 20–25 for oceans and of the order of 12.5 for enclosed water bodies; although values of unity would appear to be gross underestimates. However, it should be noted that if this less well-substantiated value ($c_1 = 1$) is indeed used, then Eq. (26) becomes

$$E_w = \rho_w \overset{*}{w}{}^3 \qquad (27)$$

where $\overset{*}{w} \approx 1.2 \times 10^{-3}$ to $1.4 \times 10^{-3}U$. It is in this form that the TKE input is often quoted without derivation and used (e.g., Kraus and Turner, 1967; Stefan and Ford, 1975a,b; de Szoeke, 1980; Ford and Stefan, 1980; Shay and Gregg, 1984), a form which therefore would appear to be possibly in error by a factor of up to 25 (for oceans).

3 TURBULENT KINETIC ENERGY BUDGET AND ENERGY TRANSFORMATIONS

A common method of attaining closure of Eqs. (1) and (2) is to utilize the turbulent kinetic energy (TKE) equation in order to provide a parameterization of the perturbation quantity (e.g., Mellor and Durbin, 1975), or, more usually, in terms of the interfacial entrainment velocity w_e (i.e., the rate of descent of the thermocline = dh/dt).

There are three major identifiable sources for TKE: at the surface as a result of wind-wave induced current shear (F_s), at the base of the upper mixed layer (see Fig. 1b), i.e., the interface located at the thermocline (E), and from convection. As can be seen from Fig. 1a there is also a sink to potential energy and via dissipation (D). A further sink is energy, F_h, lost by internal gravity waves through underlying stable waters. As Cushman-Roisin (1981) notes, this last term is usually neglected "since no acceptable parameterization has yet been proposed." Symbolically the TKE budget is thus written for the total upper mixed layer, neglecting the convective contribution, as

$$\frac{d}{dt}(\text{TKE}) = F_s + E - \frac{d}{dt}(\text{PE}) - D - F_h \qquad (28)$$

(Cushman-Roisin, 1981). (Each of these terms is discussed in detail in Section 4.1.)

However it is strongly debated which of the three main processes is most important. For example, Woods (1980) suggests that the two main sources of TKE are the wind stress and buoyant convection—these sources being located at the surface and in the ML, respectively. On the other hand, Pollard et al. (1973) have considered that the generation of TKE at the interface between the two layers in a stratified water body is the dominant process on relatively short time scales (less than f^{-1}, where f is the Coriolis parameter). The relative importance of these different source terms has been analyzed by many authors; for example, Niiler (1975) first reconciled the approaches of Kraus and Turner (1967) and Pollard et al. (1973) as asymptotic limits of one generalized theory; an interpretation echoed by, e.g., de Szoeke and Rhines (1976) and Price et al. (1978). (Details of the model equations for a full description of each of the interactions shown in Fig. 1a are deferred to Sections 3 and 4.) Further discussions (e.g., Niiler, 1977) have shown that not only are these different approaches limiting forms for the generalized TKE equation, but the processes are likely to dominate on different time scales (see full discussion in Section 6).

It has already been noted that there are three major potential source regions for TKE (neglecting transformations for the present): the surface shear layer, the interfacial shear layer, and buoyant convection. The first of these (wind input) is relatively easily evaluated for the *total kinetic energy* input at the surface, although some of the empirical constants are not convincingly determined for the possibly different energy regimes of lakes and oceans. Some of this energy is input to the wave regime and may be considered to be dissipated almost immediately in this region; while the remainder (a fraction m_1) becomes available to the TKE budget.

The interfacial shear stress layer can also be considered as a potential source region for TKE. Figure 1b illustrates the velocity shear across this surface created by a constant velocity in the mixed layer and an almost quiescent underlying layer.

The third source term for TKE is that due to convective mixing. Woods (1980) discusses in detail the importance of convective mixing to the TKE budget and emphasizes the necessary conditions for such convection to occur. The occurrence and depth of convection is governed by the magnitude of the deficit of the surface energy budget (see Section 2). When convection does occur, vertical motions are induced and these lead to turbulent kinetic energy. The magnitude of this source term is related to the surface buoyancy flux B_0, and the relative contributions of this and the wind-wave induced TKE may be assessed using the Monin-Obukhov length scale, L (e.g., Phillips, 1977). For small depths (i.e., $|z| << |L|$), the TKE budget is maintained predominantly by the Reynolds stress and the buoyancy flux plays only a minor role. Conversely when L is small, then at depths larger than L the TKE budget may be dominated by the buoyancy induced TKE. Shay and Gregg (1984) identify conditions when such a case was observed such that the contribution of convection to the TKE budget was relatively large. However, this does not imply that "convective forcing was dominant throughout most of the mixed layer" (Shay and Gregg, 1984), but that when convection occurred as a result of large-scale instabilities in the thermal profile, a larger proportion of this would be made available to the TKE budget rather than for thermal storage. Atmospheric studies of convection within a boundary layer (e.g., Tennekes, 1973, 1975; Zilintinkevich, 1975), using the time-dependent TKE equation (see below), suggest appropriate parameterizations for several characteristic cases, by expressing the time rate of change of TKE in terms of w_e, h, and the vertically averaged standard deviation of the vertical velocity in the mixed layer.

Although most of the models discussed in Sections 4 and 5 include various members of these source terms together with parameterizations for the transformations discussed in Section 3, it should also be noted from Fig. 1 that, while it is recognized that it is the turbulent KE that is responsible for thermocline erosion, the *mean* KE field may also be affected. Hence, it is also necessary to consider the temporal and spatial development of the mean kinetic energy and mean potential energy of the system and its interaction with the perturbed energy field. For example, it is the shear in the mean flow at the interface that produces TKE which is able to permit descent of the thermocline; yet when this occurs the velocity field in the mixed layer itself will of necessity be modified—as will the mean potential energy.

The derivation of the full TKE equation (as represented graphically in Fig. 1) is given in Niiler and Kraus (1977) and in part in Niiler (1975) and quoted in simplified form by many authors (e.g., Price et al., 1978; Cushman-Roisin, 1981).

The basic turbulent energy equation, as derived by Kraus (1972), is given by Niiler and Kraus (1977) as their equation 10.5, which (after correction of the typographical error) relates the rate of change of TKE, q^2, to the work of the stress on the main shearing flow, the rate of working of the buoyant force, energy transfer due to turbulence and pressure terms and the dissipation rate, ϵ:

$$\frac{1}{2}\frac{\partial}{\partial t}q^2 = -\overline{w'\mathbf{v}'}\frac{\partial \bar{\mathbf{v}}}{\partial z} + \overline{w'b'} - \frac{\partial}{\partial z}\left[\frac{1}{2}\overline{w'(w'^2 + \mathbf{v}'^2)} + \frac{\overline{w'p'}}{\rho}\right] - \epsilon \qquad (29)$$

where b' and p' are the fluctuations in the buoyancy b and pressure p. The terms in this equation are not collected together in an identical way to Eq. (28) which has as separate variables dissipation D and the rate of change of PE. In Eq. (29) these contribute to several of the terms. Their explicit appearance in Eq. (48) below will permit comparison with Eq. (28) to be made more easily.

[A quasi-equilibrium form of this TKE equation is also used in the Mellor and Yamada (1974) Level-2 model, applied to the aquatic case by Mellor and Durbin (1975). The details of this are given in Section 4.2.]

4 THERMAL STRATIFICATION MODELING

4.1 Mixed Layer (ML) Models

Observations of the phenomenon of thermal stratification in oceans and lakes suggest that once the stratification is established, the upper layer of water is quasi-homogeneous in nature. Consequently it may be reasonable to take as a simplifying assumption in the mathematical model, that there exists a homogeneous mixed layer (ML) (also called the upper quasihomogeneous layer, UQL; e.g., Filyushkin and Miropol'skiy, 1981; Shelkovnikov, 1983), whose temperature and thickness may vary throughout the seasons; although Kundu (1980) cites several sets of observational data which do not support the detail of such a simplification. André and Lacarrère (1985) suggest that a significant shear across the ML depth can persist, especially in a stress-driven (as compared to a buoyancy-driven) case.

Nevertheless, introducing the assumption that the upper layers of a water body are almost homogeneous in both velocity and temperature (apart from the obvious two interfacial shear layers, Fig. 1b) permits a single differential equation to be used for the whole of this mixed layer. Furthermore it is a common assumption (e.g., Niiler, 1975) that the waters underlying the interface are a "quiescent abyss," i.e., no energy "leaks" into these waters from above (nor also by direct radiative penetration). [Although in a shallow tidal sea, turbulence will be generated at the sea bed (Simpson and Hunter, 1974), this case will not be considered here.] This may be valid, as a first approximation, for relatively turbid waters while being compatible with the frequent assumption that all radiative energy absorption occurs in the surface sublayer (see fuller discussion elsewhere in this chapter). The severe limitations of these assumptions have recently been realized by Posmentier (1980) and Price et al. (1986); see full discussion in Chapter 6. It is also worth noting here that such restrictions have never been necessitated in the alternative (eddy diffusion) approach to ML modeling (Section 4.2).

Consequently, such assumptions of the integral mixed layer (IML) model result in a loss of information and lead to models which cannot describe in any detail whatsoever processes acting inhomogeneously within this layer (e.g., algal and nutrient recycling). [The alternative modeling approach in which the existence of a ML is not assumed to exist a priori, but in which the ML structure is predicted, is described in detail in Section 4.2. Woods (1984) also notes the advantages of resolution (at the expense of complexity and hence computational time) of eddy diffusion models and suggests that they would be most advantageous for application to the tropical ocean and in summer periods at higher latitudes.]

Furthermore, IML models cannot simulate convection accurately, especially in high winds (e.g., Price et al., 1978), largely since they cannot discriminate between thermal energies (and energy deficits) at the surface and energy absorbed differentially throughout the mixed layer. Hence, the differential absorption of shortwave radiation observed and parameterized by many authors (e.g., Jerlov, 1976; Zaneveld and Spinrad, 1980; Simpson and Dickey, 1981a,b) is replaced simply by the two values: input at layer top and penetration through the layer bottom into the "quiescent" deeper water (Fig. 1b).

Notwithstanding the inherent limitations of the integral (or bulk) mixed layer stratification model, it is instructive to describe here the steps used in development of such an integral approach. Since the Niiler and Kraus (1977) model [dubbed the "energy conservation" model (ECM) by Price et al., 1978] appears to be the most comprehensive in its inclusion of the various active processes (neglecting large-scale advection and inhomogeneities within the mixed layer), this ECM is used as the illustrative example in this chapter.

In order to derive any ML model, boundary conditions must first be derived and simplifications made to the various terms of Eq. (29) (so that the differential equation can then be integrated over the ML depth). Each of the terms in this equation is now considered in turn.

4.1.1 Buoyancy term. The buoyancy term $\overline{w'b'}$ can be derived by analogy with the heat conservation equation, such that

$$\frac{\partial b}{\partial t} + \frac{\partial}{\partial z} \overline{w'b'} = -\frac{g\alpha}{\rho c_p} \frac{\partial \phi}{\partial z} \tag{30}$$

This equation can be integrated (1) from depth z to the surface (where z is measured vertically upward from an origin at the surface) to give

$$\frac{\partial \bar{b}}{\partial t} [z]_z^0 + \overline{w'b'}|_0 - \overline{w'b'}|_z = -\frac{g\alpha}{\rho c_p} [\phi]_z^0 \tag{31}$$

and (2) from depth $-h$ to the surface:

$$\frac{\partial \bar{b}}{\partial t} [z]_{-h}^0 + \overline{w'b'}|_0 - \overline{w'b'}|_{-h} = -\frac{g\alpha}{\rho c_p} [\phi]_{-h}^0 \tag{32}$$

Noting that $\overline{w'b'}|_0 = B_0$ (surface buoyancy flux) and at the lower interface $\overline{w'b'}|_{-h} = -w_e \Delta b$, where Δb is the difference in buoyancy across the $z = -h$ interface and w_e is the rate of increase of ML depth [Eq. (3)], then elimination of $\partial \bar{b}/\partial t$ gives

$$\overline{w'b'} = B_0 \left(1 + \frac{z}{h}\right) + w_e \Delta b \frac{z}{h} + \frac{g\alpha}{\rho c_p} (\phi_{sp} - \phi) + \frac{g\alpha z}{\rho c_p h} [\phi_{sp} - \phi(-h)] \tag{33}$$

Assuming an exponential decay of penetrative shortwave radiation with depth (see earlier discussion) such that

$$\phi = \phi_{sp} \exp(\eta z) \tag{34}$$

and that at the lower interface $\phi(-h) \ll \phi_{sp}$ (an assumption which, it should be noted, restricts the applicability of the derived equation to deep MLs and/or turbid waters), then Eq. (33) is written as

$$\overline{w'b'} = B_0 + \frac{z}{h} \left(B_0 + w_e \Delta b + \frac{g\alpha}{\rho c_p} \phi_{sp}\right) + \frac{g\alpha}{\rho c_p} \phi_{sp}(1 - e^{\eta z}) \tag{35}$$

4.1.2 Shear term. Within the mixed layer itself the shear term is zero. However at the interfacial level, Niiler and Kraus (1977) present evidence that

$$\overline{w'\mathbf{v}'}|_{z=-h} = C\bar{\mathbf{v}}|\bar{\mathbf{v}}| - w_e \bar{\mathbf{v}} \tag{36}$$

where C is a generalized drag coefficient.

4.1.3 Triple correlations. At the bottom boundary this has a value of $-w_e q^2/2$ (Niiler and Kraus, 1977, eq. 10.21) and at the surface is equal to a fraction m_1

of the total input KE from the wind forcing (viz. $= m_1 \overset{*3}{w}$), such that

$$\left[\frac{1}{2} \overline{w'(w'^2 + v'^2)} + \frac{\overline{w'p'}}{\rho} \right]_{z=-h} + \frac{1}{2} w_e q^2 = 0 \tag{37}$$

$$-\left[\frac{1}{2} \overline{w'(w'^2 + v'^2)} + \frac{\overline{w'p'}}{\rho} \right]_{z=0} = m_1 \overset{*3}{w} \tag{38}$$

The importance of including these triple correlations is discussed by Warn-Varnas and Piacsek (1979) and André Lacarrère (1985).

4.1.4 Mixed layer modeling approaches.

Although Niiler (1975) continues to derive his model without integrating over the mixed layer depth, many other authors (e.g., Niiler and Kraus, 1977) at this stage note that integrating over the depth h of this layer will remove the depth dependency from the equations as applied to the ML, consequently reducing the amount of computation time required by replacing a set of layer equations by a single equation representing the bulk characteristics of the mixed layer.

Integration of the TKE equation [Eq. (29)] demands formulation of integral quantities for each of the flux terms. The buoyancy flux term results from integration of Eq. (35), with $J_0 = \phi[g\alpha/(\rho c_p)]\phi_{sp}$:

$$\int_{-h}^{0} \overline{w'b'} dz = \frac{B_0 h}{2} - \frac{1}{2} w_e \Delta b h + J_0 \left(\frac{h}{2} - \frac{1}{\eta} \right) \tag{39}$$

Similarly, for the momentum flux equation,

$$\int_{-h}^{0} \overline{w'v'} \cdot \frac{\partial \mathbf{v}}{\partial z} dz = \lim_{\Delta h \to 0} \int_{-h-\Delta h}^{-h} \overline{w'v'} \cdot \frac{\partial \mathbf{v}}{\partial z} dz \tag{40}$$

since the flux is zero within the ML itself (viz. in the range $z = 0$ to $z = -h$). Using Eq. (36) to substitute for $\overline{w'v'}$ at this depth ($z = -h$) gives

$$\int_{-h}^{0} \overline{w'v'} \cdot \frac{\partial \mathbf{v}}{\partial z} dz = \lim_{\Delta h \to 0} \int_{-h-\Delta h}^{-h} (C \, \bar{\mathbf{v}}|\bar{\mathbf{v}}| - w_e \bar{\mathbf{v}}) \cdot \frac{\partial \mathbf{v}}{\partial z} dz = \frac{1}{3} C|\bar{\mathbf{v}}|^3 - \frac{1}{2} w_e \bar{\mathbf{v}}^2 \tag{41}$$

Finally, the flux of mechanical energy is given in its integrated form by

$$\int_{-h}^{0} \frac{\partial}{\partial z} \left[\frac{1}{2} \overline{w'(w'^2 + \mathbf{v}'^2)} + \frac{\overline{w'p'}}{\rho} \right] dz = \left[\frac{1}{2} \overline{w'(w'^2 + \mathbf{v}'^2)} + \frac{\overline{w'p'}}{\rho} \right]_{-h}^{0} \tag{42}$$

Substituting from Eqs. (37) and (38) into Eq. (42) thus gives

$$\int_{-h}^{0} \frac{\partial}{\partial z} \left[\frac{1}{2} \overline{w'(w'^2 + \mathbf{v}'^2)} + \frac{\overline{w'p'}}{\rho} \right] dz = -m_1 \overset{*3}{w} - \frac{1}{2} w_e q^2 \tag{43}$$

Substitution of each of these integrated flux relationships into the TKE equation (29) will yield the *integrated TKE equation:*

$$\frac{1}{2}\frac{\partial}{\partial t}(\bar{q}^2) = \left[-\frac{1}{3}C|\bar{\mathbf{v}}|^3 + \frac{1}{2}w_e\bar{\mathbf{v}}^2 \right] + \left[\frac{B_0 h}{2} - \frac{1}{2}w_e\Delta b h + J_0\left(\frac{h}{2} - \frac{1}{\eta}\right) \right]$$

$$+ \left[m_1 \overset{*3}{w} - \frac{1}{2}w_e q^2 \right] - \int_{-h}^{0} \epsilon dz \qquad (44)$$

where

$$\bar{q} = \int_{-h}^{0} q\, dz \qquad (45)$$

Niiler and Kraus (1977) consider this equation *for the steady state* $(\partial/\partial t) \equiv 0)$, thus assuming that there is no change in overall TKE, $\bar{q}^2$, which might occur at the expense of the main KE or PE fields. Using the same nomenclature as these authors, Eq. (44) gives for this IML model:

$$\frac{1}{2}w_e(q^2 + c_i^2 - \bar{\mathbf{v}}^2) = m_1 \overset{*3}{w} + \frac{1}{2}hB_0 + \left(\frac{h}{2} - \frac{1}{\eta}\right)J_0 - \frac{1}{3}C|\bar{\mathbf{v}}|^3 - \int \epsilon dz \qquad (46)$$

where c_i is the velocity of interfacial long, internal waves, given by

$$c_i^2 = \Delta b h \qquad (47)$$

It should be noted that this corrects a typographical error in equation (10.27) of Niiler and Kraus (1977), specifically returning the factor of $(-1/3)$ into term H [see Eq. (48)].

In the model development of Niiler (1977), the time dependency is retained. Assuming the TKE density to be uniform within the mixed layer, this additional term of $0.5 \int (\partial q^2/\partial t)dz$ is shown to be approximately equal to $0.5\, c_e \overset{*2}{w} w_e$ (where c_e is an empirical constant) such that Eq. (44) is rewritten as

$$\frac{1}{2}w_e(q^2 + c_i^2 - \bar{\mathbf{v}}^2 + c_e \overset{*2}{w}) = m_1 \overset{*3}{w} + \frac{1}{2}hB_0 + \left(\frac{h}{2} - \frac{1}{\eta}\right)J_0$$

$$\qquad \text{A} \quad \text{B} \quad \text{C} \quad \text{D} \qquad \text{E} \qquad \text{F} \qquad \text{G}$$

$$-\frac{1}{3}C|\bar{\mathbf{v}}|^3 - \int \epsilon dz$$

$$\qquad \text{H} \qquad \text{I}$$

$$(48)$$

The interpretation of the terms is as follows:

A = rate of energy needed to agitate the entrained water
B = rate of work to lift entrained water and mix it
C = reduction of mean KE by entrainment
D = rate of change of TKE

E = rate of work by wind
F = rate of PE change by surface fluxes
G = rate of PE change by solar radiation
H = work from internal waves
I = dissipation

As noted previously, this equation is of a similar form to Eq. (28) in which F_s is given by term E in Eq. (48); E by term C; D by term I; F_h by term H. The rate of change of TKE in Eq. (28) is analogous to term D and of PE to terms B, F, G. Term A appears to have no direct analogue in Eq. (28) (but is shown below to be negligible).

Modeling approaches to thermal stratification which utilize the integral energy (mixed layer) concept differ in their inclusion, neglect or difference in parameterization in each of these different sources of TKE and in how the net TKE is partitioned between enhancing waves, currents and the descent of the thermocline (Fig. 1). Three basic modeling approaches can be identified and these will be discussed in detail below. These have been identified by Price et al. (1978) and others as (1) the energy conservation model (ECM); (2) the dynamic instability model (DIM); and (3) the turbulent erosion model (TEM).

As noted by Niiler and Kraus (1977) that, while the specification of the length scale, l, is the most arbitrary part of a turbulent diffusion closure model [e.g., Mellor and Durbin, 1975; see Eq. (92) below], the most arbitrary part of an IML model is in the specification of the dissipation term. Niiler and Kraus (1977) compare several formulations and advocate that a fraction m, of the wind energy, a fraction s, of the mean KE reduction as a result of entrainment, and a fraction n, of the convective flux, all contribute to the dissipation (cf. Fig. 1a). This gives rise to a simplification by substituting in Eq. (48) for the dissipation term (I) now given as

$$\int_{-h}^{0} \epsilon \, dz = (m_1 - m) \overset{*}{w}^3 + (1 - s) \frac{1}{2} w_e \bar{v}^2 + (1 - n) \frac{1}{2} h \frac{B_0 + |B_0|}{2} \tag{49}$$

where the z axis is orientated vertically upward with an origin at the water surface and the ML base is at $z = -h$. [Note the buoyancy term, $(B_0 + |B_0|)/2$, has a value of B_0 for $B_0 > 0$ and zero for $B_0 \leq 0$ and is thus one representation for the Heaviside function.] Upon substitution and algebraic simplification this gives

$$\frac{1}{2} w_e (q^2 + c_i^2 - s\bar{v}^2 + c_e \overset{*}{w}^2) = m\overset{*}{w}^3 + \frac{h}{4} [(1 + n)B_0 - (1 - n)|B_0|]$$

$$+ \left\{ \frac{h}{2} - \frac{1}{\eta} \right\} J_0 - \frac{1}{3} C|\bar{v}|^3 \tag{50}$$

The final version of the Niiler and Kraus (1977) model (but including the time dependency of TKE as suggested by Niiler, 1977) suggests that, by an order of magnitude analysis, terms A and H are negligible. Hence

$$w_e(c_i^2 - s\bar{\mathbf{v}}^2 + c_e\overset{*}{w}{}^2) = 2m\overset{*}{w}{}^3 + \frac{h}{2}[(1 + n)B_0 - (1 - n)|B_0|] + \left(h - \frac{2}{\eta}\right)J_0 \quad (51)$$

where, it should be stressed, the coefficients s, m, n and c_e must be empirically derived for each specific study and are hence "tuning coefficients" [see, e.g., Davis et al. (1981), who used this tuning capacity to represent the results of the MILE experiment using the Niiler, 1975, model].

In addition to this Niiler and Kraus (1977) model, there are two other major simplifications of the steady state Eq. (46) which have retained favor over the last decade: those of Kraus and Turner (1967) (KT model) and of Pollard et al. (1973) (PRT model) (Fig. 2). Niiler (1975) was the first to ascertain that both the KT and PRT models were implicit within the full TKE model. This idea was also developed by de Szoeke and Rhines (1976), Niiler (1977) and by Price et al. (1978). All these authors note that following simplification of Eqs. (46) and (48) to Eq. (51), this latter question can then be further simplified in order to regain the KT and PRT models.

The Kraus-Turner (KT) model (Kraus and Turner, 1967) combines the ideas of a compensation depth to determine the extent of convective mixing and include a parameterization of the descent of the thermocline by then integrating quantities over the ML depth, expressing their results nondimensionally. They assume a balance between input KE, G^*, convectively generated TKE, W^*, and dissipation, D^*:

$$W^* + G^* = D^* \quad (52)$$

Expressing these in terms of the variables of the model and integrating over the ML depth allowed Kraus and Turner (1967) to derive for the mixed layer expressions for both (1) the thermal budget:

$$\frac{dT_s}{dt} = \frac{2}{h^2}\left[(S' + B')h - \left(G' - D' + \frac{S'}{\eta}\right)\right] \quad (53)$$

where G' and D' are nondimensionalized (by dividing by $g\alpha\rho$) versions of G^* and D^*; S' is the nondimensionalized penetrative solar radiation (scale depth η); B' is the nondimensionalized buoyancy flux (S' and B' being equal to S^* and B^* divided by $\rho\,c_p$); h is the ML depth; and T_s its temperature; and for (2), the interfacial entrainment velocity, $w_e \equiv dh/dt$:

$$\Lambda\frac{dh}{dt} = \frac{1}{(T_s - T_h)h}\left[2\left(G' - D' + \frac{S'}{\eta}\right) - (S' + \eta)\,h\right] \quad (54)$$

where T_h is the temperature at a depth z and Λ is the Heaviside function given by

$$\Lambda \equiv \Lambda(w_e) = \begin{cases} 1 & \text{for } w_e > 0 \\ 0 & \text{for } w_e < 0 \end{cases} \quad (55)$$

This model can be regained from the full TKE model as follows. Consider Eq. (51) for the steady state ($c_e = 0$) in which there is no dissipation (viz. $s = 1$) and no heating (viz. $J_0 = 0$, $B_0 = 0$).

Then, we have

$$\frac{1}{4} N^2 h^2 \frac{dh}{dt} = m \overset{*3}{w} + \frac{1}{2} \frac{dh}{dt} \bar{v}^2 \tag{56}$$

where the Brunt-Väisälä frequency N is given by (Niiler, 1977, Cushman-Roisin, 1981)

$$N^2 = 2c_i^2 h^2 \tag{57}$$

The KT model results from assuming that thermocline deepening is caused by turbulence propagating from the surface and eroding the thermocline [a turbulent erosion model (TEM); Price et al., 1978, Cushman-Roisin, 1981], i.e., in Eq. (56), the left hand side is balanced only by the first term on the right hand side [from Eq. (57)]:

$$\frac{dh}{dt} = \frac{4m \overset{*3}{w}}{N^2 h^2} \quad \text{or} \quad \frac{2m \overset{*3}{w}}{c_i^2} \tag{58}$$

This is equivalent to reallocating the value of s in Eq. (51) as zero (a procedure used also by Denman, 1973). Alternatively the TEM can be derived from Eq. (28) by neglecting E and F_h such that

$$\frac{d}{dt} (\text{PE}) = F_s - D \tag{59}$$

A modification to the Garwood (1977) model is presented in a similar, simplified format by Adamec and Elsberry (1984). For an actively deepening layer, an equation analogous to Eq. (56) can be derived by retaining the surface heat flux term (viz. $c_e = 0$, $s = 1$, $J_0 = 0$, but with $n \neq 0$):

$$\frac{1}{4} N^2 h^2 \frac{dh}{dt} = m \overset{*3}{w} + \frac{hn}{2} B_0 + \frac{1}{2} \frac{dh}{dt} \bar{v}^2 \tag{60}$$

Following the argument above regarding the derivation of the KT model, the last term on the right hand side of Eq. (60) is neglected such that

$$\frac{dh}{dt} = \frac{4m \overset{*3}{w}}{N^2 h^2} + \frac{2n}{N^2 h} B_0$$

$$= \frac{2m \overset{*3}{w}}{c_i^2} + \frac{hn}{c_i^2} B_0 \tag{61}$$

Substituting for c_i^2,

$$c_i^2 = \Delta b h = \alpha g \Delta T\, h \tag{62}$$

and for B_0 in terms of the surface heat flux, Q_n from

$$B_0 = \frac{g\alpha}{\rho c_p} Q_n \tag{63}$$

permits Eq. (61) to be rewritten in the form quoted by Adamec and Elsberry (1984):

$$\frac{dh}{dt} = \frac{2m\overset{*}{w}^3/h + n(g\alpha/\rho c_p)Q_n}{\alpha g \Delta T} \tag{64}$$

When there is no entrainment-induced deepening, $dh/dt = 0$ such that

$$h = \frac{-(2m\overset{*}{w}^3\rho c_p)}{ng\alpha Q_n} \tag{65}$$

An interesting extension to the basic TEM structure has been proposed recently by Cushman-Roisin (1981) based on the schematization of mixing in terms of a field of rising and descending thermals (Fig. 7). He proposes that if (descending) thermals occupy a fraction of the cross-sectional area at any level, f_T, and (rising) antithermals a fractional area of $(1 - f_T)$, then the fractional parameter m can be related to f_T. However, by relating this parameter to von Karman's constant, k, he shows that the optimum value for m is 1.25 and consequently the value for f_T is 0.109 (in good agreement with the range of observed laboratory and field data from which f_T can be deduced). A similar concept, of intermittency, is also given by Denton and Wood (1981) described in terms of "interfacial domes."

As an alternative to the KT model, Pollard et al. (1973) proposed that Eq. (52) should be modified to include the TKE produced in the interfacial shear layer

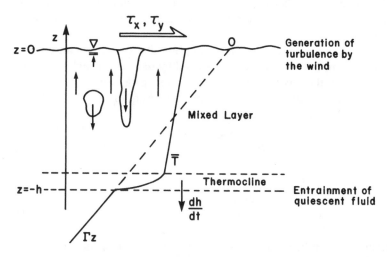

Figure 7 Schematic model of the ML model of Cushman-Roisin (1981).

at the base of the thermocline (the PRT model). They further identify W^* as the rate of change of mean potential energy $(\overline{PE})$ and G^* as $\tau \cdot \mathbf{v}$, where they stress this is the rate of working on the *mean* flow (original authors' italics), thus neglecting energy generation in excess of $\tau \cdot \mathbf{v}$ since the surface velocity u_s on which the wind acts is generally slightly larger than the mean slab velocity $\mathbf{u}$. Pollard et al. (1973) also neglect rates of change of TKE (as does the KT model) but also neglect dissipation [D^* in Eq. (52)]. In place of dissipation, they suggest the mechanical energy balance [modifying Eq. (53)] is completed in terms of alterations to the mean kinetic energy $(\overline{KE})$ of the water, a term not included in the KT model. This balance is thus written as

$$\frac{\partial}{\partial t}\overline{PE} + \frac{\partial}{\partial t}\overline{KE} = \tau \cdot \mathbf{v} \tag{66}$$

Algebraic manipulation results in an expression for w_e as

$$\frac{g\alpha h}{c_p \rho}Q + [g\alpha h(T + \Gamma h) - (u^2 + v^2)]w_e = 0 \tag{67}$$

where Q is the net surface heat flux and Γ the temperature gradient below the mixed layer. Pollard et al. (1973) note that for zero net surface heating, this reduces to

$$\frac{g\alpha h(T + \Gamma h)}{u^2 + v^2} \equiv R_i \tag{68}$$

Pollard et al. (1973) make the assumption that for a deepening thermocline ($dh/dt > 0$), the value of the Richardson number R_i is unity. Hence, they conclude, that in this initial deepening phase, the water column is marginally stable. Use of this critical Richardson number approach is equated to balancing at each time the rate of working of the stress on the mean flow against the rate of increase of potential plus kinetic energy.

Since this PRT model is based on the assumption that the dominant process in thermocline deepening is the turbulence produced by the shear instability at the interface at the bottom of the mixed layer, this "dynamic instability model (DIM)" (Price et al., 1978) can be regained from Eq. (56) by assuming $m = 0$ (still with $J_0 = 0$), but retaining s as a parameter:

$$h^2 = \frac{2s\bar{v}^2}{N^2} \tag{69}$$

Using Eq. (57), this can be rewritten as

$$\frac{c_i^2}{\bar{v}^2} = s \tag{70}$$

This is essentially a generalization of Eq. (68), in which the value of R_i (here equal to $c_i^2/\bar{v}^2$) is no longer assumed to be unity (see also discussion by Price et al., 1978).

Alternatively, setting $F_s = D = F_h = 0$ in Eq. (28), the DIM can be described by the equation

$$\frac{d}{dt}(PE) = E \tag{71}$$

Although Cushman-Roisin (1981) suggests that since the TKE adjusts quasi-instantaneously to perturbations, it is reasonable to remove the time dependency of Eq. (51) (viz. $c_e = 0$), it is instructive to consider the solution given by Niiler (1977) in which Eq. (51) was written as

$$\frac{1}{2}w_e\left(c_e\overset{*}{w}^2 + \frac{N^2h^2}{2} - |\bar{\mathbf{v}}^2|\right) = m\overset{*}{w}^3 \tag{72}$$

$$(P) \qquad (Q) \qquad (R) \qquad (S)$$

i.e., no surface heating and no radiation penetration. For this specific example, Niiler (1977) considers over what time scales the various processes included in Eq. (72) may dominate. These are illustrated in Fig. 8. Initially there is a balance between P and S. Hence integrating Eq. (72) gives $h \approx 2\,(m/c_e)\overset{*}{w}\,t$; a balance which holds for the first ~ 100 seconds of the simulation. This balance is replaced by one between terms Q and S such that $h \propto t^{1/3}$. In the third phase a balance is attained between Q and R, superseded by a balance at longer time scales, once again, between Q and S (see also discussion of time scales in Section 6).

In addition, although the Niiler and Kraus (1977) model to some degree expands and hence supersedes the Niiler (1975) model, this earlier model is often quoted and used directly in oceanic applications (e.g., Dillon and Powell, 1979).

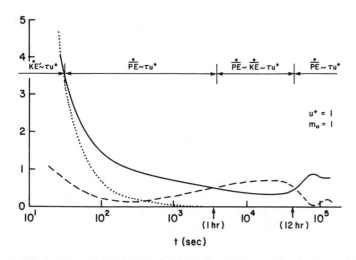

Figure 8 Numerical solution of the complete ML equation showing various asymptotic temporal regimes. Solid curve: the ratio S/Q; dashed curve: R/Q; dotted curve: P/Q. Parameter values used by de Szoeke and Rhines (1976) are $\overset{*}{w} = 1$, $m = 1$ and $c_e = 1$.

It is derived from Eq. (51) as follows: Terms F and G of Eq. (48) are amalgamated as a term for the overall rate of change of PE which is related to the heat content, θ by

$$\text{Increased PE} = \frac{1}{2} hnB_0 = -\frac{hg\alpha}{\rho^2 c_p} \frac{\partial \theta}{\partial t} \tag{73}$$

where θ is given by

$$\theta = \rho c_p \left(T + \frac{\Gamma h}{2} \right) h \tag{74}$$

Hence the steady state TKE equation can be rewritten as

$$w_e(c_i^2 - s\bar{v}^2) = 2m\overset{*}{w}^3 - g \frac{\alpha h}{\rho^2 c_p} \frac{\partial \theta}{\partial t} \tag{75}$$

It should now be noted that from the definition of c_i^2 [Eq. (47)], or from Eq. (57), we have

$$c_i^2 = \Delta bh$$

$$= \frac{\alpha}{\rho} gh(T + \Gamma h) \tag{76}$$

Niiler (1975) utilizes the transformation $\bar{U} = h\bar{v}$, $\dot{Q}_0 = \partial\theta/\partial t$, which, together with Eq. (76), permits the rewriting of Eq. (75) as

$$w_e\left[\frac{\alpha}{\rho} gh^3(T + \Gamma h) - s\bar{U}^2 \right] = 2m\overset{*}{w}^3 h^2 - \frac{g\alpha}{\rho^2 c_p} h^3 \dot{Q}_0 \tag{77}$$

Introducing the Brunt-Väisälä frequency $N^2 = \alpha g\Gamma/\rho$ and the definition of θ from Eq. (74):

$$w_e\left[\frac{\alpha}{\rho} gh^3\left(\frac{\theta}{\rho c_p h} + \frac{\Gamma}{2} h \right) - s\bar{U}^2 \right] = 2m\overset{*}{w}^3 h^2 - \frac{g\alpha}{\rho^2 c_p} h^3 \dot{Q}_0 \tag{78}$$

and hence

$$w_e\left(\frac{\alpha g}{\rho^2 c_p} h^2\theta + \frac{h^4}{2} N^2 - s\bar{U}^2 \right) = 2m\overset{*}{w}^3 h^2 - \frac{g\alpha}{\rho^2 c_p} h^3 \dot{Q}_0 \tag{79}$$

and it was this equation, but with zero surface heat flux, that was analyzed by de Szoeke and Rhines (1976) with the addition of a term $w_e c_e \overset{*}{w}^2 h^2$ on the left hand side of Eq. (79) to represent the energy needed to spin up the level of turbulence $c_e \overset{*}{w}^2$ in the increment of the mixed layer dh in time dt, i.e., the re-inclusion of term D from Eq. (48).

The three distinct ML model types thus identified are (1) Niiler and Kraus (1977) (ECM), which is also used in the original Niiler (1975) format; (2) Pollard et al. (1973) (PRT model or DIM); and (3) Kraus and Turner (1967) (KT model or TEM). Although the DIM and TEM are subsumed in the ECM, since many authors use the original KT model (e.g., Darbyshire and Edwards, 1972; Denman, 1973; Stefan and Ford, 1975a,b; Thompson, 1976; Haney and Davis, 1976; and Kim, 1976), the PRT model (e.g., Thompson, 1976; Kundu, 1981; and discussed by Kim, 1976) or the Niiler model (e.g., Dillon and Powell, 1979; Davis et al., 1981; and discussed by Kim, 1976) formulations, these original identities will be retained for discussion purposes. The implementation of these models by the cited authors will be discussed in detail in Section 5.

In the lacustrine environment, similar approaches have been used in ML modeling, although often with grosser approximations and frequently starting from first principles rather than by developing appropriate models from the oceanic ML models described above. Essentially such lake models are TEMs. Since it is intended to utilize such models to simulate seasonal time scales and longer, this is appropriate. For example, Stefan and Ford (1975a) derived their lake model by considering the input kinetic energy (from the wind) will change the potential energy of the water column or will be dissipated. This decision is accomplished by consideration of the ratio of the mechanical energy input, E_k, and the potential (buoyant) energy of the isothermal ML, E_p; where these two variables are given by

$$E_k = \tau \overset{*}{w} A_s \Delta t \tag{80}$$

$$E_p = g\Sigma V(i, k)[\rho(m + 1, k) - \rho(i, k)](m + 1 - i)\Delta z \tag{81}$$

where $V(i, k)$ the volume of layer i, $\rho(i, k)$ the density of layer i (both at time k), A_s is the surface area of the lake, Δt is a small time interval and Δz an incremental depth. Stefan and Ford (1975a) show that the ratio E_k/E_p is equivalent to the inverse of a type of Richardson number. Their critical value for this ratio was taken to be unity; although later authors have substituted this by a functional form of R_i. For example, the present version of the Massachusetts Institute of Technology stratification model (e.g., Harleman, 1982) (as well as the Waterways Experiment Station model CE-THERM; Environmental Laboratory, 1982) uses the formulation

$$\frac{\Delta E_{\text{pot}}}{\Delta E_{\text{total kinetic}}} = f'(R_i) \tag{82}$$

where $f'(R_i)$ is given by either

$$f'(R_i) = 0.057R_i \frac{29.46 - R_i^{1/2}}{14.20 + R_i} \tag{83}$$

if dissipation is considered and

$$f'(R_i) = \frac{R_i}{14.2 + R_i} \tag{84}$$

if dissipation is balanced by mechanical production of turbulent kinetic energy (Bloss and Harleman, 1979). These functions, shown in Fig. 9, suggest that at

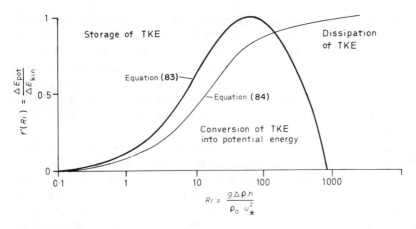

Figure 9 Dependence of the conversion of turbulent kinetic energy (TKE) into potential energy (PE) on the Richardson number (redrawn from Bloss and Harleman, 1979).

low Richardson numbers, most of the input KE is used to increase the TKE content of the ML; whereas at high R_i, most of it is dissipated and only at an intermediate value of R_i is the entrainment process efficient.

A similar approach is discussed by Sherman et al. (1978) and Imberger et al. (1978) but using four empirical constants. Spigel and Imberger (1980) synopsize this in the form of an ECM (named DYRESM), expressed in finite difference form, using the authors' own notation, by

$$\frac{1}{2}\left(C_T q^2 + \frac{\Delta\rho}{\rho_0}\, gh\right)\Delta h = \frac{C_K}{2}\, q^3 \Delta t + \frac{C_S}{2}\, U^2 \Delta h - \Lambda_L \Delta t \qquad (85)$$

$$\quad\;\; \text{(V)} \qquad \text{(W)} \qquad\quad \text{(X)} \qquad\; \text{(Y)} \qquad\;\; \text{(Z)}$$

This is seen to be directly analogous to the Niiler and Kraus (1977) steady state model [Eq. (50) with $c_e = 0$] in which the following identities can be identified: $C_S \equiv s$, $C_K \equiv 2m$, $\Lambda_L = (1/3)C|\bar{\mathbf{v}}|^3$, and $C_T = 1$; but in which the contribution of convection to the TKE budget [terms F and G in Eq. (48)] is included in the term $C_K q^3$ (term X) by redefining q by

$$q^3 = u_f^3 + \eta_*^3 w_*^3 \qquad (86)$$

where η_* is an additional empirical constant and u_f is the convective velocity defined by

$$u_f = \left(-\frac{\alpha g h Q_n}{\rho_0 c_p}\right)^{1/3} \qquad (87)$$

It is this retention of q explicitly that the authors highlight as the distinguishing characteristic of the model DYRESM (Spigel et al., 1986), coupled with the inclusion of the energetics of billowing due to Kelvin-Helmholtz instability at the base of the mixed layer, a mechanism identified in observations in Wellington

Reservoir (Western Australia) (Imberger, 1985). Spigel and Imberger (1980) also suggest that the internal wave term is negligible in lakes or, at worst, should be included in the parameterization of the coefficient C_K. They also note the inclusion in this model specification [Eq. (85)] of the KT model ($C_S = C_T = \Lambda_L = 0$) and the PRT model ($C_K = C_T = \Lambda_L = 0$).

Furthermore, the ratio of E_k to E_p is, in terms of the oceanic formulation of a TEM, essentially of the form

$$\frac{E_k}{E_p} = \frac{\rho \overset{*}{w}^3 A_s \Delta t}{V \Delta \rho g h} \tag{88}$$

For a straight sided lake, with thermocline depth h,

$$V = A_s h \tag{89}$$

such that Eq. (88) becomes

$$\frac{E_k}{E_p} \equiv \sigma = \frac{\rho \overset{*}{w}^3 \Delta t}{\Delta \rho g h^2} \tag{90}$$

Solving for $\Delta h / \Delta t$:

$$\frac{\Delta h}{\Delta t} = \frac{\rho \overset{*}{w}^3 \Delta h}{\sigma \Delta \rho g h^2} \tag{91}$$

and substituting for the Brunt-Väisälä frequency N^2

$$\frac{\Delta h}{\Delta t} = \frac{\overset{*}{w}^3}{\sigma N^2 h^2} \tag{92}$$

In comparison with the ocean TEM given by Eq. (58), Eq. (94) can be seen to be of similar form if

$$\sigma = \frac{1}{4m} \tag{93}$$

where Cushman-Roisin (1981) argues that the value of $2\ m$ is equal to the inverse of the von Karman constant, i.e., $m = 1.25$. This would imply that the critical energy ratio, σ, should be 0.4 (not 1), in accordance at least qualitatively with the ideas embodied in formulations such as Eq. (76) and Fig. 9, although it should be noted that, once again, that in this model turbulent transfer below the thermocline is neglected.

Alternative remedies to the limitations of ML models in describing long term stable stratification patterns, heating at depth and in describing the thermocline interface in terms of a (discontinuous) step function has been proposed by Posmentier (1980) and Price et al. (1986). In both these approaches, it is postulated that there is a diffusion of thermal energy through the thermocline interface which is determined by the local value of the Richardson number. Posmentier (1980) deduces that such a heat flux has a maximum for Richardson numbers ≈ 0.6 with significantly smaller values for extremely stable or nearly neutral situations.

4.2 Differential Models

Eddy diffusion models of the thermocline start with the same set of equations [Eqs. (1) and (2)] but parameterize the Reynolds' stress terms using an eddy diffusion coefficient, K. This closure can be undertaken at any one of a hierarchical set of "levels" (Mellor and Yamada, 1974). At the lowest level, a relationship of the form

Heat:
$$-\overline{w'T'} = K_H \frac{\partial T}{\partial z} \tag{94}$$

Momentum:
$$-\overline{w'\mathbf{v}'} = K_M \frac{\partial \mathbf{v}}{\partial z} \tag{95}$$

is postulated and it is this formulation that was originally applied extensively. Differences between models relate to the mode of parameterization of the value of the eddy diffusion coefficients, K_H and K_M as functions of both the ambient meteorological and oceanographic conditions and as a function of the stability. Two basic alternative formulations are current: that of the Mellor and Yamada level-2 (MYL2) scheme in which

$$K_M = lqS_M \tag{96}$$

$$K_H = lqS_H \tag{97}$$

and the classical representation of an eddy diffusion coefficient as the product of a neutral value (denoted by subscript 0) and a function of the stability (often expressed in terms of the Richardson number R_i) (e.g., Rossby and Montgomery, 1935; Munk and Anderson, 1948):

$$K_M = K_{M0}g(R_i) \tag{98}$$

$$K_H = K_{H0}f(R_i) \tag{99}$$

In these equations, $q^2/2$ is the TKE; l a turbulent length scale; and S_M, S_H, $f(R_i)$, and $g(R_i)$ are functions of stability. It should be noted that both approaches require a measure of stability relationships, but in the MYL2 approach, additional equations for the TKE and for the length scale, l, are needed in comparison with a relationship for K_{M0} and K_{H0} in the classical approach. In the MYL2 model, this requirement produces a link with the IML models since the TKE equation used by Mellor and Yamada (1974) and Mellor and Durbin (1975) is a form of the TKE equation for the steady state discussed in Section 4.1:

$$0 = K_M \left[\left(\frac{\partial u}{\partial z} \right)^2 + \left(\frac{\partial v}{\partial z} \right)^2 \right] + K_H \frac{g}{\rho} \frac{\partial \rho}{\partial z} - \frac{q^3}{c_2 l} \tag{100}$$

In this equation the term $q^3/c_2 l$ represents the dissipation, c_2 is a constant equal to 15.0 (Mellor and Durbin, 1975) and the length scale l is estimated from the Blackadar boundary layer formula (see discussion in Mellor and Yamada, 1974; Mellor and Durbin, 1975):

$$\frac{1}{l} = \frac{1}{l_0} + \frac{1}{kz} \tag{101}$$

where

$$l_0 = \gamma \frac{\int_{z=-h}^{z=0} |z| q dz}{\int_{z=-h}^{z=0} q dz} \tag{102}$$

The value of the constant γ is taken to be 0.2.

In essence, then, the Mellor and Durbin (1975) model neglects the vertical diffusion of TKE and balances the local rate of TKE production by the local rate of dissipation (Kundu, 1980).

In a modification to the Mellor and Yamada (1974) level 3 model, labeled as a level 2 1/2 model by Mellor and Yamada (1982), a more comprehensive TKE equation is used whereby the left hand side of Eq. (100) is replaced by the time dependent term

$$\frac{\partial (q/2)^2}{\partial t} - \frac{\partial}{\partial z} \left[K_q \frac{\partial (q^2/2)}{\partial z} \right] \tag{103}$$

Mellor and Durbin (1975) neglect third-order correlations, while, e.g., Garwood (1977) include a parameterization of these as part of the TKE production term. However, Warn-Varnas and Piacsek (1979) include these specifically in an extension of the MD model—essentially an MYL-5 model, which is compared with the Mellor and Durbin (MYL-2) model by Warn-Varnas et al. (1981). Although this resultant model requires tuning, they demonstrate the need to include such terms. They suggest that such a modification can, in some cases, increase the ML depth by a factor of three and result in higher entrainment rates.

A third level Mellor and Yamada (1974) model is also utilized by Worthem and Mellor (1980) in an application of a stratification model to the upper tropical ocean. In this case GATE observations indicate the existence of a relatively large velocity shear in the upper ocean layer; of large-scale upwelling; large amplitude internal waves; a large diurnal variations in thermodynamic forcing (e.g., in solar radiation over the daylight hours).

Marchuk et al. (1977) present a variant on the Mellor and Durbin differential model, which utilizes a simpler closure model for eddy diffusion more akin to the parallel developments in lake stratification modeling (see Section 5.2). Their value of the eddy diffusion coefficient for momentum is related to the rate of turbulent decay, D_T, and the TKE per unit mass, q_m^2, by

$$K_M = 0.08 \frac{q_m^2}{D_T} \tag{104}$$

and the value for K_H is related to K_M by means of a Munk-Anderson formulation for the turbulent Prandtl number.

Lower order closure models have also been frequently applied in both oceans and lakes. Pacanowski and Philander (1981), in their investigation of tropical oceans, utilize values of $f(R_i)$ and $g(R_i)$ given by

$$f(R_i) = (1 + a_3 R_i)^{-(b_3+1)} \tag{105}$$

$$g(R_i) = (1 + a_3 R_i)^{-b_3} \tag{106}$$

and conclude that such a representation gives improved results compared with constant values for the eddy diffusivities, with values of $K_0 \sim 5 \times 10^{-3} \text{ m}^2 \text{ s}^{-1}$, $a_3 = 5$ and $b_3 = 2$.

A similar formulation was adopted by James (1977) in a study of the annual temperature cycle in a frontal region of the Celtic Sea. He utilized the Munk and Anderson (1948) formula directly to represent the effects of stratification but modified the neutral value as a discontinuous function of wind speed, while noting that the maximum diffusivity occurred at mid-depths. This subsurface maximum for the value of K_{H0} has been noted by others (e.g., Kundu, 1980; Weber, 1981; 1983) and investigated theoretically by Henderson-Sellers (1985b). This last author developed, from boundary layer concepts, an analytical representation of the neutral value, K_{H0}, which depicted such behavior:

$$K_{H0} = \frac{k \overset{*}{w_s} z}{P_0} \exp(-k^* z) \tag{107}$$

where P_0 is the neutral value of the turbulent Prandtl number and k^* is related to the reciprocal of the Ekman depth (Smith, 1979). In this approach the non-neutral value is then given by

$$K_H = K_{H0}(1 + 37 R_i^2)^{-1} \tag{108}$$

(Henderson-Sellers, 1982). This eddy diffusion profile has a subsurface maximum which compares well with observations (Filatov et al., 1981) as well as with the MYL-5 calculated profile (Warn-Varnas et al., 1981) cf. MYL-2 calculation (Mellor and Durbin, 1975) and the calculations of Kundu (1980) depicted in Fig. 10. These formulations [Eq. (107) and (108)] are embodied in the University of Salford Eddy Diffusion model, U.S.E.D.* (Henderson-Sellers, 1985b). In this model, the dynamics and thermodynamics are effectively decoupled by means of Eq. (107) so that Eq. (2) is solved in the form

$$A(z) \frac{\partial T}{\partial t} = \frac{\partial}{\partial z} \left[A(z)(\alpha_H + K_H) \frac{\partial T}{\partial z} \right] + \frac{\partial [A(z)\phi(z)]/\partial z}{\rho_w c_p} \tag{109}$$

where the molecular diffusivity of heat, α_H, is included so that for low wind speed or large depth situations, there is still a minimal diffusion occurring. There is no need to specify the TKE, as in the Mellor and Durbin (1975) diffusion model, since the form of closure chosen for K_H is sufficient. Inclusion of a full surface energy budget (Henderson-Sellers, 1986) and part penetration of short-

---

*More recently recorded as the model EDD1 (Eddy Diffusion Dimension 1).

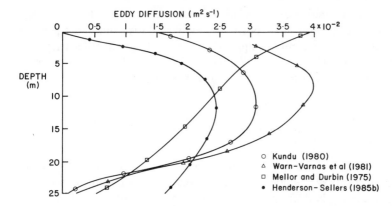

Figure 10 Profiles of eddy diffusion coefficient showing subsurface maximum.

wave radiation in the model U.S.E.D. ensures it is realistic, numerically stable over long time periods and able to simulate convection successfully (see also later discussions).

4.3 "Diffusive" Mixed Layer Models

It is interesting to note that the realization that the process of molecular diffusion must be included in stratification models has also been demonstrated for ML models by Denton and Wood (1981). By considering diffusion at low Péclet numbers, they combine many of the attributes of ML and eddy diffusion models. In contrast to Fig. 1b, which shows a discontinuity in temperature, in their model Denton and Wood (1981) assume that temperature is continuous but that the temperature gradient is discontinuous. This avoids the awkward concept of a discrete temperature and density jump at the interface and permits an illuminating discussion of penetrative convection. Essentially, this model utilizes Eq. (109) for constant $A(z)$ and in which the heat penetration term is not considered:

$$\frac{\partial T}{\partial t} = \frac{\partial}{\partial z} (\alpha_H + K_H) \frac{\partial T}{\partial z} \qquad (z < -h) \tag{110}$$

together with a simple expression for the total heat budget of the mixed layer [similar to that often used in simple oceanic climate models; see Section 5.4 and especially Eq. (118)]:

$$\frac{d}{dt}(hT_s) = Q_n - Q_e + w_e T_s \tag{111}$$

where Q_e is the interfacial heat flux and Q_n is the net surface heat input. The assumption of discontinuity of temperature profile across the interface permits an expansion of dT/dt, which, on rearrangement, gives an expression for w_e

$$w_e = -\frac{(dT_s/dt) - (\partial T/\partial t)_i}{(\partial T/\partial z)_i} \tag{112}$$

where subscripts s and i indicate values immediately above and below the interface. [Note Denton and Wood use z positive downward, such that the equations presented here have additional minus signs associated with gradient terms.] Substitution of Eqs. (110) ($[\partial T/\partial t]_i$) and (111) (for dT_s/dt) into Eq. (112), together with a gradient representation for Q_e, gives an expression for the rate of change of ML depth w_e as

$$w_e = \frac{-Q_n}{h(\partial T/\partial z)_i} + \frac{K_H + \alpha_H}{h} + (K_H + \alpha_H)\left(\frac{\partial^2 T/\partial z^2}{(\partial T/\partial z)}\right)_i + \left(\frac{\partial K_H}{\partial z}\right)_i \quad (113)$$

The first term on the right hand side of this equation describes changes in ML depth due to the heat flux Q_n (nonpenetrative convection), whereas the other terms are the new terms arising both from turbulent *and* molecular diffusion at the thermocline (interfacial) level and from the assumptions relating to the temperature profile. The model is closed by using a functional form for K_H at the interfacial depth, $(K_H)_i$, given by Eq. (99). To satisfy data from both high and low Péclet number cases, they suggest that $f(R_i)$ is given by

$$f(R_i) = 1.18(1 + 0.41R_i^{3/2})^{-1} \quad (114)$$

which agrees well with Turner's (1968) experiments with a salinity difference across an interface but not with the boundary layer data of Ueda et al. (1981) upon which Eq. (106) was largely based. The depth profile for K_H is normalized by the interfacial value $(K_H)_i$ and found, from experimental data, to be exponential, in agreement with the theoretical arguments of Henderson-Sellers (1985b) [Eq. (107)].

Denton and Wood (1981) point out that from this time dependent analysis, which includes both turbulent and molecular diffusion terms, it is possible to provide a more realistic model of temporal variation of mixed layer depth which has been tested with experimental data sets.

In Price et al.'s (1986) extension of the DIM of Price et al. (1978), applied by them to a study of diurnal mixing, a critical value of the bulk Richardson number, R_b, of 0.65 is used to ensure mixed layer stability; together with a critical gradient Richardson number, R_g, of 0.25 for shear flow stability and a requirement that $\partial \rho/\partial z \leq 0$ for static stability. These two Richardson numbers are given by

$$R_b = -\frac{g\Delta\rho h}{\rho_0(\Delta v)^2} \quad (115)$$

$$R_g = -\frac{g\partial\rho/\partial z}{\rho_0(\partial v/\partial z)^2} \quad (116)$$

Such modifications permit mixing across and beneath the thermocline boundary (assumed in many previous models to be a (near-) discontinuity and corrects the misconception, expressed earlier, that there is "no leakage of energy into the (hypolimnetic) deeper waters." In terms of simulated temperature profiles it has the effect of "smoothing" the interfacial values (Fig. 11), a capability that has never been denied in the eddy diffusion models (cf. Figs. 6 and 16). It is with

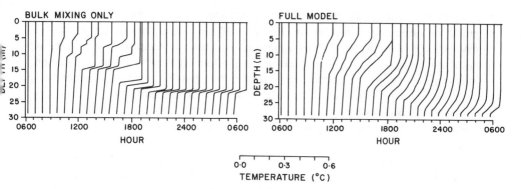

Figure 11 Mixing experiments run using the numerical model of Price et al. (1986) with bulk wind mixing process only for comparison to the full model (right); idealized forcing is used in this sensitivity analysis (after Price et al., 1986 copyright by the American Geophysical Union).

these aims in mind that this review has hoped to elucidate these and other areas of potential cross fertilization.

5 APPLICATION OF STRATIFICATION MODELS

It is not the intention here to undertake a comparative evaluation of simulations undertaken by the various versions of the TKE model since this has already been done adequately before (e.g., de Szoeke and Rhines, 1976; Thompson, 1976; Camp and Elsberry, 1978; Price et al., 1978; Martin, 1985). Rather we wish to emphasize the applications of these models in their various formats, especially with respect to different types and sizes of water bodies.

However it is worth describing here the results of sensitivity testing of the shortwave penetrative term for both lakes and oceans since, although such testing has been undertaken largely with eddy diffusion models, the conclusions are as relevant to bulk mixed layer models and equally for oceans and lakes.

Testing of the importance of accurate specification of the extinction coefficient has been presented by Hurley Octavio et al. (1977) and Henderson-Sellers (1985a), in which it was stressed that the stratification pattern in lakes was highly dependent on the turbidity, as quantified by the value of η in Eqs. (6), (7), (9) or (10), a conclusion reinforced using a later version of the University of Salford Eddy Diffusion model U.S.E.D. (Henderson-Sellers, 1985b). Oceanic sensitivity studies have been largely undertaken using the Mellor and Yamada level 2 1/2 model by Simpson and Dickey (1981a, 1981b), Dickey and Simpson (1983). They note that many oceanic models ignore such differential absorption or parameterize it inappropriately by ignoring surface absorption or, as in many of the ML studies (e.g., Niiler and Kraus, 1977), assuming total surface absorption. These authors conclude that there is little difference between the various parameterizations of shortwave energy penetration at high wind speeds, but that at wind speeds lower than about 10 m s^{-1} it is vital to include an accurate representation of the depth dependency in the penetration of the shortwave radiation. These difference regimes are attributed (Simpson and Dickey, 1981b) to the supposition that at low wind speeds, the TKE budget is dominated by shear production, dissipation and diffusion, while at high wind speeds, shear production is balanced by dissipation. In this second paper, they compare a further two possible parameterizations and conclude that the two term exponential and the arctangent model give nearly equivalent results (as supported by the results shown here in Fig. 4) although they suggest that the former is easier to interpret and evaluate and is more efficient computationally.

Within the context of the importance of the full specification of the shortwave radiation profile, as discussed above and earlier with regard to the calculation of a compensation depth, it is important to note that in an integral or bulk ML in which the total radiation absorbed in the ML is assumed to be evenly distributed with depth or, equivalently, at the surface (e.g., Kitaigorodskii, 1979), then the static instability caused by surface heat loss (as identified by Woods, 1980) *cannot* be represented. Consequently any model which does not allow for such radiation penetration and the possibility of surface instabilities must be viewed with caution.

Often such assumptions are linked with an intention to derive an *analytical* solution. In order to do this several more (over) simplifying assumptions are often made. Although bulk ML models are usually derived to incorporate surface heating, many authors exemplify the use of their model by immediately setting the surface heating term either to positive definite or zero (e.g., de Szoeke and Rhines, 1976; the majority of examples in Niiler and Kraus, 1977; Niiler, 1977; Kitaigorodskii, 1979; Kundu, 1980; Spigel and Imberger, 1980; Cushman-Roisin, 1981). This difference in solution approach should be stressed. Although the correlation is far from perfect, it has been the case, until recently, that most ocean models were solved analytically (see, e.g., discussions of Niiler and Kraus, 1977) whereas most lake models were derived in finite difference form (e.g., Stefan and Ford, 1975a; Spigel and Imberger, 1980; Thompson and Imberger; 1980). However, several recent implementations of oceanic ML models have been directed toward eventual incorporation in a GCM (Section 5.4) and have been recast in finite difference form accordingly (e.g., Thompson, 1976; Elsberry and Garwood, 1980; Semtner, 1984).

The one-dimensional model discussed in this chapter, common to both oceanic and lacustrine environments, has been frequently analyzed. In this case, the processes which could invalidate such an assumption differ between the types of water body. In the ocean, lateral advection of, especially, heat into the water column is of concern whereas in a lake, where surface processes are assumed to be homogeneous across the lake surface, inhomogeneous processes such as seiches, local upwelling, spatially distinct inflows and outflows may require parameterization. Thompson and Imberger (1980) examine the validity of the 1-D assumption in terms of the Wedderburn number, given by

$$W = \frac{g\Delta\rho d^2}{\rho \overset{*}{w}{}^2 L^*} \tag{117}$$

(e.g., Imberger and Hamblin, 1982), where $\Delta\rho$ is the density difference between the two layers, d is the depth of the diurnal mixed layer and L^* the length of the lake in the wind direction. For values of $W > 3$, the 1-D assumption is found to hold, provided that the model contains parameterization of the shear production of TKE (Patterson et al., 1984).

Another assumption common to both one-dimensional models of lake and oceanic applications is, perhaps surprisingly, that water density is a function of temperature alone (e.g., Zholudev, 1983). Suspended and dissolved material can of course alter this relationship and in specific case studies this has been taken into account (e.g., for lakes Stefan et al., 1982; for oceans Aikman, 1984; Price et al., 1986). Saline effects in point ocean models are usually neglected but included in 3-D simulation models, although as Bretherton (1982) and Haney (1985) point out, in the global ocean this can lead to significant errors at high latitudes where the coefficient of thermal expansion of sea water is relatively small.

As noted in Section 1, perhaps the greatest contrast in ML models for lakes and oceans is in the necessity to include for lakes variations, with depth, of the cross-sectional area, $A(z)$. Discussion of the appropriateness of constant area lake models to describe real lakes in Bedford and Babajimopoulos (1977) concludes that in the more approximate model with constant area, parameter values (such as eddy diffusion coefficients) must be tuned significantly. The contrast is readily illustrated in Fig. 12 for simulations of a typical U.K. lake with (a) realistic values for $A(z)$ and (b) a constant value for $A(z)$. Both models have the same surface area, so that the volume of the constant A case is much larger. Consequently at depth, the greater volume of water to be heated results in a smaller temperature rise so that summer hypolimnetic temperatures are several degrees cooler. The "funneling effect" of real bathymetry results effectively in heat reaching the bottom of the lake more quickly. Since the constant area case (Fig. 12b) is closely akin to the water column model necessitated in any point ocean model, although not in a full 3-D simulation, the contrast between the two parts of Fig. 12 essentially reflects the differences between a lake (Fig. 12a) and a point ocean (Fig. 12b) at the same latitude, wherein all other dynamic effects (currents, wave field, advection, etc.) have been neglected.

A useful graphical method of model comparison is given by Garwood (1979) (Fig. 13). The figure compares five possible model variations. In each 3-D plot, the vertical parameter, P^*, is the nondimensionalized mixed layer entrainment expressed as a function of stability, $H^* = h/L_1$, and a neutral depth parameter, $Z^* = h/L_2$ (both nondimensional). Garwood (1979) discussed the various characteristics of these five selected models (but notes other variants are possible, e.g., Camp and Elsberry, 1978, who combine (b) and (c).

The time scale of scientific interest may, as has already been noted, help to determine the model selection; so that, for example, a model of the PRT type may be useful on a short but not a long time scale. For simulations over years or decades, especially those on climatological time scales, it is vital that any unperturbed (control) simulation is algorithmically and numerically unconditionally stable before comparisons are made in, say, a doubling CO_2 experiment. The algorithmic stability has been discussed briefly in the literature. Niiler (1977) highlights models which have been used to simulate a single seasonal cycle only and identifies only one model, that of Warren (1972), which appears to have been proven to be stable over several years (although Haney and Davies, 1976, and Henderson-Sellers, 1978, also present results from long term, unconditionally stable models). It would appear that the common characteristics of these stable models in, in fact, the inclusion of a complete description of the surface energy budget which allows interactive feedbacks to operate between atmosphere and ocean.

A further useful test of the validity of any stratification model for use on long time scales can be assessed by its ability to reproduce the observed hysteresis in the curve of heat content versus surface temperature (Gill, 1974). This is endorsed by, e.g., Bretherton (1982), who recommends, as a convenient verification framework, plots of heat stored against surface temperature such as that shown in Fig. 14, based on OWS Echo observations from Gill and Turner (1976). He demonstrates that a simple two layer model with fixed ML depth cannot exhibit such hysteresis and that the agreement between model and observations can be assessed

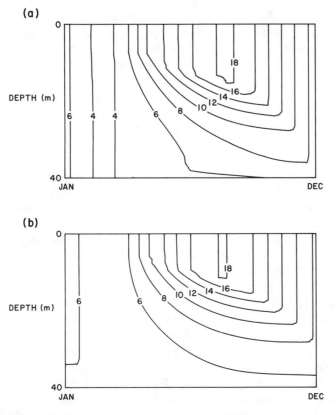

Figure 12 Thermal structure for typical U.K. lake with the assumptions of (a) realistic bathymetry and (b) constant lake area with depth.

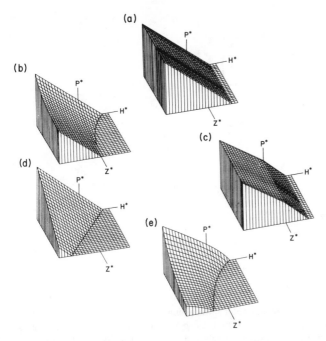

Figure 13 Nondimensional solutions for mixed layer entrainment, P^*, as a function of the stability, H^*, and the neutral depth parameter, Z^*, for the models of (a) Kraus and Turner (1967), Tennekes (1973), Denman (1973), Niiler (1975); (b) Elsberry et al. (1976); (c) Gill and Turner (1976); (d) Alexander and Kim (1976), Kim (1976); (e) Garwood (1977). (After Garwood, 1979, published by the American Geophysical Union.)

using such techniques (see also Semtner, 1984). For the lake simulations de-scribed in Section 5.2, it can be seen from Fig. 15 that using the eddy diffusion model U.S.E.D. (Henderson-Sellers, 1985b), this hysteresis phenomenon is well simulated, as a consequence of its dynamical thermocline parameterization. This hysteresis effect is also noted in the diurnal cycle of temperature by Price et al. (1986).

5.1 Oceans

The original applications of thermal stratification models to the oceans were those based on eddy diffusion models (e.g., Munk and Anderson, 1948). However in the 1960s and 1970s these largely lost favor as a result of the (then) inadequate theory for calculating functional forms for both the neutral values and the stability functions $[f(R_i), g(R_i)]$ which could be applied universally without prior calibra-tion. During this period the various versions of IML models described in Section 4.1 were used to simulate thermal stratification on various time scales. As has been pointed out already, the DIM is most applicable to time scales of the order of one inertial period ($\sim$1 day), whereas the TEM can be used for longer periods.

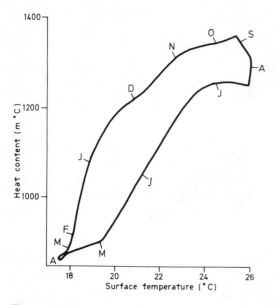

Figure 14 Hysteresis between heat content and water surface temperature. Observations from Ocean Weather Station Echo. (Reprinted with permission from *Deep-Sea Research*, 23, A. E. Gill and J. S. Turner, A comparison of seasonal thermocline models with observations, copyright 1976).

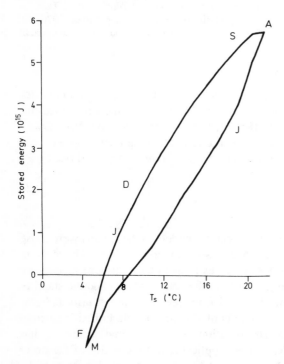

Figure 15 Hysteresis between heat content and water surface temperature. Model simulations for typical U.K. lake (Fig. 12*a*) using eddy diffusion model U.S.E.D.

One major exception to this prevalence of ECMs in ocean modeling is the group of differential models spawned following the initial exposition by Mellor and Durbin (1975), although it should be noted (as described below in Section 5.2) that in lake applications there is a greater balance between the use of IML and differential models. More recently, the need to provide thermocline models on appropriate time and space scales has emerged in terms of the demand for a relatively cheap (computationally) thermal model for incorporation into a linked atmosphere-ocean general circulation climate model (e.g., Semnter, 1984). Indeed this highlights that no single model is likely to be adequate for all time and space scales; and recent investigations stress the simplicity versus reality of simulation and the appropriateness for the problem under consideration.

Although some of the applications have already been mentioned in Section 4 during the model derivations, this section summarizes the major case studies and theoretical studies in ocean stratification modeling. One of the first applications was, of course, in the original paper of Kraus and Turner (1967). Their results were derived from simplistic, saw-toothed forcing as well as with observations from Ocean Weather Ship Papa; whereas Gill and Turner (1974) used a sinusoidal forcing with a modified KT model. In the initial presentation of the DIM (Pollard et al., 1973), only temperature variations within a time scale of a few days or less were investigated; and it was left to later authors (e.g., de Szoeke and Rhines, 1976) to reconcile these two approaches in terms of applicable time scales. Ocean Station Papa observations were again used in the 12-day simulation of Denman and Miyake (1973) who used the model of Denman (1973) which was itself based on the Kraus-Turner TEM. This same model has also been applied to a study of the seasonal cycle of the Arabian Sea surface temperature by Shetye (1986) who suggests that use of this parameterization of the thermodynamics can explain SST observations except during the southwest monsoon season.

Thompson (1976) presented algorithms for the numerical solution of three ML models: a modified PRT model, the Denman (1973) version of the KT model and a simplistic model assuming the ML to be of fixed depth (a function of latitude only). He compared simulation results with Ocean Station N observations and concluded that all three could be made to be computationally efficient. The fastest is the constant depth model, but this is unlikely to give sufficient resolution or accuracy for many requirements. Thompson (1976) suggested that the modified PRT model may be a possible contender for inclusion in a coupled atmosphere-ocean model, although since the time scale of applicability of this model appears to be days rather than years (e.g., de Szoeke and Rhines, 1976), the TEM model-type would seem to be a better candidate; see also discussion in Section 5.4 plus details of papers cited therein). This conclusion appears to be corroborated by details of a study discussed by Woods (1984) in which random fluctuations were imposed on a series of ML models. The resulting temperature fluctuations suggested that the discrepancies in Thompson's (1976) were more likely to be due to model inadequacy rather than permitting a model comparison. Additionally, following publication of an erratum (Thompson, 1977), Garwood and Camp (1977) questioned these conclusions and suggest that modifications to the DKT model by, e.g., Elsberry et al. (1976), Kim (1976) and Garwood (1977), may remedy the deficiencies in that original formulation by which excessive winter deepening is predicted.

Applications of differential models to the oceanic environment are found less frequently, but are by no means rare. Following the study of Mellor and Durbin (1975), this type of modeling approach has been utilized by, e.g., Marchuk et al. (1977), Kundu (1980) (but cf. comment by Deardorff, 1980), Klein (1980), Klein and Coantic (1981) and in the U.S. Navy's operational forecast model (Clancy and Pollak, 1983). A comparison of the Mellor and Durbin (1975) model with the ML model of Gill and Turner (1976) is presented by Miyakoda and Rosati (1984) who conclude that for SST anomalies, both models give reasonable results. They also stress the need to use appropriate scalings in both time and space.

Klein and Coantic (1981) utilize a modification to the level 3 model of Mellor and Yamada (1974) to investigate the response of the ocean to sudden changes in wind stress. Their numerical scheme allows fine spatial resolution ($\sim$1 m) but large time steps ($\sim$1 h). They too emphasize the importance of turbulent diffusion at high wind speeds. They suggest from their model results that the impact of breaking surface waves is minimal, while a comparison with observations was successful without any tuning. A simpler (level 2) model was used also in a later study of nutrient dynamics by Klein and Coste (1984).

More recently, these models have been applied to investigate the diurnal cycling of temperature within an oceanic mixed layer. Dickey and Simpson (1983) concentrated on the effects of the specification of the optical characteristics of the water (see also discussion by Woods et al., 1984). Using the shortwave penetration models described above, together with an MYL2 1/2 model, the effects of the differential radiation absorption (with depth) was investigated, especially with respect to influences on the biomass, suggesting that such diurnal effects need to be included in global climate models. A similar conclusion was reached by Woods (1980) in his study of the diurnal variations in compensation depth. Fuller studies of various aspects of the diurnal thermal stratification cycle are presented by Shay and Gregg (1986) using observational data from both the Bahamas and warm-core rings in the Gulf Stream, by Price et al. (1986) using observational data from R/P FLIP and by Spigel et al. (1986) using data from Wellington Reservoir in Australia (Imberger, 1985). Price et al. (1986) deduce the existence of a 'diurnal jet': an enhanced near-surface current caused primarily by diurnal variations in ML depth and moderated by changes in wind stress. Price (1985, p.c.) suggests that this effect is likely to be less marked in lakes. In their reservoir study, using the model DYRESM [see Eq. (85)], Spigel et al. (1986) identify billowing, occurring at the base of the mixed layer, as being of importance to the diurnal thermodynamics; their observations also showing the existence of a transient thermocline similar in character to the "diurnal jet" identified in the R/P FLIP data.

5.2 Lakes

As with oceans, the first stratification simulations were made using eddy diffusion models, often in which the value of K_H was held (unrealistically) to be constant independent of ambient stability (see, e.g., Harleman, 1982). Not surprising such simulations were, in general, inadequate. Incorporation of a varying value of eddy diffusion was made in the early 1970s (e.g., Orlob and Selna, 1970), whereby a decaying exponential was assumed in the epilimnion together with a constant value below the thermocline. Further modifications were made by Sundaram and Rehm (1971, 1973) who used a stability dependent formulation and a modified hypolimnetic value. Harleman (1982) notes that by the mid 1970s it was considered, at that time, that this approach could not progress without more explicit reference to wind-induced mixing and this led to the investigation of the possible use of ML models for lake applications.

The incorporation of an IML submodel into lake temperature and water quality models assumed initially that the solution would be accomplished numerically (unlike oceanic point models where often an analytic solution is sought) and the formulations presented in the literature are all couched in terms of finite difference grid notation. Stefan and Ford (1975a) used their IML lake model to simulate temperature profiles in several lakes in the Minneapolis-St. Paul locality and investigated the model sensitivity to variables such as lake bathymetry, the order of the entrainment and heating calculations, selection of the value for finite difference grid length, etc. Although this model type has been much used in lake simulations, it is important to note that since there is no heat transfer below the thermocline by diffusion (only radiative penetration) errors can accumulate under certain climatological conditions/localities. This suggests that finally some compromise between IML and eddy diffusion models may be most appropriate for lake modeling (and by analogy also for ocean modeling).

This type of approach was also adopted by the MIT group (e.g., Hurley Octavio et al., 1977) when wind mixing was introduced explicitly into the existing numerical scheme. Harleman (1982) reviews the development and application of an IML model to lakes in which an homogeneous slab overlies a quiescent layer in which there is a non-zero, yet constant, thermal gradient.

Although a ML model, DYRESM (Imberger and Patterson, 1981), has been utilized in an interbasin comparison without retuning (Patterson et al., 1984). This is accomplished by setting the tuning coefficients to standardized values; although the range of values reported in Sherman et al. (1978) and Fischer et al. (1979) is relatively large, needing specific tuning for sheltering effects (see also Ivey and Patterson, 1984); although only two lakes (one in British Columbia and one in Western Australia) were used in this study.

More recently, however, the eddy diffusion model described in Section 4 [using Eqs. (107) and (108)] has been utilized in comparative studies for lakes in different parts of the world without the need to retune any internal model coefficients. Successful simulations have been undertaken for lakes in the U.K. (e.g., Henderson-Sellers, 1985b) and for South Africa and Canada (Fig. 16). Using a similar model, based on the solution of Eq. (109), Babajimopoulos and Papadopoulos (1986) simulated the stratification observed in Lake Ostrovo (Vegoritis) in Greece. However, in their parameterization of Eq. (99), based on that of McCormick and Scavia (1981), they retained three parameters as tuning coeffi-

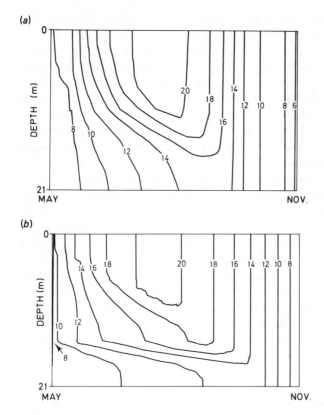

Figure 16 (*a*) Observations and (*b*) simulation of 5 year average for Clearwater Lake, Ontario. (Simulations using the eddy diffusion model U.S.E.D.)

cients and found their model to be sensitive to the value of β [Eq. (8)], which they took to be 1. More recently an eddy diffusion parameterization has been applied to cooling lakes by Adams et al. (1987) who take a pragmatic view that although data illustrate the depth dependency of K_H, "a characteristic depth-independent value . . . was sought in order to limit the number of fitting parameters in the numerical model."

It is interesting to note that for large lakes, such as the Laurentian Great Lakes, both eddy diffusion models (Lam and Schertzer, 1987) and IML models (Ivey and Patterson, 1984) have been applied successfully, although Ivey and Patterson (1984) identify the need, for lakes of this scale, to include Coriolis effects, as do Garwood et al. (1985a,b) for the oceanic case.

Two major advantages of this eddy diffusion approach over the IML model approach, clearly evident in this present comparison, are that (1) there is no need to utilize empirical tuning coefficients (see also Martin, 1985) and (2) the ML is predicted and not assumed a priori (see also e.g. Mellor and Durbin, 1975; Prangsma and Kruseman, 1984). Perhaps the sole advantage of using IML models is their better computational efficiency, largely at the expense of detail within the ML; although Krenkel and French (1982) suggest that the use of such a bulk parameterization obviates the need to consider the physics of mixing processes such as eddies, surface and internal waves, etc.

5.3 Storm Events

The simulation of storm events is an important test of stratification models. Both integral mixed layer and differential stratification models have been applied to this phenomenon. Elsberry and Garwood (1978) use an IML to investigate upper ocean temperature anomalies and Marchuk et al. (1977) apply their differential model to the simulation of the response of the ocean to a storm event.

D'Asaro (1985) uses data collected as part of the Storm Transfer and Response Experiment (STREX) in November 1980 to compare two one-dimensional ML models. The first model was that of Niiler and Kraus (1977), calibrated by Davis et al. (1981). This model includes both the effect of surface fluxes and subsurface shears. The model, when forced by observed surface wind stress and heat flux, predicted a ML deepening and temperature change consistent with the observations. The second model used assumed ML deepening to depend only on a bulk Richardson number computed from changes in velocity and density across the ML base. This model, which omits the surface flux, could not predict the observed ML deepening. The model comparisons emphasize the importance of the surface flux during storm events in determining the rate of ML deepening.

Observations at Ocean Station P on August 23rd 1977 by Dillon and Caldwell (1978) suggest that during storms the turbulent mixing processes described in Section 4 are inadequate since large scale disturbances at the ML may sweep large masses of cool water into the warmer isothermal layer of the ML (Fig. 17). Camp and Elsberry (1978) note that one-dimensional mixing models simulate a major fraction (but not all) of the upper ocean response to atmospheric storms, partly due to the fact that the majority of the mixing effects due to surface forcing are concentrated within a small percentage of the time (Elsberry and Camp, 1978)— a result of the cubic functional form for wind mixing [e.g., Eq. (26)]. Numerical experiments reported by Elsberry and Garwood (1980) emphasize that during a storm event rapid deepening occurs, *"especially if there is upward heat flux."*

Such a simulation, in which additional mixing is produced synergistically by a large surface energy budget deficit as well as an increased TKE at the surface, is exemplified in Fig. 18. This shows a simulation of a north temperate lake from the eddy diffusion model U.S.E.D. (Henderson-Sellers, 1985b) in which there was a sudden increase in wind speed, during the period Julian dates 180–220 inclusive, from its climatological value of about 3 m s^{-1} to 20 m s^{-1}. The wind speed was held constant at this value throughout the selected period and then allowed to revert to its climatological value. As can be seen from Fig. 18, the lake was found to be overturned quickly following the increase in wind speed, due to both an increase in net surface energy loss and deeper penetration of wind-induced turbulent mixing. This agrees, at least qualitatively, with the observation

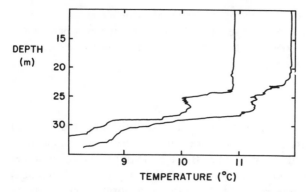

Figure 17 Temperature profiles showing large-scale disturbance at mixed layer base which are interpreted by Dillon and Caldwell (1978) as being incipient catastrophic events. Reprinted by permission from *Nature*, Vol. 276, pp. 601–602. Copyright © 1978 Macmillan Journals Limited.

of Haney (1980) that during a wind event, wind mixing is incapable alone of creating the observed mixing and changes in surface fluxes must also be taken into account.

The more extreme meteorological phenomenon of a hurricane was simulated by Martin (1982) in a study of hurricane Eloise—observations also simulated by Price (1981) using a 3-D, three-layer ocean model. Martin (1982) utilized the level 2 Mellor and Yamada (1974) model (Fig. 19) and demonstrated, firstly, by omitting vertical advection and, secondly, by ignoring wind direction changes,. the importance of including these two variables.

5.4 Coupled Ocean-Atmosphere GCMs

One important, more specialized application of thermal stratification modeling in oceans is in coupled ocean-atmosphere general circulation climate models (GCMs)

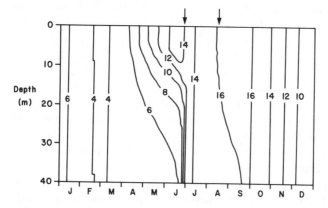

Figure 18 Overturn in lake following large storm event. The climatological wind speed is replaced instantaneously by a value of 20 m s^{-1} on Julian day 180 (which lasts until day 220).

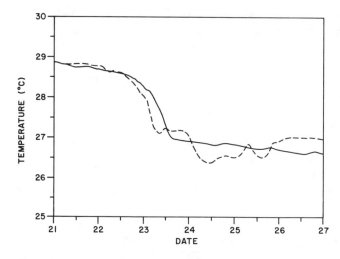

Figure 19 Comparison of model-predicted (solid) and observed (dashed) SST for the case where surface stress, heat flux and vertical motion are computed explicitly, essentially using bulk formulae (after Martin, 1982, published by the American Geophysical Union).

(e.g., Foreman, 1986). Such models are fully three-dimensional as opposed to the previous point models. Although few such simulations have been undertaken (partly because of the cost of extensive computer time needed; see discussion in Bryan, 1984), many authors have indeed oriented their analyses taking the inherent spatial and temporal constraints of GCMs into account. Marchuk et al. (1977) suggested that the inherent mathematical simplicity of the integral approach is well-suited to GCM applications; whereas Garrett (1979) investigated the implementation of an eddy diffusion (differential) model.

While many early ML models investigated ML deepening over relatively short time scales (frequently less than 6 months), for GCM applications it has long been realized that an inherent year-to-year numerical stability is necessary, e.g., Haney and Davies (1976). These authors present climatological simulations for the last year of a 100-year simulation by which time a stable and repeating thermal structure is evident. They identify, in addition to the Monin-Obukhov length scale, the importance of a second length scale given by $\overset{*}{w}/f$ (first identified by Rossby and Montgomery, 1935) such that the mixed layer depth, h, for their climatological ocean circulation model is given by the minimum of the Monin-Obukhov length scale, L, and this second length scale (i.e., there is no explicit calculation of ML depth from TKE or other considerations).

A similar concern for the applicability of ML models to longer period simulations is expressed by de Szoeke (1980) who extends the Niiler (1975) model to encompass horizontal variations in wind stress. This permits a detailed study of upwelling/downwelling regions to be undertaken by means of the inclusion of the wind stress curl. However he notes deficiencies in the approach with regard to the role of surface heating restabilizing the upper mixed layer, suggesting it to be an area ripe for further developments.

Wells (1979) suggests that the modified KT models (e.g., Kim, 1976; Garwood, 1977) are appropriate for middle latitudes since the sea surface temperatures (SST) are governed largely by the local exchanges of heat and mechanical energy in these latitudes; whereas in equatorial latitudes, lateral advection of energy needs to be included (e.g., Webster and Lau, 1977; Pacanowski and Philander, 1981). He describes results of coupling a simple KT model (as modified by Gill and Turner, 1976) into a southern hemisphere spectral GCM for fixed January. Although advection is included within the mixed layer it is neglected below this depth. However, no simultaneous coupling of the ocean-atmosphere system was used. Instead the atmospheric model was integrated for 4 days with a fixed SST, which was then itself recalculated every fourth day. Results from this model were compared with a simulation in which the atmosphere forces the ocean but no feedback effect of changing SST on the atmosphere was included. Conclusions regarding the differences between these two model runs were limited to simulation periods of 72 days, despite the fact that simulation spin-up periods are usually considered to be years for the atmosphere and centuries for the ocean (e.g., Haney and Davies, 1976; Manabe, 1983). One further problem is that of incompatible time scales (Elsberry and Garwood, 1980): typically an ML model uses a time step of 1 hour (cf. the advective component of an oceanic GCM with time steps of several hours).

Of more directed relevance to ocean general circulation models is the analysis of Haney and Davies (1976), who used a modified KT model to simulate the seasonal cycle at Ocean Station N using simplified sinusoidal forcing.

Parallel development of ocean circulation models and mixed layer models have led to the need to embed the latter within the former to produce an efficient three-dimensional ocean circulation model. Initial attempts (e.g., Haney, 1980) utilized crude representations of mixed layer depth, while recommending the integration of these two model types (see also Elsberry and Garwood, 1980). The embedding of the Garwood (1977) ML model within the Haney (1980) circulation model is described by Elsberry and Garwood (1980); see Fig. 20. Some simplification of the embedded ML approach is proposed by Haney (1985), based on the "dynamic adjustment" of Adamec et al. (1981) in which ML deepening is controlled by a local gradient Richardson number, thus providing a link with the differential model approach (see also discussion in Section 4.3 of Price et al.'s (1986) parameterization).

Recently, Meehl (1984) has reviewed the coupling of ocean-atmosphere models in terms of a hierarchy of oceanic components. He suggests that even oversimplistic oceanic representations (e.g., (1) the "swamp" representation of, e.g., Manabe and Wetherald, 1980, Washington and Meehl, 1983, with no heat storage capacity; and (2) fixed depth ocean surface layer models where there is a heat capacity but no dynamics, e.g., Manabe and Stouffer, 1980; Spelman and Manabe, 1984; Washington and Meehl, 1984) may be useful in helping to understand the processes acting in air-sea interaction. He identifies the advantage of utilizing a fixed depth ML as permitting the inclusion of a full seasonal cycle in the atmospheric GCM (not possible with the oceanic "swamp") (see also Hansen et al., 1983; Meehl and Washington, 1985). As a modification to the constant depth ML, Meehl

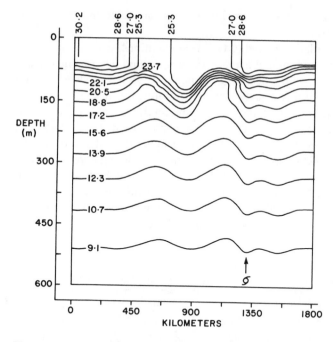

Figure 20 Ocean temperature distribution predicted using an embedded ML model within an ocean circulation model. The cross section is aligned along the track of a model storm (after Elsberry and Garwood, 1980, *Bulletin of the American Meteorological Society,* American Meteorological Society, Boston).

(1984) then utilizes a model in which ML depth is allowed to vary in a prespecified way; while acknowledging that more complete (and hence more complex) models of oceanic dynamics and thermodynamics exist, either in terms of a full computation of the oceanic surface layer (as discussed in this review) or in terms of a three-dimensional oceanic circulation model (e.g., Washington et al., 1980; Spelman and Manabe, 1984; Cox and Bryan, 1984). Meehl (1984) concludes that use of prespecified ML depth and calculated sea surface temperatures is inadequate for the simulation of the annual cycle of zonal heat storage, especially in the tropics, recognizing the need for a fully interactive coupling between atmosphere and ocean. Preliminary results of such a coupling are presented by Semtner (1984), using the Held and Suarez (1978) atmospheric model. He compares three oceanic representations: a simple diffusive convective two-layer model; a two-layer model, but including a prognostic model of the ML; and a three-layer primitive equation model. His conclusions are based upon only a 12-year integration for his third model type, which he acknowledges is inadequate for full equilibrium. Nevertheless he feels it appropriate to comment that, as might be expected, with this hierarchy accuracy/usefulness increases with increasing complexity of oceanic ML model, but so does computational time required. However no sea-ice is included, an omission partially rectified by Pollard et al. (1983); but in this oceanic model, the atmospheric forcing is prescribed. Once again, by testing several models with differing degrees of complexity, it is concluded that fixed depth ML models are inadequate.

For ocean general circulation model (OGCM) applications, appropriate values for the ML depth, either as an annual average or prespecified geographically and seasonally, remain dubious. Since one major concern of global modelers relates to heat storage terms, use of a two layer schematization, even if using well-simulated ML depths, will necessarily underestimate the total thermal capacity of the ocean by neglecting the heat storage in the water at depths below the thermocline. Such arguments are detailed by Meehl (1984) who notes that ML depths from various sources (e.g., Fig. 21) need to be modified (as a modeling approximation) and replaced by an "effective mixed layer depth" (Fig. 22). Such effective depths are used extensively (e.g., Manabe and Stouffer (1980) who use a value of 68 m as an annual value; Harvey and Schneider (1985); Wigley and Schlesinger (1985) who undertake simulations with mean ML depths of 70 and 110 m). This method

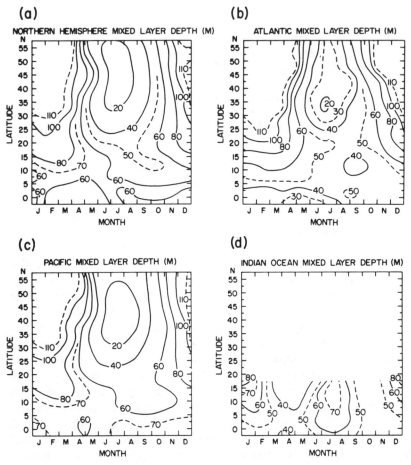

Figure 21 Annual cycle of zonal mean mixed layer depths in meters from various sources (after Meehl, 1984, *Journal of Physical Oceanography*, American Meteorological Society, Boston).

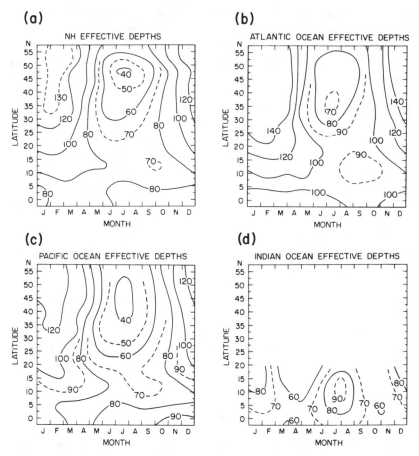

Figure 22 Annual cycle of zonal means of effective ocean surface-layer depths in meters (after Meehl, 1984, *Journal of Physical Oceanography,* American Meterological Society, Boston).

of obviating the restrictions of the implicit step function profile of a slablike ML model can be contrasted with the addition of an intermediate layer by Harvey and Schneider (1985); the exponential slab ML of Pollard et al. (1983); and the Richardson-number-induced "smoothing" of Price et al. (1986) (see Section 4.3)—again illustrating one of the limitations of the (computationally more efficient) bulk modeling approach (cf. differential models).

The coupling of an oceanic boundary layer model and an atmospheric boundary layer model is described by Davidson and Garwood (1984). However they only examine short time period (12–24 h) and conclude that over this time period, there is little difference between the results of the coupled model and the separate uncoupled models. Taking into account the known relaxation times of the two fluids, as discussed above, this result is then not perhaps too surprising. Although it sheds no light on the coupling of atmospheric and oceanic models for climate studies, the parameterizations of the two boundary layers employed might have some relevance to such studies.

Although the effects on climate of increasing CO_2 have been extensively commented on in the literature, as noted above, full ocean-atmosphere simulations, with a model which includes sea-ice thermodynamics (Bryan and Manabe, 1985), have been, to date, rarely undertaken while providing an attainable scientific goal (Washington and VerPlank, 1986), largely as a consequence of the enormous computational demands such a study would entail. For example, two experiments with a swamp ocean, with annually averaged solar forcing, coupled with an atmospheric GCM (Washington and Meehl, 1986) suggest that sensitivity of the modeled climate impact depends on both the ice parameterization and the basic state of the control run utilized. Other attempts to include the oceanic effects in such a study have recently been made using a simplified analysis of ocean thermodynamics by Wigley and Schlesinger (1985). They utilize a simple column energy balance approach, representing the ocean-atmosphere system by only four "compartments" or "boxes": two atmospheric (one over land, one over ocean), an oceanic mixed layer of fixed depth and a deeper diffusive ocean. The heating rate of the ML is calculated by assuming a constant depth in which the temperature increment due to some perturbation, ΔT, increases to (1) the change in the thermal (surface) forcing, ΔQ; (2) the atmospheric feedback, expressed in terms of a climate feedback parameter, λ, and, additionally to many fixed depth ML ocean models, leakage of energy is permitted into the underlying waters. This energy, ΔF acts as a surface boundary condition to a diffusive ocean, in which the turbulent diffusion coefficient K_H is assumed to be a constant. The differential equations describing the rates of heating in the two "layers" of this box-diffusion model are thus,

Mixed layer:
$$C_m \frac{d\Delta T}{dt} = \Delta Q - \lambda \Delta T - \Delta F \tag{118}$$

(which is essentially Eq. (111) applied to a fixed ML depth (viz. $w_e \equiv 0$) with the additional atmospheric feedback term, $-\lambda \Delta T$)

Deeper waters:
$$\frac{\partial \Delta T_0}{\partial t} = K_H \frac{\partial^2 \Delta T_0}{\partial z^2} \tag{119}$$

assuming a constant value of K_H. This latter equation may be evaluated at depth z (measured vertically upward from zero at the interface) or calculated numerically using a finite difference grid, with grid spacing Δz. In either case, the heat source at the surface is the energy "leaking" out of the ML, ΔF, which thus acts as a surface boundary condition to the lower level differential equation. However, Wigley and Schlesinger (1985) use a simpler parameterization by assuming that at the interface there is continuity between ML temperature ΔT and deeper layer temperature, evaluated at the interfacial level, $\Delta T_0(0, t)$, i.e.,

$$\Delta T_0(0, t) = \Delta T(t) \tag{120}$$

With this formulation, the value of ΔF can be calculated from

$$\Delta F = \gamma^* \rho c_p K_H \left(\frac{\partial \Delta T_0}{\partial z} \right)_{z=0} \tag{121}$$

and used in Eq. (118). In this last equation, γ^* is the parameter utilized to average over land and ocean and has a value between 0.72 and 0.75, ρ is the water density, and c its specific heat.

Wigley and Schlesinger (1985) utilize both the numerical model described by Eqs. (118) and (119) and an analytical model based on Eq. (121) to evaluate different atmospheric forcing, related to possible impacts of increasing atmospheric carbon dioxide using two possible analytical representations for such forcing. The models are used primarily to conclude that GCM modeling predictions and observational evidence are not contradictory, but additionally illustrates the growing need of ML modelers to include some representation of "leakage" of energy to waters below the thermocline—a modification also discussed by Price et al. (1986). In addition, Harvey (1986) evaluates the importance of advection in such simulations in a comparison of the simulated transient climate responses to increasing atmospheric carbon dioxide using both a box-diffusion (BD) and a box-advection-diffusion (BAD) model.

However, the simplifications of the annual timestep box-diffusion model [Eqs. (118), (119)] may invalidate its applications for climate simulations. In a recent study (Henderson-Sellers, 1987) using an eddy diffusion model based on numerical solution of Eq. (109), it was suggested that the SST anomalies (from increasing atmospheric carbon dioxide) calculated by the box-diffusion model may be overestimates, by a factor of ~ 3. Further investigation of the impact of CO_2 on sea surface temperatures are obviously needed as a vital component of future integrated climate impact studies.

6 DISCUSSION AND RECOMMENDATIONS

Parallel development of thermal stratification models for lakes and oceans has resulted in strong similarities between the basic formulations. Although contrasting phenomena are present in the real world in these two physical situations (e.g., waves, seiches, boundary effects), since the assumptions utilized by researchers in both limnology and oceanography have been aimed largely at eliminating such dynamic problems, the resulting one-dimensional thermal stratification models can be used to mutual benefit. With the caveats expressed in Section 1 regarding the validity of the modeling assumptions, it becomes possible to discuss the model development independently of the type of water body to which it will be eventually applied. Here we have traced such development and variations upon the two basic themes: integral and differential models.

The advantages of the former are that they are simpler in conception and solution, since a single equation is utilized to represent spatial averages over the depth of the upper ML. Consequently less computer time is required for solution. Indeed earlier models of this type were designed for even simpler solution using analytical methods by identifying temporal ranges over which a variety of approximations could result in an easy algebraic solution. To do this, however, the loss of spatial resolution in the ML reduces the ability of the integral ML models in certain aspects: e.g., storm events, lower level thermodynamics, the application to phytoplankton distributions; although recent modifications of, e.g., Denton and Wood (1981) and Price et al. (1986), in allowing energy leakage through the thermocline are beginning to bridge the gap to differential models.

The other area of concern relates to the use of empirical tuning coefficients. These are more prevalent in the integral approach due largely to the difficulty of parameterizing the TKE dissipation rate. In integral ML models developed from the ECM of Niiler and Kraus (1977), there are usually three (sometimes more) coefficients to be evaluated in this way, which, as noted by Garwood (1979) makes a fair intermodel comparison difficult to achieve. The empiricism in differential models is of a different kind in that it relates to the formulation of the eddy diffusion coefficients. Since most workers have attempted to delineate these a priori, then they cannot be considered as tuning coefficients since they are simply a parameterization of mixing developed intentionally to be independent of the case study of concern.

Additionally, as noted before, ML models assume the existence of a mixed layer whereas the differential models predict its existence. Since the two models are beginning to be able to be shown to be in good agreement (e.g., preliminary results of Martin, 1985), and furthermore since their future implementation is likely to be as part of a larger model (e.g., the U.S. Navy's ocean forecast model and coupled ocean-atmosphere general circulation models), it is perhaps apposite that the best features of both should be amalgamated into a new generation of definitive stratification models which would, ideally, be easily and cheaply implemented numerically (i.e., perhaps using the ML ideas for the upper layers) while retaining the spatial resolution of the differential models.

The second major resolution problem relates to the temporal domain, as discussed in part earlier. Several workers have discussed the time scales for which certain approximate (usually analytical) solutions are appropriate, e.g., the PRT-type model for short time periods and the KT-type formulation for longer periods. Such ideas are encapsulated by de Szoeke and Rhines (1976), who show that each of the forcing mechanisms is dominant over different time scales such that in a numerical simulation inclusion of all the terms should be ensured; while in analytical solution, approximations can be made to neglect individual terms over specific time scales. The time scales of validity are however themselves functions of the values of the empirical parameters selected. Figure 23 can be compared with Fig. 8 to illustrate the ranges of applicability for two extreme combinations of parametric values. In each case it is observed, from a full numerical solution, that initial linear deepening lasts for approximately 100 seconds, followed by a deepening proportional to $t^{1/3}$ as the wind-driven turbulence entrains lower water, changing the potential energy of the column. As the flow meanwhile accelerates, interfacial shear production of turbulence dominates and deepening is proportional to $t^{1/2}$. At later times the Coriolis force causes the mean current to deviate from the wind direction, such that, after about one-half of one sidereal day, shear production is superseded once again by the "turbulent erosion" mechanism in which deepening is again proportional to $t^{1/3}$. This analysis substantiates the use of the TEM for periods longer than 1 day, but also suggests that a comprehensive numerical model should include the shear production mechanism for completeness,

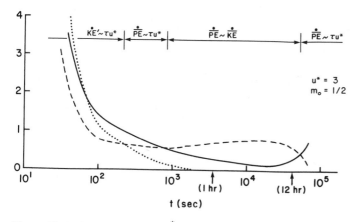

Figure 23 As for Fig. 8 but with $\overset{*}{w} = 3$ and $m = 0.5$ (after de Szoeke and Rhines, 1976).

especially during the initial "spin-up" period. This modeling approach was later extended to three dimensions by de Szoeke (1980) in order to evaluate the effects of horizontal variability of wind stress.

Often the time scale of interest dictates the model type to be used. For short period studies (e.g., Prangsma and Kruseman, 1984; Davidson and Garwood, 1984), it is perhaps possible to make assumptions regarding the temporal scale of thermocline development, which is much longer than that of the forcing processes and hence regard the ML depth as quasi-constant over the time period of interest (e.g., Prangsma and Kruseman, 1984). However for longer simulations (e.g., climatic time scales, Section 5.4), many of the ML models become unstable and are unable to depict a seasonal *cycle*. For example, models without dissipation will produce a thermocline that, over such time scales, continues to deepen (Niiler, 1977). However, the long term stability of the eddy diffusion model U.S.E.D. (Henderson-Sellers, 1985b) is readily demonstrable (see Fig. 24). This stability seems to be partly a result of the inclusion of a full surface energy budget and partly a result of the fact that an eddy diffusion formulation obviates many of the assumptions needed by an ML model.

Indeed many groups are discovering that for more realistic and stable solutions from ML models in both lakes and oceans (which have to date relied on a simple mixed layer deepening term while the underlying waters remain undisturbed both dynamically and thermodynamically), it is necessary to add a turbulent diffusion term for deeper waters (Waide, 1982, p.c.). Of course in many (but not all) turbulent diffusion models, such a below-thermocline diffusion term has always been part of the model schematization. (A discussion of some of the aspects of deep water turbulence can be found in Davies, 1985.)

Further developments of thermal stratification models will be dictated by two main considerations. Firstly, for incorporation into large and expensive coupled ocean-atmosphere GCMs, computational speed is likely to be favored at the expense of a high degree of accuracy. Two examples for illustration are: (1) the development at Oregon State University (OSU) of a coarse resolution synchro-

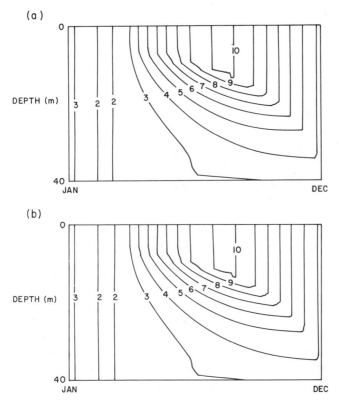

Figure 24 Stability of eddy diffusion models is demonstrated in terms of year 10 (upper) and year 100 (lower) simulation of a typical U.K. lake (isopleths are in °C).

nously coupled model which has only two atmospheric layers and six layers in the ocean, but an hourly temporal resolution (Gates et al., 1985; Han et al., 1985). In addition, it is intended that this model will utilize an embedded Kraus-Turner ML model. Preliminary results (Gates et al., 1985) highlight the need to simulate the SEB realistically. (2) Foreman (1986) describes a 5-day synchronous coupling scheme and identifies several areas of error in his 10 month simulation using a modified Cox-Bryan model, but without sea ice. (He does not, however, describe the type of stratification model used in this project.) Such aims (to build coupled ocean-atmosphere GCMs) will encourage the development of one-dimensional models which themselves then become spatially dependent upon their grid location within a larger model. Discontinuities of thermocline depth between neighboring grid cells is likely to be a source of difficulty in the numerical schematization if oversimple ML models are utilized. Secondly it is likely that processes other than those in the vertical direction will be investigated. This line of research has been developed in lake applications by Imberger and co-workers and in oceanic situations, large-scale advection of heat is often discussed, but infrequently included in the model explicitly.

It is also likely that analytical models will be largely superseded by numerical models, partly for the reasons discussed above, while the analysis itself should assist in the selection of appropriate simplifying parameterizations in the numerical code in order to minimize computation time.

REFERENCES

Adamec, D. and Elsberry, R. L., 1984, Sensitivity of mixed layer predictions at Ocean Station Papa to atmospheric forcing parameters, *J. Phys. Oceanogr.*, 14, 769–780.

Adamec, D., Elsberry, R. L., Garwood, R. W. Jr. and Haney, R. L., 1981, An embedded mixed layer-ocean circulation model, *Dyn. Atmos. Oceans*, 5, 69–96.

Adams, E. E., Wells, S. A. and Ho, E. K., 1987, Vertical diffusion in a stratified cooling lake, *J. Hyd. Eng.*, 113, 293–307.

Aikman, F., III, 1984, Pycnocline development and its consequences in the Middle Atlantic Bight, *J. Geophys. Res.*, 89, 685–694.

Alexander, R. C. and Kim, J.-W., 1976, Diagnostic model study of mixed layer depths in the summer North Pacific, *J. Phys. Oceanogr.*, 6, 293–298.

André, J. C. and Lacarrère, P., 1985, Mean and turbulent structures of the oceanic surface layer as determined from one-dimensional, third-order simulations, *J. Phys. Oceanogr.*, 15, 121–132.

Babajimopoulos, C. and Papadopoulos, F., 1986, Mathematical prediction of thermal stratification of Lake Ostrovo (Vegoritis), Greece, *Water Resour. Res.*, 22, 1590–1596.

Bedford, K. W. and Babajimopoulos, C., 1977, Vertical diffusivities in areally averaged models, *Proc. ASCE, J. Env. Eng. Div.*, 103(EE1), 113–125.

Blanc, T. V., 1985, Variation of bulk-derived surface flux, stability, and roughness results due to the use of different transfer coefficient schemes, *J. Phys. Oceanogr.*, 15, 650–669.

Bloss, S. and Harleman, D. R. F., 1979, Effect of wind-mixing on the thermocline formation in lakes and reservoirs, *MIT Tech Rept*, 249, MIT, Mass.

Bretherton, F. P., 1982, Ocean climate modelling, *Prog. Oceanogr.*, 11, 93–129.

Brezonik, P. L., 1978, Effect of organic color and turbidity of Secchi disk transparency, *J. Fish. Res. Bd. Can.*, 35, 1410–1416.

Bryan, K., 1984, Accelerating the convergence to equilibrium of ocean-climate models, *J. Phys. Oceanogr.*, 14, 666–673.

Bryan, K., 1975, Three-dimensional numerical models of the ocean circulation, pp 94–106 in *Numerical Models of Ocean Circulation* (ed. R. O. Reid), National Academy of Sciences.

Caldwell, D. R., 1983, Small-scale physics of the ocean, *Rev. Geophys. Space. Phys.*, 21, 1192–1205.

Camp, N. T. and Elsberry, R. L., 1978, Oceanic thermal response to strong atmospheric forcing. II Simulation with mixed layer models, *J. Phys. Oceanogr.*, 8, 215–224.

Churchill, J. and Pade, B. H., 1981, Acoustically and visually tracked drogue measurements of nearsurface water velocities in Lake Huron, plus observations of a coastal upwelling, *Woods Hole Oceanographic Institution Report*, WHOI-81-91, Woods Hole, Mass.

Clancy, R. M. and Pollak, K. D., 1983, A real-time synoptic ocean thermal analysis/forecast system, *Prog. Oceanogr.*, 12, 383–424.

Cox, M. D., 1975, A baroclinic numerical model of the world ocean: preliminary results, pp 107–120 in *Numerical Models of Ocean Circulation* (ed R. O. Reid), National Academy of Sciences.

Cox, M. D. and Bryan, K., 1984, A numerical model of the ventilated thermocline, *J. Phys. Oceanogr.*, 14, 674–687.

Csanady, G. T., 1979, A developing turbulent surface shear layer model, *J. Geophys. Res.*, 84, 4944–4948.

Cushman-Roisin, B., 1981, Deepening of the wind-mixed layer: a model of the vertical structure, *Tellus*, 33, 564–582.

Dake, J. M. K. and Harleman, D. R. F., 1969, Thermal stratification in lakes, analytical and laboratory studies, *Water Resour. Res.*, 5, 484–495.

Darbyshire, J. and Edwards, A., 1972, Seasonal formation and movement of the thermocline in lakes, *Pure and Applied Geophysics*, 93, 141–150.

D'Asaro, E. A., 1985, Upper ocean temperature structure, inertial currents, and Richardson numbers observed during strong meteorological forcing, *J. Phys. Oceanogr.*, 15, 943–962.

Davidson, K. L. and Garwood, R. W., 1984, Coupled oceanic and atmospheric mixed layer model, *Dyn. Atmos. Oceans*, 8, 283–296.

Davies, A. M., 1985, A three dimensional modal model of wind induced flow in a sea region, *Prog. Oceanogr.*, 15, 71–128.

Davies, A. M., 1986, Application of a spectral model to the calculation of wind drift currents in a stratified sea, *Cont. Shelf Res.*, 5, 579–610.

Davis, R. E., de Szoeke, R. and Niiler, P. P., 1981, Variability in the upper ocean during MILE. Part II: Modeling the mixed layer response, *Deep-Sea Research*, 28A, 1453–1475.

Deardorff, J. W., 1968, Dependence of air-sea transfer coefficients on bulk stability, *J. Geophys. Res.*, 73, 2549–2557

Deardorff, J. W., 1980, Comments on "A numerical investigation of mixed-layer dynamics," *J. Phys. Oceanogr.*, 10, 1695–1696.

Denman, K. L., 1973, A time-dependent of the upper ocean, *J. Phys. Oceanogr.*, 3, 173–184.

Denman, K. L. and Miyake, M., 1973, Upper layer modification at Ocean Station Papa: observations and simulation, *J. Phys. Oceanogr.*, 7, 185–196.

Denton, R. A. and Wood, I. R., 1981, Penetrative convection at low Péclet number, *J. Fluid Mech.*, 113, 1–21.

de Szoeke, R. A., 1980, On the effects of horizontal variability of wind stress on the dynamics of the ocean mixed layer, *J. Phys. Oceanogr.*, 10, 1439–1454.

de Szoeke, R. A. and Rhines, P. B., 1976, Asymptotic regimes in mixed-layer deepening, *J. Mar. Res.*, 34, 111–116.

Dickey, T. D. and Simpson, J. J., 1983, The influence of optical water type on the diurnal response of the upper ocean, *Tellus*, 35B, 142–154.

Dillon, T. M. and Caldwell, D. R., 1978, Catastrophic events in a surface mixed layer, *Nature*, 276, 601–602.

Dillon, T. M. and Powell, T. M., 1979, Observations of a surface mixed layer, *Deep-Sea Research*, 26A, 915–932.

Dillon, T. M., Richman, J. G., Hansen, C. G. and Pearson, M. D., 1981, Near-surface turbulence measurements in a lake, *Nature*, 290, 390–392.

Effler, S. W., Wodka, M. C. and Field, S. D., 1984, Scattering and absorption of light in Onondaga Lake, *J. Env. Eng.*, 110, 1134–1145.

Elder, R. A. and Wunderlich, W. O., 1968 *Evaluation of Fontana Reservoir field measurements*, ASCE Specialty Conference on Current Research into the Effects of Reservoirs on Water Quality, Portland, Oregon, January 1968.

Elsberry, R. L. and Camp, N. T., 1978, Oceanic thermal response to strong atmospheric forcing I. Characteristics of forcing events, *J. Phys. Oceanogr.*, 8, 206–214.

Elsberry, R. L. and Garwood, R. W., Jr., 1978, Sea-surface temperature anomaly generation in relation to atmospheric storms, *Bull. Amer. Meteor. Soc.*, 59, 786–789.

Elsberry, R. L. and Garwood, R. W., Jr., 1980, Numerical ocean prediction models—goal for the 1980s, *Bull. Amer. Meteor. Soc.*, 61, 1556–1566.

Elsberry, R. L., Fraim, T. S. and Trapnell, R. N., 1976, A mixed layer model of the oceanic thermal response to hurricanes, *J. Geophys. Res.*, 81, 1153–1162.

Environmental Laboratory, 1982, *CE-QUAL-R1: A numerical one-dimensional model of reservoir water quality*, User's Manual, Instruction Report E-82-1, U.S. Army Corps of Engineers, Waterways Experiment Station, CE, Vicksburg, Miss.

Filatov, N. N., Rjanzhin, S. V. and Zaycev, L. V., 1981, Investigation of turbulence and Langmuir circulation in Lake Ladoga, *J. Great Lakes Res.*, 7, 1–6.

Filyushkin, B. N. and Miropol'skiy, Yu.Z., 1981, Seasonal variability of the upper thermocline and self-similarity of the temperature profiles, *Oceanology,* 21, 299–304.

Fischer, H. B., List, E. J., Koh, R. C., Imberger, J. and Brooks, N. H., 1979, *Mixing in Inland and Coastal Waters,* Academic, 483pp.

Ford, D. E. and Stefan, H. G., 1980, Thermal predictions using integral energy model, *Procs. ASCE., J. Hyd. Div.,* 106(HY1), 39–55.

Foreman, S. J., 1986, Ocean and atmosphere interact!, *Meteorol. Mag.,* 115, 358–361.

Garratt, J. R., 1977. Review of drag coefficients over oceans and currents. *Mon. Weath. Rev.,* 105, 915–929.

Garrett, C., 1979, Mixing in the ocean interior, *Dyn. Atmos. Oceans,* 3, 239–265.

Garwood, R. W., Jr., 1977, An oceanic mixed layer model capable of simulating cyclic states, *J. Phys. Oceanogr.,* 7, 455–468.

Garwood, R. W., Jr., 1979, Air-sea interaction and dynamics of the surface mixed layer, *Rev. Geophys. Space Phys.,* 17, 1507–1524.

Garwood, R. W. and Camp, N. T., 1977, Comments on "Climatological numerical models of the surface mixed layer of the ocean," *J. Phys. Oceanogr.,* 7, 469–471.

Garwood, R. W., Jr., Gallacher, P. C. and Muller, P., 1985a, Wind direction and equilibrium mixed layer depth: general theory, *J. Phys. Oceanogr.,* 15, 1325–1331.

Garwood, R. W., Jr., Muller, P. and Gallacher, P. C., 1985b, Wind direction and equilibrium mixed layer depth in the tropical Pacific Ocean, *J. Phys. Oceanogr.,* 15, 1332–1338.

Gates, W. L., Han, Y.-J. and Schlesinger, M. E., 1985, The global climate simulated by a coupled atmosphere-ocean general circulation model: preliminary results, in *Coupled Ocean-Atmosphere Models* (ed. J. C. J. Nihoul), Elsevier, Amsterdam, 131–151.

Geernaert, G. L., Katsaros, K. B. and Richter, K., 1986, Variation of the drag coefficient and its dependence on sea state, *J. Geophys. Res.,* 91, 7667–7679.

Gill, A. E., 1974, The relationship between heat content of the upper ocean and the sea-surface temperature, *NORPAX Highlights,* 2(3).

Gill, A. E. and Turner, J. S., 1974, Mixing models for the seasonal thermocline, *NORPAX Highlights,* 2(5), 9–12.

Gill, A. E. and Turner, J. S., 1976, A comparison of seasonal thermocline models with observations, *Deep-Sea Res.,* 23, 391–401.

Gray, W. G. (ed.), 1986, *Physics-Based Modeling of Lakes, Reservoirs, and Impoundments,* Amer. Soc. Civil Engineers, New York, 308pp.

Gulliver, J. S. and Stefan, H. G., 1982, Lake phytoplankton model with destratification, *Proc. ASCE, J. Env. Eng. Div.,* 108(EE5), 864–882.

Han, Y-J., Schlesinger, M. E. and Gates, W. L., 1985, An analysis of the air-sea-icea interactions simulated by the OSU-coupled atmosphere-ocean general circulation model, in *Coupled Ocean-Atmosphere Models* (ed. J. C. J. Nihoul), Elsevier-Amsterdam, 167–182.

Haney, R. L., 1971, Surface thermal boundary condition for ocean circulation models, *J. Phys. Oceanogr.,* 1, 241–248.

Haney, R. L., 1980, A numerical case study of the development of large-scale thermal anomalies in the central North Pacific Ocean, *J. Phys. Oceanogr.,* 10, 541–556.

Haney, R. L., 1985, Midlatitude sea surface temperature anomalies: a numerical hindcast, *J. Phys. Oceanogr.,* 15, 787–799.

Haney, R. L. and Davies, R. W., 1976, The role of surface mixing in the seasonal variation of the ocean thermal structure, *J. Phys. Oceanogr.,* 6, 504–510.

Hansen, J., Russell, G., Rind, D., Stone, P., Lacis, A., Lebedeff, S., Ruedy, R. and Travis, L., 1983, Efficient three-dimensional global models for climate studies: models I and II, *Mon. Wea. Rev.,* 111, 609–662.

Harleman, D. R. F., 1982, Hydrothermal analysis of lakes and reservoirs, *Procs. ASCE., J. Hyd. Div.,* 108(HY3), 302–325.

Harvey, L. D. D., 1986, Effect of ocean mixing on the transient climate response to a CO_2 increase: analysis of recent model results, *J. Geophys. Res.,* 91, 2709–2718.

Harvey, L. D. D. and Schneider, S. H., 1985, Transient climate response to external forcing on 10^0–10^4 year time scales 2. Sensitivity experiments with a seasonal, hemispherically veraged coupled atmosphere, land, and ocean energy balance model, *J. Geophys. Res.*, 90, 2207–2222.

Held, I. M. and Suarez, M. J., 1978, A two-level primitive equation atmospheric model designed for climatic sensitivity experiments, *J. Atmos. Sci.*, 35, 106–229.

Henderson-Sellers, B., 1977, The thermal structure of small lakes: the influence of a modified wind speed, *Water Resour. Res.*, 13, 791–793.

Henderson-Sellers, B., 1978, The longterm thermal behaviour of a freshwater lake, *Procs. ICE (Part 2)*, 65, 921–927.

Henderson-Sellers, B., 1982, A simple formula for vertical eddy diffusion coefficients under conditions of nonneutral stability, *J. Geophys. Res.*, 87, 5860–5864.

Henderson-Sellers, B., 1984, *Engineering Limnology*, Pitman, London, 356pp.

Henderson-Sellers, B., 1985a, The relative significance of various physico-chemical factors in limnological eutrophication studies, in *Scientific Basis for Water Resources Management (Procs. Jerusalem Symposium, September 1985)*, IAHS Publ. no 153, 147–159.

Henderson-Sellers, B., 1985b, New formulation of eddy diffusion thermocline models, *Appl. Math. Model.*, 9, 441–446.

Henderson-Sellers, B., 1986, Calculating the surface energy balance for lake and reservoir modeling: a review, *Rev. Geophys.*, 24, 625–649.

Henderson-Sellers, B., 1987, Modelling sea surface temperature rise resulting from increasing atmospheric carbon dioxide concentrations, *Climatic Change* (in press)

Huang, N. E., Bliven, L. F., Long, S. R. and DeLeonibus, P. S., 1986, A study of the relationship among wind speed, sea state, and the drag coefficient for a developing wave field, *J. Geophys. Res.*, 91, 7733–7742.

Huber, W. C. and Harleman, D. R. F., 1969, *Laboratory and analytical studies of thermal stratification in reservoirs*, MIT Tech Report 112, MIT, Cambridge, Mass.

Hurley Octavio, K. A., Jirka, G. H. and Harleman, D. R. F., 1977, Vertical transport mechanisms in lakes and reservoirs, MIT Technical Report No. 227, Cambridge, Mass.

Imberger, J., 1985, The diurnal mixed layer, *Limnol. Oceanogr.*, 30, 737–770.

Imberger, J. and Hamblin, P. F., 1982, Dynamics of lakes, reservoirs, and cooling ponds, *Ann. Rev. Fluid Mech.*, 14, 153–187.

Imberger, J. and Parker, G., 1985, Mixed layer dynamics in a lake exposed to a spatially variable wind field, *Limnol Oceanogr.*, 30, 473–488.

Imberger, J. and Patterson, J., 1981, A dynamic reservoir simulation model: DYRESM 5, in *Transport Models for Inland and Coastal Waters*, Academic, 310–360.

Ivey, G. N. and Patterson, J. C., 1984, A model of the vertical mixing in Lake Erie in summer, *Limnol Oceanogr.*, 29, 553–563.

James, I. D., 1977, A model of the annual cycle of temperature in a frontal region of the Celtic Sea, *Estuarine and Coastal Marine Science*, 5, 339–353.

Jerlov, N. G., 1976, *Marine Optics*, Elsevier, New York.

Kim, J-W., 1976, A generalized bulk model of the oceanic mixed layer, *J. Phys. Oceanogr.*, 6, 686–695.

Kitaigorodskii, S. A., 1979, Review of the theories of wind-mixed layer deepening, in *Marine Forecasting* (ed. J. C. J. Nihoul), Oceanography Series 25, Elsevier, Amsterdam, pp 1–33.

Klein, P., 1980, A simulation of the effects of air-sea transfer variability on the structure of marine upper layers, *J. Phys. Oceanogr.*, 10, 1824–1841.

Klein, P. and Coantic, M., 1981, A numerical study of turbulent processes in the marine upper layers, *J. Phys. Oceanogr.*, 11, 849–863.

Klein, P. and Coste, B., 1984, Effects of wind-stress variability on nutrient transport into the mixed layer, *Deep-Sea Res.*, 31, 21–37.

Kondo, J., Sasano, Y. and Ishii, T., 1979, On wind-driven current and temperature profiles with diurnal period in the oceanic planetary boundary layer, *J. Phys. Oceanogr.*, 9, 360–372.

Kraus, E. B., 1972, *Atmosphere-Ocean Interaction*, Clarendon Press, Oxford.

Kraus, E. B. and Rooth, C., 1961, Temperature and steady state vertical heat flux in the ocean surface layers, *Tellus*, 13, 231–238.

Kraus, E. B. and Turner, J. S., 1967, A one-dimensional model of the seasonal thermocline II. The general theory and its consequences, *Tellus*, 19, 98–105.

Krenkel, P. A. and French, R. H., 1982, State-of-the-art of modeling surface water impoundments, *Water Sci. Technol.*, 14, 241–261.

Kundu, P. K., 1980, A numerical investigation of mixed-layer dynamics, *J. Phys. Oceanogr.*, 10, 220–236.

Kundu, P. K., 1981, Self-similarity in stress-driven entrainment experiments, *J. Geophys. Res.*, 86, 1979–1988.

Lam, D. C. L. and Schertzer, W. M., 1987, Lake Erie thermocline model results: comparison with 1967–1982 data and relation to anoxic occurrences, *J. Gt. Lakes Res.*, 13, 757–769.

Large, W. G. and Pond, S., 1981, Open ocean momentum flux measurements in moderate to strong winds, *J. Phys. Oceanogr.*, 11, 324–336.

McAlister, E. D. and McLeish, W., 1969, Heat transfer in the top millimeter of the ocean, *J. Geophys. Res.*, 74, 3408–3414.

McCormick, M. J. and Scavia, D., 1981, Calculation of vertical profiles of the lake-averaged temperature and diffusivity in Lakes Ontario and Washington, *Water Resour. Res.*, 17, 305–310.

Madsen, O. S. 1977. A realistic model of the wind-induced Ekman boundary layer. *J. Phys. Oceanogr.*, 7, 248–255.

Manabe, S., 1983, Carbon dioxide and climatic change, in *Theory of Climate* (ed. B. Saltzman), Academic Press, New York, 39–84.

Manabe, S. and Bryan, K., 1985, CO_2-induced change in a coupled ocean-atmosphere model and its paleoclimatic implications, *J. Geophys. Res.*, 90, 11689–11707.

Manabe, S. and Stouffer, R. J., 1980, Sensitivity of a global climate model to an increase of CO_2 concentration in the atmosphere, *J. Geophys. Res.*, 85, 5529–5554.

Manabe, S. and Wetherald, R. T., 1980, On the distribution of climate change resulting from an increase in CO_2 content of the atmosphere, *J. Atmos. Sci.*, 37, 99–118.

Marchuk, G. I., Kochergin, V. P., Klimok, V. I. and Sukhorukov, V. A., 1977, On the dynamics of the ocean surface mixed layer, *J. Phys. Oceanogr.*, 7, 865–875.

Martin, P. J., 1982, Mixed-layer simulation of buoy observations taken during hurricane Eloise, *J. Geophys. Res.*, 87, 409–427.

Martin, P. J., 1985, Simulation of the mixed layer at OWS November and Papa with several models, *J. Geophys. Res.*, 90, 903–916.

Meehl, G. A., 1984, A calculation of ocean heat storage and effective ocean surface layer depths for the Northern Hemisphere, *J. Phys. Oceanogr.*, 14, 1747–1761.

Meehl, G. A. and Washington, W. M., 1985, Sea surface temperatures computed by a simple ocean mixed layer coupled to an atmospheric GCM, *J. Phys. Oceanogr.*, 15, 92–104.

Mellor, G. L. and Durbin, P. A., 1975, The structure and dynamics of the ocean surface mixed layer, *J. Phys. Oceanogr.*, 5, 718–728.

Mellor, G. L. and Yamada, T., 1974, A hierarchy of turbulent closure models for planetary boundary layers, *J. Atmos. Sci.*, 31, 1791–1806.

Mellor, G. L. and Yamada, T., 1982, Development of a turbulence closure model for geophysical fluid problems, *Rev. Geophys. Space Phys.*, 20, 851–875.

Miyakoda, K. and Rosati, A., 1984, The variation of sea surface temperature in 1976 and 1977 2. The simulation with mixed layer models, *J. Geophys. Res.*, 89, 6533–6542.

Monismith, S. G., 1985, Wind-forced motions in stratified lakes and their effect on mixed-layer shear, *Limnol Oceanogr.*, 30, 771–783.

Munk, W. H. and Anderson, E. R., 1948, Notes on a theory of the thermocline, *J. Mar. Res.*, 7, 276–295.

Nieuwstadt, F. T. M. and van Dop, H., (eds.), 1982, *Atmospheric Turbulence and Air Pollution Modelling*, Reidel, Dordrecht, 358pp.

Niiler, P. P., 1975, Deepening of the wind-mixed layer, *J. Mar. Res.*, 33, 405–422.

Niiler, P. P., 1977, One-dimensional models of the seasonal thermocline *in The Sea* (Vol 6: Marine Modeling), ed. I. N. McCave, J. J. O'Brien and J. H. Steele, J. Wiley, N.Y., 97–115.

Niiler, P. P. and Kraus, E. B., 1977, One-dimensional models of the upper ocean, Chapter 10 in *Modelling and Prediction of the Upper Layers of the Ocean*, (ed. E. B. Kraus), Pergamon.

Orlob, G. T. and Selna, L. G., 1970, Temperature variations in deep reservoirs, *Procs. ASCE, J. Hyd. Div.*, 96(HY2), 391–410.

Pacanowski, R. C. and Philander, S. G. H., 1981, Parameterization of vertical mixing in numerical models of typical oceans, *J. Phys. Oceanogr.*, 11, 1443–1451.

Patterson, J. C., Hamblin, P. F. and Imberger, J., 1984, Classification and dynamic simulation of the vertical density structure of lakes, *Limnol Oceanogr.*, 29, 845–861.

Paulson, C. A. and Simpson, J. J., 1977, Irradiance measurements in the upper ocean, *J. Phys. Oceanogr.*, 7, 952–956.

Phillips, O. M., 1977, *The Dynamics of the Upper Ocean* (2nd ed), CUP

Pollard, D., Batteen, M. L. and Han, Y. J., 1983, Development of a simple oceanic mixed-layer and sea ice model, *J. Phys. Oceanogr.*, 13, 754–768.

Pollard, R. T., Rhines, P. B. and Thompson, R. O. R. Y., 1973, The deepening of the wind-mixed layer, *Geophys. Fluid Dynam.*, 3, 381–404.

Posmentier, E. S., 1980, A numerical study of the effects of heat diffusion through the base of the mixed layer, *J. Geophys. Res.*, 85, 4883–4887.

Prangsma, G. J. and Kruseman, P., 1984, Aspects of mixed layer modelling applied to JASIN data, *Dyn. Atmos. Oceans*, 8, 321–341.

Price, J. F., 1981, On the upper ocean response to a moving hurricane, *J. Phys. Oceanogr.*, 11, 153–175.

Price, J. F., Mooers, C. N. K. and van Leer, J. C., 1978, Observation and simulation of storm-induced mixed-layer deepening, *J. Phys. Oceanogr.*, 8, 582–599.

Price, J. P., Weller, R. A. and Pinkel, R., 1986, Diurnal cycling: observations and models of the upper ocean response to diurnal heating, cooling, and wind mixing, *J. Geophys. Res.*, 91, 8411–8427.

Rossby, C. C. and Montgomery, B. R., 1935, The layer of frictional influence in wind and ocean currents, *Papers in Physical Oceanography*, 3, 1–101.

Schwab, D. J., 1978, Simulation and forecasting of Lake Erie storm surges, *Mon. Weath. Rev.*, 106, 1476–1487.

Semtner, A. J., Jr., 1984, Development of efficient, dynamical ocean-atmosphere models for climatic studies, *J. Clim. Appl. Meteor.*, 23, 353–374.

Shay, T. J. and Gregg, M. C., 1984, Turbulence in an oceanic convective mixed layer, *Nature*, 310, 282–285 + corrigendum, 311, 84.

Shay, T. J. and Gregg, M. C., 1986, Convectively driven turbulent mixing in the upper ocean, *J. Phys. Oceanogr.*, 16, 1777–1798.

Shelkovnikov, N. K., 1983, A study of entrainment at the interface in a two-layer liquid, *Soviet Met. Hydrol.*, 1983, No. 1, 40–45.

Sherman, F. S., Imberger, J. and Corcos, M. G., 1978, Turbulence and mixing in stably stratified water, *Ann. Rev. Fluid Mech.*, 10, 267–288.

Shetye, S. R., 1986, A model study of the seasonal cycle of the Arabian Sea surface temperature, *J. Mar. Res.*, 44, 521–542.

Simpson, J. H. and Hunter, J. R., 1974, Fronts in the Irish Sea, *Nature*, 250, 404–406.

Simpson, J. J. and Dickey, T. D., 1981a, The relationship between downward irradiance and upper ocean structure, *J. Phys. Oceanogr.*, 11, 309–323.

Simpson, J. J. and Dickey, T. D., 1981b, Alternative parameterizations of downward irradiance and their dynamical significance, *J. Phys. Oceanogr.*, 11, 876–882.

Smith, I. R., 1979, Hydraulic conditions in isothermal lakes, *Freshwater Biol.*, 9, 119–145.

Smith, S. D., 1980, Wind stress and heatflux over the ocean in gale force winds, *J. Phys. Oceanogr.*, 10, 709–726.

Smith, S. D., 1981, Comment on "A new evaluation of the wind stress coefficient over water surfaces," *J. Geophys. Res.*, 86, 4307.

Spelman, M. J. and Manabe, S., 1984, Influence of oceanic heat transport upon the sensitivity of a model climate, *J. Geophys. Res.*, 89, 571–586.

Spigel, R. H. and Imberger, J., 1980, The classification of mixed-layer dynamics in lakes of small to medium size, *J. Phys. Oceanogr.*, 10, 1104–1121.

Spigel, R. H., Imberger, J. and Rayner, K. N., 1986, Modeling the diurnal mixed layer, *Limnol. Oceanogr.*, 31, 533–556.

Stefan, H. and Ford, D. E., 1975a, Temperature dynamics in dimictic lakes, *Procs. ASCE., J. Hyd. Div.*, 101(HY1), 97–114.

Stefan, H. and Ford, D. E., 1975b. Mixed layer depth and temperature dynamics in temperate lakes, *Verh. Internat. Verein. Limnol.*, 19, 149–157.

Stefan, H. G., Dhamotharan, S. and Schiebe, F. R., 1982, Temperature/sediment model for a shallow lake, *Procs. ASCE, J. Env. Eng. Div.*, 108(EE4), 750–765.

Sundaram, T. R. and Rehm, R. G., 1971, Formation and maintenance of thermoclines in temperate lakes, *Am. Inst. Aeronaut. Astronaut. J.*, 9, 1322–1330.

Sundaram, T. R. and Rehm, R. G., 1973, The seasonal thermal structure of deep temperate lakes, *Tellus*, 25, 157–167.

Tennekes, H., 1973, A model for the dynamics of the inversio above a convective boundary layer, *J. Atmos. Sci.*, 30, 558–567.

Tennekes, H., 1975, Reply to "Comments on 'A model for the dynamics of the inversion above a convective boundary layer'" by S. S. Zilintinkevich, *J. Atmos. Sci.*, 32, 992–995.

Thompson, R. O. R. Y., 1976, Climatological numerical models of the surface mixed layer of the ocean, *J. Phys. Oceanogr.*, 6, 496–503.

Thompson, R. O. R. Y., 1977, Erratum of "Climatological numerical models of the surface mixed layer of the ocean," *J. Phys. Oceanogr.*, 7, 157.

Thompson, R. O. and Imberger, J., 1980, Response of a numerical model of a stratified lake to wind stress, in *Stratified Flows. Proc. Int. Symp. Trondheim*, 562–570.

Tucker, W. A. and Green, A. W., 1977, A time-dependent model of the lake-averaged vertical temperature distribution of lakes, *Limnol Oceanogr.*, 22, 687–699.

Turner, J. S., 1968, The influence of molecular diffusivity on turbulent entrainment across a density interface, *J. Fluid Mech.*, 33, 639–656.

Ueda, H., Mitsumoto, S. and Komori, S., 1981, Buoyancy effects on the turbulent transport processes in the lower atmosphere, *Quart. J. Roy. Meteor. Soc.*, 107, 561–578.

Warn-Varnas, A. C. and Piacsek, S. A., 1979, An investigation of the importance of third-order correlations and choice of length scale in mixed layer modelling, *Geophys. Astrophys. Fluid Dyn.*, 13, 225–243.

Warn-Varnas, A. C., Dawson, G. M. and Martin, P. J., 1981, Forecast and studies of the oceanic mixed layer during the Mile experiment, *Geophys Astrophys Fluid Dynam.*, 17, 63–85.

Warren, B. A., 1972, Insensitivity of subtropical model water characteristics to meteorological fluctuations, *Deep Sea Res.*, 19, 1–20.

Washington, W. M. and Meehl, G. A., 1983, General circulation model experiments on the climatic effects due to a doubling and quadrupling of carbon dioxide concentrations, *J. Geophys. Res.*, 88, 6600–6610.

Washington, W. M. and Meehl, G. A., 1984, Seasonal cycle experiment on the climate sensitivity due to a doubling of CO_2 with an atmospheric general circulation model coupled to a simple mixed layer ocean model, *J. Geophys. Res.*, 89, 9475–9503.

Washington, W. M. and Meehl, G. A., 1986, General circulation model CO_2 sensitivity experiments: snow-sea ice albedo parameterizations and globally averaged surface air temperature, *Climatic Change*, 8, 231–241.

Washington, W. M. and VerPlank, L., 1986, A description of coupled general circulation models of the atmosphere and oceans used for carbon dioxide studies, *NCAR Technical Note NCAR/ TN-271 + EDD*, 29pp.

Washington, W. M., Semtner, A. J., Meehl, G. A., Knight, D. J. and Mayer, T. A., 1980, A general circulation model experiment with a coupled atmosphere, ocean and sea ice model, *J. Phys. Oceanogr.*, 10, 1887–1908.

Weber, J. E., 1981, Ekman currents and mixing due to surface gravity waves, *J. Phys. Oceanogr.*, 11, 1431–1435.

Weber, J. E., 1983, Steady wind- and wave-induced currents in the open ocean, *J. Phys. Oceanogr.*, 13, 524–530.

Webster, P. J. and Lau, K. M. W., 1977, A simple ocean-atmosphere climate model. Basic model and a simple experiment, *J. Atmos. Sci.*, 34, 1063–1084.

Wells, N. C., 1979, A coupled ocean-atmosphere experiment: the ocean response, *Quart. J. Roy. Meteor. Soc.*, 105, 355–370.

Wieringa, J., 1974, Comparison of three methods for determining strong wind stress over Lake Flevo, *Boundary-Layer Meteorol.*, 7, 3–19.

Wigley, T. M. L. and Schlesinger, M. E., 1985, Analytical solution for the effect of increasing CO_2 on global mean temperature, *Nature,* 315, 649–652.

Williams, D. T., Drummond, G. R., Ford, D. E. and Robey, D. L., 1981, Determination of light extinction coefficients in lakes and reservoirs, *Proc. ASCE Symp. on Surface Water Impoundments,* Minneapolis, Minnesota, Vol. 2 (ed. H. G. Stefan), 1329–1335.

Woods, J. D., 1980, Diurnal and seasonal variation of convection in the wind-mixed layer of the ocean, *Quart. J. Roy. Meteor. Soc.,* 106, 379–394.

Woods, J. D., 1984, The upper ocean and air-sea interaction in global climate, pp 141–187 in *The Global Climate* (ed. J. T. Houghton), Cambridge Univ. Press, Cambridge, 233pp.

Woods, J. D. and Strass, V., 1986, The response of the upper ocean to solar heating. II: The wind-driven current, *Quart. J. Roy. Meteor. Soc.,* 112, 29–42.

Woods, J. D., Barkmann, W. and Horch, A., 1984, Solar heating of the ocean—diurnal, seasonal and meridional variation, *Quart. J. Roy. Meteor. Soc.,* 110, 633–656.

Worthem, S. and Mellor, G., 1980, Turbulence closure model applied to the upper tropical ocean, *Deep Sea Res.,* 26 (GATE SUPPL), 237–272.

Wu, J., 1973. Prediction of near-surface drift currents from wind velocity. *Procs. ASCE., J. Hyd. Div.,* 99(HY9): 1291–1302.

Wu, J., 1980, Wind-stress coefficients over sea surface nera neutral conditions—a revisit, *J. Phys. Oceanogr.,* 10, 727–740.

Wu, J., 1983, Sea-surface drift currents induced by wind and waves, *J. Phys. Oceanogr.,* 13, 1441–1451.

Wu, J., 1985a, On the cool skin of the ocean, *Boundary-Layer Meteorol.,* 31, 203–207.

Wu, J., 1985b, Parameterization of wind-stress coefficients over water surfaces, *J. Geophys. Res.,* 90, 9069–9072.

Zaneveld, J. R. V. and Spinrad, R. W., 1980, An arctangent model of irradiance in the sea, *J. Geophys. Res.,* 85, 4919–4922.

Zaneveld, J. R. V., Kitchen, J. C. and Pak, H., 1981, The influence of optical water type of the heating rate of a constant depth mixed layer, *J. Geophys. Res.,* 86, 6426–6428.

Zholudev, V. D., 1983, Similarity theory for the ocean boundary layer, *Oceanology,* 23, 17–23.

Zilintinkevich, S. S., 1975, Comments on "A model for the dynamics of the inversion above a convective boundary layer," *J. Atmos. Sci.,* 32, 991–992.

FOUR

NUMERICAL FLOW AND HEAT TRANSFER UNDER IMPINGING JETS: A REVIEW

S. Polat, B. Huang, A. S. Mujumdar, and W. J. M. Douglas

ABSTRACT

Numerical studies related to the prediction of transport processes under laminar and turbulent jets impinging on flat surfaces are reviewed in light of the experimental data available under a limited range of the relevant parameters. Brief summaries are presented of the numerical techniques commonly employed and the turbulence models used to obtain results of engineering interest. Effects of various flow and geometric parameters on the flow, heat/mass transfer in the impingement and wall jet zones of single/multiple, round/slot jets impinging normally/ obliquely on a stationary/moving surface which may be permeable or impermeable are discussed. Deficiencies in current numerical models are identified via comparison with relevant experimental data. It is concluded that the two-equation and algebraic stress models, which allow the integration of the variables up to the wall, provide the best hope for engineering calculations for the turbulent impinging jets in the near future. Prediction of the flow, heat and mass transfer under a single, semi-confined turbulent jet may be employed as a good test for new turbulence models.

NOMENCLATURE

A a constant in Eq. (37)

A_D, A_μ constants in one-equation turbulence model used by [6], Eqs. (8) and (9) respectively

C_1, C_2 constants in two-equation turbulence model, Eq. (12), and constants in algebraic stress model, Eqs. (20) and (21)

C_D a constant in one-equation turbulence model, Eq. (6)

C_P specific heat

C_w a constant in algebraic stress model, Eq. (22)

C_μ a constant in two-equation turbulence model, Eq. (11)

C'_μ a constant in one-equation turbulence model, Eq. (7)

C_{s1} a constant in algebraic stress model, Eq. (17)

H nozzle-to-surface spacing

i turbulence intensity at the jet exit

k turbulent kinetic energy

L_D length scale of turbulent energy dissipation

L_μ length scale of turbulence

l_m Prandtl mixing length

M cross flow parameter, cross flow/jet mass flow ratio

P pressure, Eq. (3)

P_s static pressure

Pe   Péclet number

$\dot{q}_w$ heat flux at the wall, Eq. (37)

R_N nozzle radius

R_v turbulent Re number at the edge of viscous sublayer

r radial position

r_T radial position where angular velocity, W, is at its maximum

Re_j jet Reynolds number

Re_T local turbulence Reynolds number ($=k^{1/2}y/v_t$)

S spacing between the centerlines of the jet and the adjacent exhaust port

S_Φ source term for variable Φ

T temperature

t time

U velocity in x direction

u fluctuating velocity in x direction

V velocity in y direction

W angular velocity

w nozzle width

y_0 position of the centerline of an inclined jet in [14]

Y^+ dimensionless distance ($=yk^{1/2}C_\mu^{1/4}/v$)

α thermal diffusivity, Eq. (4), and a constant in Eqs. (17), (19)–(21)

Γ_Φ general exchange coefficient

ϵ turbulent energy dissipation

θ angle of inclination

κ von Karman constant

λ a length scale constant, Eq. (28)

μ viscosity

v eddy viscosity

ρ density

σ	Prandtl number
Φ	general variable
ψ	stream function
ω	vorticity

Subscripts

e	at position e
E	at position E
j	inlet
P	at position P
v	at the edge of viscous sublayer
w	at the wall
avg	average
max	maximum

1 INTRODUCTION

Rapid heating, cooling, or drying of a variety of industrial products with large surface area are often provided using arrays of round or slot impinging jets. Annealing of metal and plastic sheets, tempering of glass, cooling of electric components, turbine blades, and drying of textiles, veneer, and paper are among the major industrial applications of impinging jets.

In the design of an impinging jet system for a given thermal application, the designer is immediately faced with the task of specifying a rather large number of geometric and flow parameters, e.g., jet type (round/slot), jet configuration (array geometry), nozzle-to-target surface spacing, location of exhaust ports, etc. Further in many applications effects of induced or imposed cross flow, surface motion, angle of impingement, nozzle design, large temperature differences between the jet and the impingement surface, etc., may be significant. In view of the large number of design parameters it is clear that a purely experimental approach to the study of transport processes under impinging jets is unlikely to lead to a satisfactory solution to the problem. Fortunately with the advent of the high speed computers and robust numerical techniques for solving transport equations, it is now possible to supplement (not supplant) the experimental studies so as to permit (eventually) extrapolation, interpolation or even true prediction of complex impingement transport phenomena.

The objective of this chapter is to review recent developments in the numerical prediction of impingement transport processes. Essentially all prediction schemes are based on obtaining solution of the governing conservation equations for mass, momentum, energy and species (for mass transfer). Both laminar and turbulent flow domains are considered. In the latter, an added complication is the solution of one or more additional equations to model turbulence to varying degrees of sophistication and accuracy. Results of selected configurations are presented while most of the relevant literature is cited in an annotated tabular form for easy reference by the interested reader. The emphasis here is on the computational results and their validity rather than on the numerical schemes employed although often the latter governs the former. Areas for further studies are identified in light of the state of the art in modeling impingement flows.

To assist in the presentation and discussion of relevant results, a brief summary is given in the following section of the basic flow characteristics of impinging jets.

2 FLOW CHARACTERISTICS OF IMPINGING JETS

The flow pattern of plane or axisymmetric impinging jets can be subdivided into three characteristic regions: (1) free jet region, (2) stagnation flow region, and (3) wall jet region. Figure 1 displays these regions schematically.

In addition, the free jet region may also show three characteristic regions: (1) potential core, (2) developing flow, and (3) developed flow region depending on the nozzle-to-impingement surface spacing.

The potential core region is the part of the flow where the axial velocity remains almost equal to the nozzle exit velocity. The end of the potential core is determined by the rate of growth of two mixing layers originating at the edges of the nozzle. In the developing flow region, the axial velocity starts to decay and jet spreads to the surrounding. Eventually lateral profiles of the axial velocity approaches to a bell shape. In the developed flow region, similar axial velocity profiles exist at different jet lengths.

Depending on the nozzle-to-surface spacing, the free jet region of an impinging jet may display one or more of the above regions.

The impingement or stagnation region is characterized by an increased static pressure as a result of the sharp decrease in mean axial velocity. Upon impingement the flow deflects and starts to accelerate along the impingement surface.

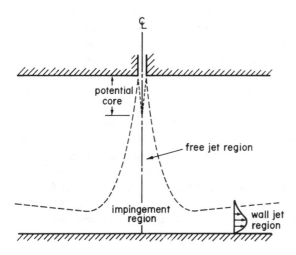

Figure 1 Flow regions of an impinging jet.

Gutmark et al. [1] and Saad [2] observed among others that the presence of the impingement plate was not felt beyond about 0.25 H away from the surface. The end of the impingement region is the location where the pressure gradient at the impingement surface becomes zero. This was reported to be about 0.5 H by Saad for single and noninteracting multiple jets (i.e., $S/H > 0.75$) for $4 < H/w < 16$. In the wall jet region, the boundary layer grows along the impingement surface. In the case of semi-confined impinging jet, if the confinement and the impingement surfaces are sufficiently long, the wall jet boundary layer reaches the confinement surface and a recirculating bubble resides in the region enveloped by the free jet and the wall jet regions. Beyond this point the flow resembles a channel flow for a two-dimensional jet.

The above discussion of flow regimes is valid for the case of normal impingement. For oblique impingement the stagnation point is difficult to define. When there is superimposed cross flow and/or the impingement surface is in motion, under certain flow and geometric conditions, no stagnation point or region may exist.

According to Vickers [3] jets with Re_j higher than 1000 may be considered turbulent. However, McNaughton and Sinclair [4] report four characteristic jet patterns for circular free jets:

1. The dissipated laminar jet, $Re_j < 300$
2. Fully laminar jets, $300 < Re_j < 1000$
3. A transition or semiturbulent jet, $1000 < Re_j < 3000$
4. A fully turbu0lent jet, $Re_j > 3000$

Gardon and Akfirat [5] regarded the slot jets with $Re_j > 2000$ as turbulent.

In general, whether an originally laminar free jet is still laminar before impact depends on many factors such as Re_j, the velocity profile at the nozzle exit, the nozzle-to-surface spacing, whether the jet is confined or not, etc. All these factors affect the level of mixing at the outer jet boundaries which determines how fast the laminar jet transforms into a turbulent one.

When the jet is laminar at impact, the flow remains laminar due to acceleration under the favorable pressure gradient existing in the impingement region. Beyond this region transition to turbulent flow takes place under the influence of free stream turbulence created by outer mixing layers of free jet region. This transition occurs at larger distances away from the stagnation point when Re_j and H/w are decreased. No direct verification has yet been carried out to test this flow model proposed originally by Gardon and Akfirat [5] on the basis of their heat transfer measurements.

3 GOVERNING EQUATIONS

The governing equations which describe the laminar flow field can be written in terms of their primitive form (i.e., velocity-pressure) or in stream function-vorticity form. For brevity, only the primitive variables form is shown below for the generalized transport equation. By selecting the appropriate generalized transport quantity Φ and the general exchange coefficient, Γ_Φ, the Navier-Stokes equations, the continuity equation, as well as the energy equation can be represented by the same equation. The term on the right hand side is the source term which counterbalances the convective and diffusive transfers as described by the left hand side of the equation.

$$\frac{\partial}{\partial x_i}(\rho U_i \Phi) - \frac{\partial}{\partial x_i}\left(\Gamma_\Phi \frac{\partial \Phi}{\partial x_i}\right) = S_\Phi \tag{1}$$

The exchange coefficients and source terms for momentum and energy equations are shown in Table 1 for the laminar two-dimensional flow case. The governing equations for an incompressible turbulent jet impinging on a flat surface are similar to that of laminar ones except

$$\mu \rightarrow \mu_{\text{eff}} = \mu + \mu_t \tag{2}$$

$$P \rightarrow P_s + \frac{2}{3}\rho k \tag{3}$$

$$\alpha \rightarrow \alpha + \alpha_t = \frac{\mu}{\sigma} + \frac{\mu_t}{\sigma_{h,t}} \tag{4}$$

The extra terms that appear due to turbulence have to be computed in order to be able to solve these equations. This is done using the so-called turbulence models. A turbulent Prandtl number of 0.9 is usually used in the energy equation unless otherwise stated.

4 TURBULENCE MODELS

A brief introduction to turbulence models is in order here to aid further discussions. Models of varying complexity have been proposed but the one-equation,

Table 1 Exchange coefficients and source terms of governing equations for a two-dimensional laminar flow

Equation	Φ	Γ_Φ	S_Φ
Continuity	1	0	0
x Momentum	U	μ	$-\dfrac{\partial P}{\partial x} + \dfrac{\partial}{\partial x}\left(\mu \dfrac{\partial U}{\partial x}\right) + \dfrac{\partial}{\partial y}\left(\mu \dfrac{\partial V}{\partial x}\right)$
y Momentum	V	μ	$-\dfrac{\partial P}{\partial y} + \dfrac{\partial}{\partial x}\left(\mu \dfrac{\partial U}{\partial y}\right) + \dfrac{\partial}{\partial y}\left(\mu \dfrac{\partial V}{\partial y}\right)$
Energy	T	$\alpha = \dfrac{\mu}{\sigma}$	0

two-equation, and algebraic stress models are the ones applied to impinging jet flows. Hence we will confine our attention only to these models.

4.1 One-Equation Models

These are the simplest turbulence models which account for the transport and history effects of turbulence employing a transport equation for a suitable velocity scale, commonly $k^{1/2}$. In most one-equation models the following transport equation for k is solved.

$$\frac{\partial k}{\partial t} + U_i \frac{\partial k}{\partial x_i} = \frac{\partial}{\partial x_i} \left(\frac{\nu_t}{\sigma_k} \frac{\partial k}{\partial x_i} \right) + G - \epsilon \tag{5}$$

where

$$G = \nu_t \left(\frac{\partial U_i}{\partial x_j} + \frac{\partial U_j}{\partial x_i} \right) \frac{\partial U_i}{\partial x_j}$$

The dissipation rate ϵ is determined from

$$\epsilon = C_D \frac{k^{3/2}}{L_D} \tag{6}$$

where L_D is the length scale of turbulence dissipation.

The turbulent viscosity is computed using Kolmogorov-Prandtl expression

$$\nu_t = C'_\mu \sqrt{k}\, L_\mu \tag{7}$$

where L_μ is the length scale of turbulence. In one-equation models L_D and L_μ are usually determined from simple empirical relations. Wolfshtein [6] and Russell and Hatton [7] used

$$\frac{L_D}{y} = 1 - \exp\left(-A_D \mathrm{Re}_T\right) \tag{8}$$

$$\frac{L_\mu}{y} = 1 - \exp\left(-A_\mu \mathrm{Re}_T\right) \tag{9}$$

with the constants $A_\mu = 0.016$; $A_D = 0.263$; $C'_\mu = 0.22$; $C_D = 0.416$; and $\sigma_k = 1.53$.

Equations (8) and (9) reduce $L_D = L_\mu = y$ at large y values and

$$\frac{L_D}{y} = A_D \mathrm{Re}_T \qquad \frac{L_\mu}{y} = A_\mu \mathrm{Re}_T \tag{10}$$

at small y's (y is the distance from wall).

On the other hand, Lampinen [8] used $L_D = L_\mu = L$. Near a wall, $L = 0.21y$ was used with a maximum value of $L = 0.0852H$ where H is the height of the channel formed between the confinement and the impingement surface. He used the following values for the constants: $C'_\mu = 1.0$; $C_D = 0.0915$; $\sigma_k = 1.53$.

These models are only valid at high local turbulence numbers, Re_j; therefore cannot be used in the viscous sublayer. Hence this region has to be bridged by so-called wall functions (see Section 5.1).

4.2 Two-Equation Models of Turbulence

The empirical determination of the turbulence length scale is the major limitation of the one-equation models. The turbulence length scale also has transport and history effects as in the case of the turbulent velocity scale, $k^{1/2}$. In order to account for these processes, models have been suggested to use a transport equation for the turbulent length scale. However, the dependent variable of this equation is not necessarily the length scale itself; any combination with k will suffice since k is already known from the solution of the k equation. Various combinations of k and L have been suggested but $\epsilon = C_D k^{3/2}/L$ has become the most popular. Here the eddy diffusivity is computed using the following expression:

$$\nu_t = \frac{C_\mu k^2}{\epsilon} \tag{11}$$

In this model, in addition to Eq. (5), the following equation for ϵ is solved.

$$\frac{\partial \epsilon}{\partial t} + U_i \frac{\partial \epsilon}{\partial x_i} = \frac{\partial}{\partial x_i} \left(\frac{\nu_t}{\sigma_\epsilon} \frac{\partial \epsilon}{\partial x_i} \right) + C_1 \frac{G\epsilon}{k} - C_2 \frac{\epsilon^2}{k} \tag{12}$$

The terms in the equation represent physical processes similar to those in the k equation. The widely used empirical constants are [9]: $C_\mu = 0.09$; $C_1 = 1.44$; $C_2 = 1.92$; $\sigma_k = 1.00$; $\sigma_\epsilon = 1.30$.

The model described so far is applicable only to flows with sufficiently high turbulence Re numbers. Therefore near walls, the so-called wall functions, are used to bridge the wall to the main flow (see Section 5.1).

Jones and Launder [9] proposed a low-Re number version with which the calculations can be carried right to the wall. In this model viscous diffusion of k and ϵ is included in Eqs. (5) and (12). Further constants C_μ and C_2 are replaced by functions of Re_T in the following manner:

$$C_\mu = 0.09 \exp \frac{-2.5}{(1 + Re_T/50)} \tag{13}$$

$$C_2 = 2.0 \, [1.0 - 0.3 \exp(-Re_T^2)] \tag{14}$$

To the right-hand side of Eqs. (5) and (12) the following terms

$$-2\nu \left(\frac{\partial k^{1/2}}{\partial y} \right)^2 \tag{15}$$

$$2.0 \, \nu\nu_t \left(\frac{\partial^2 U}{\partial y^2}\right)^2 \tag{16}$$

are added respectively. The reasons for including the extra term, Eq. (15), to the k equation are computational rather than physical to let ϵ go to zero at the wall. This term is equal to the dissipation rate in the immediate vicinity of the surface. The extra term, Eq. (16), to ϵ equation is included in order that the distribution of kinetic energy within the viscosity-affected region should be in reasonable accord with experiment. Indeed its presence is one of the less satisfactory features of this model.

The limitations of these models are the following:

1. They are based on the eddy viscosity/diffusivity concept which is not valid under all flow conditions.
2. More importantly, the eddy viscosity and diffusivity are assumed to be isotropic; that is, the same values are taken for the various $\overline{u_i u_j}$'s and $\overline{u_i \Phi}$'s.
3. Additional effects such as buoyancy, streamline curvature, etc., are not inherent to those models and have to be included separately when it is deemed necessary.

4.3 Algebraic Stress Models

To overcome some of the limitations of the $(k - \epsilon)$ model, the transport equations for Reynolds stresses, $\overline{u_i u_j}$ and analogous equations for scalar fluxes, $\overline{u_i \Phi}$ may be introduced to the turbulence model. These equations can be derived in exact forms but they contain higher-order correlations that have to be approximated in order to obtain a closed set of equations. A particular advantage of deriving the exact equations is that terms accounting for buoyancy, rotation and other effects are introduced automatically. Such models employing transport equations for individual turbulent stresses and fluxes are called Turbulent Stress/Flux Equation Models.

The implementation and solution of such a number of complex equations is not an easy task and is also expensive. Hence, for practical applications, it would be desirable to use simplified models whenever possible. For this reason, the so-called algebraic stress/flux models have been developed by simplifying the differential transport equations for individual stresses/fluxes such that they reduce to algebraic equations while retaining most of their basic features.

In one model that is proposed by Rodi [10], the diffusion terms are assumed to be proportional to those in the k-equation multiplied by the ratio $\overline{u_i u_j}/k$. The following equation then emerges for turbulent viscosity,

$$\mu_t = \frac{2}{3}(1 - \alpha)\left[\frac{C_{s1} - 1 + \alpha G/\epsilon}{(C_{s1} - 1 + G/\epsilon)^2}\right]\frac{\rho k^2}{\epsilon} \tag{17}$$

where $\alpha = 0.6$ and $C_{s1} = 2.486$ are values recommended by Hanjalic and Launder [11].

In deriving the above equation, the surface integral (and hence the wall effects) in the pressure-stress correlation is neglected but the stress-redistribution part of the pressure-strain correlation, which is important in the stagnation region of an impinging jet, is included.

Another model due to Ljuboja and Rodi [12] includes the wall effects by including the surface integral appearing in the modeled pressure-strain correlation term. The following turbulent viscosity expression results:

$$\mu_t = F_D G_1 G_2 \frac{\rho k^2}{\epsilon} \tag{18}$$

$$F_D = 2(1 - \alpha) \frac{C_{s1} - 1 + \alpha G/\epsilon}{3C_{s1}(C_{s1} - 1 + G/\epsilon)} \tag{19}$$

$$G_1 = \frac{1 + 1.5\alpha C_2' f/(1 - \alpha)}{1 + 1.5C_1' f/C_{s1}} \tag{20}$$

$$G_2 = \frac{1 - 2\alpha C_2' \rho f/(C_{s1}\epsilon - \epsilon + \alpha G)}{1 + 2C_2' f/(C_{s1} - 1 + G/\epsilon)} \tag{21}$$

$$f = \min\left(1.0, \frac{k^{3/2}}{C_w y \epsilon}\right) \tag{22}$$

with constants $C_{s1} = 2.2$; $\alpha = 0.55$; $C_1' = 0.75$; $C_2' = 0.45$; $C_w = 4.4$.

The algebraic expressions (17) or (18) together with the k and ϵ equations form an extended $(k - \epsilon)$ model which has been applied to a single turbulent impinging jet problem by Looney and Walsh [13].

5 BOUNDARY CONDITIONS

We will consider, in this section, the boundary conditions needed for both laminar and turbulent jets. Unless stated otherwise, we restrict attention to single, normally impinging jets and stationary, impermeable planar targets. The only difference between laminar and turbulent flow boundary conditions is that for the latter simultaneous solution of the turbulence model equations requires boundary conditions for k and ϵ.

Reference is made to Fig. 2 in conjunction with Table 2 in describing the boundary conditions.

A. Boundary I (Nozzle Exit)
Laminar: The usual velocity profiles used are either flat or fully developed.
Flat velocity profile:

$$U = U_j \qquad V = 0 \tag{23}$$

Fully developed profile:

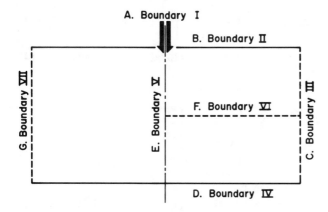

Figure 2 Position of the boundaries described in Table 2.

Table 2 Possible boundary conditions for impinging jets

A. Boundary I
    A1.   Flat jet inlet velocity profile
    A2.   Developed jet inlet profile
    A3.   Inclined jet
B. Boundary II
    B1.   Without confinement
    B2.   With confinement
C. Boundary III
    C1.   No flow (discharge is at the nozzle level)
    C2.   Free discharge (no cross flow)
    C3.   Cross flow
    C4.   Multiple jets (symmetry boundary)
D. Boundary IV
    D1.   Plane target surface
    D2.   Impermeable surface
    D3.   Smooth surface
    D4.   Stationary surface
    D5.   Curved surface
    D6.   Moving surface
    D7.   Permeable surface
    D8.   Rough surface
    D9.   Mass transfer at the surface
    D10. Large temperature driving force
    D11. Free surface
    D12. Phase change at the surface
E. Boundary V
    E1.   Symmetry
F. Boundary VI
    F1.   Well-developed, unbounded submerged turbulent jet profil
    F2.   Partially developed unbounded turbulent jet profile
G. Boundary VII
    G1.   Cross flow
    G2.   No cross flow

$$U = 2U_{max}\left[1 - \left(\frac{r}{R_N}\right)^2\right] \tag{24}$$

To describe the inlet velocity profile of an inclined slot jet, Jaussaud [14] used the following equations:

$$U = \frac{3}{2}U_{avg}\left[1 - \left(\frac{y - y_0}{y_R}\right)^2\right]\cos\theta \tag{25}$$

$$V = \frac{3}{2}U_{avg}\left[1 - \left(\frac{y - y_0}{y_R}\right)^2\right]\sin\theta \tag{26}$$

where θ is the angle of inclination and $y_R = W/2\cos\theta$.

For an axisymmetric nozzle with swirl, Huang [15] used a modified Rankine vortex type profile to specify the angular velocity profile:

$$\text{for } 0 < r < r_T \qquad W = C \cdot r \qquad \text{where } C \text{ is constant}$$

$$\text{for } r_T < r < R_N \qquad W = 1 - \frac{1 - r_T/r}{1 - r_T/R_N} \tag{27}$$

Turbulent Jet: For uniform (or flat) velocity profiles at the nozzle exit, the following boundary conditions are used.

$$U = U_j$$

$$k = k_j = iU_j$$

$$\epsilon = \epsilon_j = \frac{k_j^{3/2}}{\lambda(w/2)} \tag{28}$$

where i is the turbulence intensity and λ is the length scale constant (usually a value of 0.3 is used).

When the velocity profile at the nozzle exit is nonuniform, the following nozzle exit conditions may apply.

$$U = f(r)$$

$$k = C_\mu^{-1/2}l_m^2\left(\frac{\partial U}{\partial r}\right)^2$$

$$\epsilon = C_\mu^{1/2}k\frac{\partial U}{\partial r} \tag{29}$$

where l_m is the mixing length. Amano [16] used a 5.5 power velocity profile for an axisymmetric jet, i.e.,

$$U = U_{max}\left(1 - 0.3412\left(\frac{r}{r_N}\right)^{5.5}\right) \tag{30}$$

Temperature or enthalpy at the jet inlet is usually taken to be uniform.

B. Boundary II

Laminar and Turbulent: In the case of semi-confined jets, at the confinement plate, the no-slip condition is used, where all velocity components are zero, and the temperature set to that of the jet.

In the case of mass transfer study [17], nonevaporating surface is specified.

Without confinement, this boundary can only be approximated. Various forms are used by [18–20] for laminar flow. The usual boundary conditions here for turbulent flow is zero gradients for all variables.

C. Boundary III (Outflow)

Laminar and Turbulent: The outflow region is usually approximated using one of the following assumptions. If this boundary is sufficiently far from the jet axis, a fully developed flow between parallel plates can be specified. However, as it is not economical to carry the computation of the much larger domain, some other assumptions are needed when the outflow boundary is not sufficiently away from the jet axis. The usual approach is to assume a developing flow which is parallel to the impingement plate (i.e., the axial velocity = 0), and the gradients of other variables set to 0.

When using the primitive variables, the stream-wise velocity at this boundary is adjusted through an overall mass balance of the entire flow field and then the appropriate pressure correction is applied.

Law [21] used the developing velocity profile as given by [22] as the outflow boundary condition. Similarity approach was used by Deshpande [18] for laminar case and by Looney and Walsh [13] for their turbulent jet simulation.

For laminar multiple jet simulation [20], the outflow boundary condition becomes: V (velocity) = 0, i.e., no flow across the boundary; gradients of axial velocity and temperature are zero. The same approach may be used for turbulent multiple jets by assuming gradients of turbulence variables are also zero at this boundary.

D. Boundary IV (Impingement Plate)

Laminar: The no-slip condition is imposed at this boundary. In the case of a permeable surface, the axial velocity is nonzero and its sign depends on whether suction or blowing is applied. For diffusion at the surface, the axial velocity is modified accordingly [17]. The temperature at the impingement plate is usually set at a constant value.

Turbulent: In addition to the above conditions, the k and ϵ or their axial gradients may be set to zero.

E. Boundary V (Axis of Symmetry)

Laminar and Turbulent: At the axis of symmetry, the V velocity is set to 0, and the gradients of all other variables are set to 0.

F. Boundary VI

Turbulent: The profiles of the dependent variables at a level near the wall is specified either by using similarity solutions of plane or axisymmetric free jets or by solving free jet case iteratively and using the results as the inlet conditions to the impinging jet simulation problem.

G. Boundary VII

Laminar and Turbulent: In the case of cross flow, a velocity profile is introduced at this boundary. Depending on the velocity profile, the local velocities can be adjusted to have a predetermined cross flow to jet mass flow rate ratio, M. When $M = 0$ this boundary resembles a wall boundary.

5.1 Wall Functions

When low Re versions of a turbulence model are used, the iterations are carried up to the wall where the previously given boundary conditions are applicable. However in the case of a high-Re version, since the model cannot handle transition or low Re regions which occur near a wall, a bridging technique in the form of suitable functions must be employed.

The most commonly used wall function is the well known "logarithmic law of the wall," which assumes constant shear stress near the wall up to the turbulent region of the flow, is the following:

$$\frac{U_P}{(\tau_w/\rho)^{1/2}} = \frac{1}{\kappa} \ln \frac{Ey_P(\tau_w/\rho)^{1/2}}{\nu} \tag{31}$$

with $E = 9$ for smooth walls and $\kappa = 0.41$.

Assuming local equilibrium where $k = (\tau_w/\rho)C_\mu^{-1/2}$, the wall shear stress may be evaluated using the following relation:

$$\frac{U_P k_P^{1/2} C_\mu^{1/4}}{\tau_w/\rho} = \frac{1}{\kappa} \ln \frac{Ey_P k_P^{1/2} C_\mu^{1/4}}{\nu} \tag{32}$$

Since $\tau_w/\rho\nu = \partial U/\partial y$ this relation is substituted whenever a $\partial U/\partial y$ term appears in the source term of a dependant variable (i.e., U and k since no boundary condition for the velocity vertical to the wall is required). Usually ϵ is calculated from the following equilibrium relation

$$\epsilon = \frac{C_\mu^{3/4} k^{3/2}}{\kappa y} \tag{33}$$

where y is the distance away from the wall.

Chieng and Launder [23] proposed an improved version of this wall treatment which later has been applied to an axisymmetrical impinging jet problem by [16] and [24–26]. Since an improvement is reported it is summarized here.

Here the near-wall flow is treated as if it were composed of two layers, i.e., laminar inside the viscous sublayer and fully turbulent beyond this point. A parabolic distribution of the turbulent kinetic energy inside the viscous sublayer and a linear variation of it beyond this region is assumed. The turbulent shear stress is zero within the viscous sublayer and it goes through an abrupt increase at the edge of the sublayer and varies linearly from then on. The turbulent shear stress is again evaluated using Eq. (32). This model is summarized in Fig. 3. In numerical computation grid is so adjusted that the point P is always outside the viscous sublayer region (i.e., $Y^+ > 11.5$).

The mean rate of generation of turbulent kinetic energy near a wall is given by [16] as

$$\overline{G} = \frac{\tau_w(U_e - U_v)}{y_e} + \frac{\tau_w(\tau_e - \tau_w)}{\rho C_\mu^{1/4} k_v^{1/2} y_e} \left(1 - \frac{y_v}{y_e}\right) \tag{34}$$

The dissipation of turbulence energy in the viscous sublayer is constant and equal to $\epsilon = 2v(\partial k^{1/2}/\partial y)^2$ [27]. On introducing parabolic variation of k in this region, the dissipation inside the viscous sublayer can be expressed as

$$\epsilon = \frac{2vk_v}{y_v^2} \tag{35}$$

In the fully turbulent region, Spalding's [28] assumption is used for ϵ [Eq. (33)]. Then by integrating and averaging in the turbulent portion of the flow, the mean dissipation rate is found to be

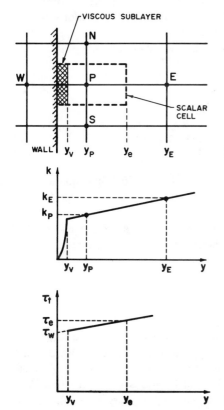

Figure 3 Near wall model used by [16,24–26].

$$\bar{\epsilon} = \frac{2k_v^{3/2}}{y_v R_v} + \frac{C_\mu^{3/4}}{y_e \kappa} \left[\frac{2}{3}(k_e^{3/2} - k_v^{3/2}) + 2a(k_e^{1/2} - k_v^{1/2}) + b \right] \tag{36}$$

where

$$a = k_P - \frac{k_P - k_E}{y_P - y_E} y_P$$

and

$$b = \begin{cases} a^{3/2} \ln \left[\dfrac{(k_e^{1/2} - a^{1/2})(k_v^{1/2} + a^{1/2})}{(k_v^{1/2} - a^{1/2})(k_e^{1/2} + a^{1/2})} \right] & \text{for } a > 0 \\[12pt] 2(-a)^{3/2} \left\{ \tan^{-1} \left[\dfrac{k_e^{1/2}}{(-a)^{1/2}} \right] - \tan^{-1} \left[\dfrac{k_v^{1/2}}{(-a)^{1/2}} \right] \right\} & \text{for } a < 0 \end{cases}$$

In a recent paper, Amano and Sugiyama [26] report another modification to this model in which each dissipation or generation term in the ϵ-equation is computed in accordance with the k equation instead of using equilibrium conditions to compute ϵ near the wall. They claim an improvement in the prediction of the stagnation point heat transfer. Due to limited space, the model equations are not given here; these can be found in the reference given.

In general, as the boundary condition for the energy equation, the following equation is used as a wall function:

$$\frac{(T_P - T_W)C_p \rho C_\mu^{1/4} k_P^{1/2}}{\dot{q}_w} = \frac{\sigma_{h,t}}{\kappa} \ln \left[\frac{E y_P (C_\mu^{1/2} k_P)^{1/2}}{\nu} \right]$$

$$+ \sigma_{h,t} \frac{\pi/4}{\sin \pi/4} \left(\frac{A}{\kappa} \right)^{1/2} \left(\frac{\sigma_{h,1}}{\sigma_{h,t}} - 1 \right) \left(\frac{\sigma_{h,t}}{\sigma_{h,1}} \right)^{1/4} \tag{37}$$

with constants for smooth impermeable walls; $E = 9$, $A = 26$. This equation was established by Jayatillaka [29] following an investigation of several experimental data.

The above wall functions are used for rigid, impermeable, stationary and smooth walls and require additional modifications whenever the other type boundary conditions (i.e., D5–D12 on Table 2) appear.

6 NUMERICAL TECHNIQUES

Most impingement transfer processes have been simulated using finite difference methods applied to the primitive equations. The flow field is discretized using orthogonal grid which is necessarily nonlinear to minimize computational error. At regions where gradients are large as well as at locations where the boundary conditions need to be adequately specified, higher grid density is required. For impingement jet simulation, it means higher grid lines near the impingement and confinement surface where the gradients are large, and also next to the jet axis. Coordinate transformations are also sometimes used [18,21,49].

The control volume approach with staggered grid [30] seems to be preferred by most researchers dealing with impinging jets. This is because the location of the vector components situated midway between the scalar grid points is claimed to improve the accuracy of solution. Control volume integration reduces the original equations into a set of algebraic equations which can then be solved numerically.

Even though the central differencing scheme is simple, it is seldom used for recirculating flow. Instead, upwind differencing or hybrid upwind differencing scheme has become the norm. In the hybrid method, a one-dimensional transport equation is solved, which results in an exact solution involving exponentiations. It can be used directly [21], but an approximation to reduce computation cost is usually used, which suppresses the diffusion term if the local Péclet number is outside the range of -2 to 2. Therefore, in the hybrid method, the central differencing is used when the convective and diffusive terms are comparable ($|Pe|$ < 2), but the upwind differencing scheme is used when the convective component is much larger than the diffusive one ($|Pe| > 2$).

To achieve stability and improve on efficiency, successive overrelaxation is normally incorporated into the finite difference scheme. The computation procedure usually involves the use of tridiagonal matrix algorithm (TDMA) to solve the set of finite difference equations along one grid line, and then proceed to the next line until the whole field is covered. The procedure is then repeated for each variable. This iterative step is repeated until a converged solution is obtained. Judicious ordering of the sequence of variable calculations in the loop can also reduce computational effort substantially.

The covergence criteria commonly specified require (a) overall mass balance is satisfied, (b) each of the sum of local residuals must reduce to a chosen level (for example, sum of the absolute local mass residuals be less than 1% of total mass entering the flow field), or (c) the difference of the point values of each variable at a reference location between two successive iterations must be less than a predefined small percentage.

7 LAMINAR IMPINGING JETS

Compared with turbulent jets, the simulation of laminar impinging jets is much simpler and straightforward. For relatively short nozzle-to-surface spacings the flow field is considered to be laminar up to a jet Reynolds number of approximately 2500. There is no direct evidence confirming this transition Re for impinging jets, confined or unconfined. Two commonly used configurations are round and slot jets impinging normally on a stationary isothermal impermeable planar surface. Although an arbitrary jet inlet velocity profile can be specified in numerical simulation, parabolic and flat profiles for the axial velocity are used in most studies. To study the effect of swirl, Huang [15] imposed a swirl velocity profile based on the modified Rankine vortex. Simulation results for an inclined slot jet in the presence of cross flow have been reported by Jaussaud et al. [14]. Most studies with the exception of that of Mujumdar et al. [17] considered constant physical property flows.

The jet may be confined by a plate at the nozzle exit parallel to the impingement plate, or it may be a free jet. The confinement plate renders specification of the appropriate boundary condition straightforward. For a free jet two possible ways of specifying the jet inlet and inflow (induced by the jet) boundary have been studied. In one the jet nozzle protrudes into the flow domain under consideration, thus requiring an additional boundary condition on the outside wall of the nozzle [18] and [20]. The other is for a jet nozzle upstream of the flow domain. This approach requires some form of approximation of the velocity profile at the top boundary [19].

A similar problem exists at the outflow boundary. In most cases it is difficult to specify the outflow boundary condition exactly. By extending the impingement and confinement surfaces long enough downstream one can assume a fully developed channel flow for which analytical solutions are obtainable. Deshpande and Vaishnav [18] used a similarity solution for this boundary. The developing velocity profile from Sparrow et al. [22] was used by Law [21] as the outflow boundary condition for a semi-confined slot jet.

For two-dimensional multiple jets appropriate boundaries are the planes of symmetry across which no flow occurs and all gradients vanish. This is true of course for normal impingement and exhaust slots located between neighboring jets.

At the impingement plate the no-slip condition holds. By defining a nonzero axial (jet direction) velocity at the plate, impingement with through flow (suction or injection) can be simulated [15,31,32]. Coupled heat and mass transfer (evaporation) has been studied by Mujumdar et al. [17] and Jaussaud et al. [14] for a single round jet and for a slot jet with cross flow, respectively.

7.1 Discussion

As expected predictions of transport processes under laminar jets yield good agreement with the few experimental data available in the literature for single round and slot jets.

Several researchers have investigated the effect of the jet velocity profile on the flow field and heat transfer at the impingement plate. The differences between flat and parabolic axial velocity profiles have been examined by [15,18,21,31,32]. The effect of imposing a swirl velocity component [15] and inclining the jet inlet [14] has also been studied.

The velocity profiles manifest their effect via marked changes in the spreading rate of the jet. For a parabolic axial velocity profile at the inlet, the momentum of the jet is concentrated at the axis of symmetry, therefore penetrating farthest onto the impingement plate. This results in a central peak in the heat or mass transfer distribution along the plate. In the case of a flat axial velocity profile, due to the high velocity gradient at the edge of the "flat," a large shear stress develops and causes a rapid outward transfer of momentum. The jet thus spreads faster compared with one with a parabolic inlet profile. van Heiningen [32] also noted a slight contraction of the parabolic slot jet just below the nozzle as a result of the upper confinement plate recirculation. As reported by [15,18,21,31,32], uniform velocity profiles result in stagnation (and even average) transfer rates that are lower than those for a parabolic velocity profile, the jet Reynolds number being held constant. As expected the effect of jet velocity profile diminishes with increased spacing between the nozzle and impingement surfaces.

Huang et al. [15] found that introduction of swirl in the entering jet causes it to spread much more rapidly than without swirl. This results in a decrease in stagnation heat transfer. If the swirl number (defined as the ratio of axial flux of angular momentum to the axial flux of axial momentum) of the jet exceeds about 0.43, the decay of the center-line axial velocity is so rapid that the adverse pressure gradient built up near the stagnation region overcomes the axial momentum of the jet. The result is a small recirculation "bubble" situated just above the impingement surface right next to the stagnation point. It causes a drastic reduction of the stagnation point heat transfer as the bubble acts as an insulator between the jet and the plate.

Similar results are reported by Jaussaud et al. [14] for a rather different flow condition. With a mild cross flow and an inclined jet (against cross-wind direction) it is expected that an adverse pressure gradient will build up slightly downstream of the intersection of the jet center line and the impingement surface; this situation is equivalent to forcing the jet to spread faster than the case of a normally impinging jet. The velocity field obtained by Jaussaud clearly shows a recirculation bubble in the stagnation region.

In summary, a jet inlet velocity profile that enhances the spreading of the jet will lower the stagnation point transfer rate by reducing the central axial momentum of the jet. Further, if the spread is rapid, a recirculation bubble may form in the stagnation zone which causes a sharp drop in the local heat transfer rate.

For a semi-confined jet, the existence of the confinement plate causes the formation of a recirculation bubble just below the plate. At low Reynolds numbers, the bubble is situated near the jet nozzle. As the Reynolds number increases, the bubble extends toward the outflow region. If the confinement plate is not sufficiently large, the bubble will eventually burst through the outflow boundary. Under these conditions, an inflow situation occurs. However, as noted above, such a boundary condition at the outflow is physically unrealistic to specify and numerically problematic.

The impact of this recirculation bubble on the heat transfer rate at the impingement plate is felt via the formation of a constricted channel at the thickest section of the bubble. Because the equivalent channel area is small, the fluid accelerates in this region, thus causing a small increase in the local transfer coefficient. For a jet without enhanced spreading, this constriction is situated right under the nozzle edge (in the case of flat inlet velocity profile it moves slightly downstream [32]). Its effect on local transfer coefficient is not noticeable because the transfer rate decreases rapidly in this region. However, for enhanced spreading of the jet this constriction moves sufficiently downstream that the increase in the local heat transfer coefficient is predictable [14,15].

Interestingly, for a parabolic jet velocity profile, Law [21] reported formation of a second recirculation bubble on the impingement plate but further downstream than the one described above, i.e., the one driven by the jet itself. It occurs at Re between 100 and 400, and is more pronounced for larger nozzle-to-plate spacings. However, no secondary bubble was detected for a flat jet velocity profile. There is reasonable correlation between the stream-wise location of the bubble center and the location of the local minimum of Sherwood number. This is probably due to the retardation of transport process by the bubble. It should be noted that no other workers reported this recirculation bubble under similar flow conditions.

Deshpande and Vaishnav [18] studied the impingement of a free jet. They reported the movement of the recirculation bubble from near the nozzle toward the wall jet region, over the range of Re from less than 5 to above 100.

At higher Reynolds numbers, most of these recirculation zones are either suppressed or move further downstream. The work reported by Ravuri and Tabakoff [19] on an unconfined laminar round jet covers Reynolds numbers from 4900 to 38,000. However, with a Reynolds number much higher than approximately 2500, the flow should be turbulent, thus the use of laminar simulation is not appropriate.

Mikhail et al. [20] studied the heat transfer under a row of laminar slot jets. As can be seen in Fig. 4, the stagnation region heat transfer is not influenced by the presence of neighboring jets. The wall jet region is also similar to that of a single jet except near the separation point, where the flow detaches from the plate towards the exhaust. In this region, i.e., just below the exhaust slot, the Nu increases again reaching a peak at the line of symmetry.

The effect of an inclined slot jet in cross flow on coupled heat and mass transfer on an impingement surface was studied by Jaussaud et al. [14]. With the

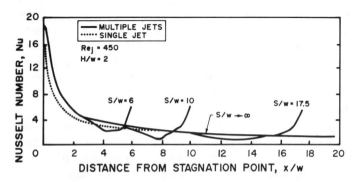

Figure 4 Nusselt number distribution at the impingement surface for various S/w values of laminar multiple impinging jets [20].

nozzle inclined against or into the cross-flow, two recirculation bubbles develop, one on either side of the nozzle. At low cross flows, $M < 0.4$ (M is defined as the cross flow to jet flow ratio), the maximum Nu values as well as their locations change with the angle of inclination of the jet. At an angle of approximately $-10°$ (against the cross flow), the Nu was noted to be optimal. However, at higher cross flows (i.e., $M = 2.0$), jet inclination has only a marginal effect.

The application of suction and blowing was studied by van Heiningen et al. [32], Saad et al. [31], and Huang et al. [15]. With uniform suction applied at the impingement surface, the convective flow towards the surface is increased, which results in a higher Péclet number. At the same time the thermal boundary layer is thinner, creating a larger temperature gradient near the wall. The heat transfer along the impingement surface thus increased with suction; the opposite is true for blowing.

Evaporation under a laminar impinging round jet was studied numerically by Mujumdar et al. [17]. It was found that the mass tranfer at the impingement surface results in a lower heat transfer rate, and the Reynolds analogies were shown to be inappropriate if appreciable mass fluxes exist. It was also shown that using a definition of mass transfer coefficient based on a log-mean molar driving potential, the Sherwood number becomes less sensitive to the influence of mass transfer and temperature difference between the jet and the impingement surface.

The effect of Reynolds number on stagnation Stanton number is usually expressed as $St_s = Const. \times Re^n$. The values of n as derived from laminar boundary layer theory for a flat inlet velocity profile is -0.5. Tabulated below are the values of n from numerical simulations as well as experimental results, together with the applicable Reynolds number ranges and H/w. Some of these results are derived from mass transfer data using the heat/mass transfer analogy.

In Fig. 5 the experimental and numerical results for the stagnation Stanton number are plotted against Reynolds number.

From Table 3 it can be seen that the values of n are close to -0.5 for flat inlet velocity profile whereas there is larger scatter in n values computed for parabolic profiles. Figure 5 shows that the stagnation transfer coefficients for a flat inlet velocity profile are lower than those for a parabolic inlet velocity profile at the same Reynolds number.

In the case of experimental results from Gardon and Akfirat [5], it was shown by van Heiningen [35] that the actual inlet profile is developing, thus the stagnation St numbers lie between those computed for flat and parabolic profiles.

The underprediction of heat transfer at higher Re ($n = -0.64$) by Saad et al. [31] is probably due to too coarse a grid near the impingement surface.

The deviation of Law's [21] results for parabolic velocity profile is quite large ($n = -0.23$ at H/w of 4 vs. -0.50) and could be related to the reported recirculation bubble attached on the impingement plate further downstream, which was not observed by other workers under similar flow conditions.

As described above, numerical studies of laminar impingement flow have covered most two-dimensional flow configurations. The results are fairly consistent among various numerical predictions as well as with the very limited ex-

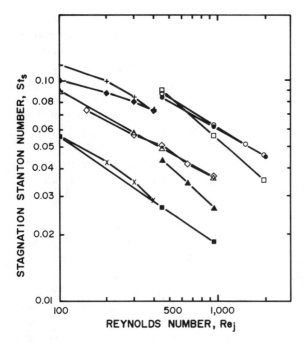

Figure 5 St, as a function of Re_j for various studies of laminar impinging jets.

Sym-bol	Ref.	Jet inlet velocity profile
○	[15]	parabolic
●	[17]	parabolic
◆	[21]	parabolic, $H/w = 4$
+	[21]	parabolic, $H/w = 2$
×	[21]	flat, $H/w = 4$
□	[31]	parabolic
■	[32]	flat
△	[32]	parabolic
▲ exp.	[5]	developing
◇ exp.	[34]	parabolic

perimental data. There are no major difficulties encountered with either the mathematical or the numerical model.

8 TURBULENT IMPINGING JETS

Table 4 shows a summary of the studies considered in this review. It covers (a) plane jet studies and (b) axisymmetric jet studies. The second column of this table shows the combination of boundary conditions used in each study. All these studies for incompressible turbulent impinging jets are performed for two-dimensional cases using either a one-equation (k) or a two-equation $(k - \epsilon)$ model of turbulence. Among these studies, only Looney and Walsh [13] tested algebraic stress model (ASM) of turbulence as well as the standard $(k - \epsilon)$ model and found that the ASM produced results at the stagnation region which differed markedly from experiments. The one-equation, two-equation, and algebraic stress model equations are summarized in Section 4.

The flow at the stagnation region of an impinging jet is pressure driven and therefore good predictions of mean flow and pressure distribution are usually reported no matter what turbulence model is used. However, the real engineering interest is usually in prediction of the heat and mass transfer rates under impinging jets. This is where the results of different numerical studies show greatest variation between themselves and with the experimental results.

The only way to verify the validity of the results of a numerical simulation, in the case of turbulent impinging jets, is to compare them to the results of experiments where the boundary conditions of the flow domain are the same as those of the numerical study. However, all experimental results are subject to an uncertainty level themselves which is rarely quantified and reported by the authors. Experimental parameters such as turbulence level profiles at nozzle exit or jet velocity profile are generally not specified. A flow domain which is assumed to be two dimensional may have three dimensional effects due to peculiarities of the apparatus. In case of the impingement Nu or Sh distributions, resolution of the measurements affects the accuracy of results particularly in the impingement region where Nu and Sh profiles change rapidly. Hence, discrepancies between numerical simulation and experimental ones should not always be attributed solely to the former as is generally the case.

8.1 Discussion

Table 4 is a summary of the relevant studies. A majority of the studies used the $(k - \epsilon)$ model of turbulence together with Boussinesq viscosity model. Some studies [6–8] used a one-equation model which required empirical specification of the turbulence length scales; this is inappropriate in view of the complicated structure of impinging jets. Higher order models, namely $(k - \epsilon)$ and algebraic stress models have been used in the more recent studies. Jones and Launder's [9] two-equation turbulence models (high and low Re versions) seem to have gained popularity due to their success with the prediction of recirculating flows. These models are relatively easy to understand and implement, and economical to use. Since these models solve a differential equation for dissipation of turbulent kinetic

Table 3 Comparison of n values (St_s Re_j^n) of various laminar impinging jet studies

	Reference	Re range	H/w	n	Note[a]
Numerical	[19]	4900–38,040		−0.69	1
	[35]	100–1000	4	−0.49	3
	[32]	100–1000	4	−0.4	
	[31]	950–1960	1.5–8	−0.64	
	[15]	450–1950	1.5	−0.44	
	[17]	450–2000	6	−0.42	4
	[21]	100–400	4	−0.48	2, 3
	[21]	100–400	2	−0.35	2
	[21]	100–400	4	−0.23	2
	[33]		>3	−0.50	3
Experimental	[5]	450–950	0.5–5	−0.67	5
	[34]	150–950	4	−0.38	6

[a]1, Computed from reported values of h; 2, computed from reported values of Sh; 3, flat inlet velocity profile; 4, with negligible mass transfer; 5, developing inlet velocity profile; 6, semi-confined nozzle which protrudes below the confinement plate.

Table 4 Impinging jet studies

Study	Boundaries	Turbulence model	Equations solved in
	a. Two-dimensional plane impinging jet studies		
[6]	C2[a], D1, D2, D3, D4, E1, F1	one-equation model	$\psi - \omega$
[7]	C2, D1, D2, D3, D4, E1, F1	one-equation model	$\psi - \omega$
[8]	A1, A3, B2, C1, C3, D2, D6, G1	one-equation model	$\psi - \omega$
[13]	C2, D1, D2, D3, D4, E1, F1, F2	$(k - \epsilon)$ high Re model & two ASM's	primary variables
[35]	A1, B2, C2, D1, D2, D3, D4, E1	$(k - \epsilon)$ high Re model	primary variables
[36]	A1, B2, C2, D1, D2, D3, D4, E1	$(k - \epsilon)$ high Re model	primary variables
[38]	no information	$(k - \epsilon)$ high Re model	primary variables
[40]	C2, D1, D2, D3, D4, E1, F1	$(k - \epsilon)$ high Re model	$\psi - \omega$
[41]	C2, D1, D2, D3, D4, E1, F1	$(k - \epsilon)$ high Re model	$\psi - \omega$
[42,43]	A2, B2, C2, D1, D2, D3, D4, D6, E1, G1, G2	$(k - \epsilon)$ high Re model	primary variables
[49]	A2?, B2, C2, D1, D2, D3, D4, E1, F1	$(k - \epsilon)$ low Re model	$\psi - \omega$
	b. Two-dimensional axisymmetric impinging jet studies		
[16]	A1, A2, B1, C1, C2, D1, D2, D3, D4, E1	$(k - \epsilon)$ high Re model	primary variables
[24]	A1, A2, B1, C1, D1, D2, D3, D4, E1	$(k - \epsilon)$ high Re model	primary variables
[25]	A1, A2, B1, C1, C2, D1, D2, D3, D4, E1	$(k - \epsilon)$ high Re model	primary variables
[26]	A1, B1, C2, D1, D2, D3, D4, E1	$(k - \epsilon)$ high Re model	primary variables

[a]Refer to Table 2.

energy, the need to determine the turbulence length scale empirically is eliminated.

It can easily be seen from Table 5 that a wide range of Re_j (450–180,000) and H/w (2–67.5) is covered by the numerical studies considered. Among these [6,7,13,26,35,36,38,42,43] studied impingement heat transfer due to turbulent slot jets. Heat transfer under impinging axisymmetric jets was studied by [26] and more recently by [39].

In the following discussion the main interest will be laid on the turbulent heat transfer predictions because of its relative importance in industrial applications

Table 5 Summary of the results of the turbulent impinging jet studies

Study	Re_j	H/w	A	B	C	D	E	F	G	H	Remarks
a. Two-dimensional plane impinging jet studies											
[6]	450–22,000	8						x	x		i, ii
	11,000	8								x	iii
	43,000	40		x	x	x	x				
[7]	—	—									iv
	30,000	20	x	x	x	x	x	x	x		
[13]	42,000	15	x	x	x	x	x	x	x		v
	22,000	43	x	x	x	x	x	x	x		
	50,000	30					x				
[35]	5,200	6		x	x			x			
	8,100	6		x	x			x			
	10,400	2.6		x	x	x		x			
	20,750							x			
[36]	5,000	6				x		x	x		
	10,000	2.6, 6	x	x				x	x	x	
	20,000	2.6, 6						x	x		
[38]	7,100	6, 12						x			
[40,41]	5,650	67.5	x	x	x		x				
	25,000	40	x	x							
	43,000	40	x	x	x	x	x				
[42,43]	11,000	8					x	x			
	11,000	16				x		x			
	11,000	12					x				vi
	22,000	8				x	x	x			
	22,000	12					x				
[49]	11,000	8				x					
	43,000	40				x					vii
	100–130,000	2	x			x			x		
b. Two-dimensional axisymmetric impinging jet studies											
[16,24]	170,000	4–12	x	x		x	x				
and	180,000	18					x				
[25]	100,000	2–20					x				
[26]	20,000	4						x			
	20,000	7						x			
	20,000	10						x			

A = Axial centerline velocity decay; B = maximum velocity development along the impingement surface; C = lateral velocity profile at the wall jet region; D = pressure distribution along the impingement surface; E = skin friction distribution at the impingement surface; F = Nu/Sh/St/h distribution along the impingement surface; G = turbulent kinetic energy development along the jet axis; H = turbulent kinetic energy distribution at the wall.

Remarks

In addition to above, also reported
   i. Axial velocity decay for $H/w = 8$, $Re_j = 950$
  ii. St_s vs. Re_j and St_s vs. σ
 iii. Contours of ψ, ω, k, T, P, P_s
  iv. a) Contours of turbulent kinetic energy, shear stress, and turbulent diffusivity
      b) Turbulent kinetic energy and shear stress profiles in the wall jet region
   v. a) Turbulent kinetic energy and shear stress profiles in the wall jet region
      b) Nu vs. H/w for $6000 < Re_j < 50,000$
  vi. a) Shear stress and Nu profiles for $0 < M < 0.57$ and $V_s/U_j = 0$
      b) Shear stress and Nu profiles for $M = 0.57$ and $0 < V_s/U_j < 0.36$ in the case of $Re_j = 11,000$ and $22,000$ and $H/w = 8$
 vii. ψ, ω, k, ϵ contours for $H/w = 40$ and $Re_j = 43,000$

181

and due to the more challenging nature of the problem. The flow predictions are also paid attention wherever applicable since the correct velocity field description is essential for the accurate prediction of the temperature field.

Stagnation point heat transfer. Heat and mass transfer at the impingement point depend on where the position of the impingement surface within the various regions of the jet (i.e., potential core, developing and developed flow regions) is. This was clearly shown by Gardon and Akfirat [5] who studied the heat transfer under vertically impinging unconfined jets over wide ranges of H/w and Re_j. From a close investigation of the stagnation point heat transfer rates measured by these authors, the two regimes of heat transfer were recognized for turbulent jets as can be seen in Fig. 6. These are

1. For initially turbulent jets ($Re_j > 2000$) and for short nozzle-to-surface spacings ($H/w < 8$–10), stagnation point Nu's depend on the initial jet turbulence levels as well as Re_j and H/w values.
2. For $Re_j > 2000$ and $H/w > 14$, stagnation point Nu's are correlated to within 5% by $Nu_s = 1.2\ Re_j^{0.58}\ (H/w)^{-0.62}$ for $2000 < Re_j < 50,000$ and $14 < H/w < 60$. For $H/w > 14$ the effect of initial turbulence is completely masked by the mixing induced turbulence at the outer regions of a jet.

Figure 6 also shows Saad's [2] experimental data for $Re_j = 10,000$ and $4 < H/w < 16$ for two different nozzle sizes (or inlet turbulence levels). Although the absolute values are different, Saad's results for confined slot jets also show the significant effect of the inlet turbulence levels on stagnation point heat transfer when $H/w < 16$. As expected, this effect increases as H/w decreases.

In Fig. 6 the stagnation point Nu for slot jets predicted by various authors is compared with available experimental data. van Heiningen [35] and Polat et al. [36] made predictions for the heat transfer in the case where impingement surface was within the potential core of a plane jet, i.e., in regime (1). In both cases, H/w was equal to 2.6 and 6, and Re_j varied from 5000 to 20,000. van Heiningen found a minimum instead of a maximum at the stagnation point which may be attributed to his specification of the grid distribution near the symmetry line. He reported that a few grid points next to the wall near the symmetry line fell into the viscous sublayer region (i.e., $Y^+ < 11.5$). Since the high Re version of ($k - \epsilon$) model was used together with wall functions, Y^+ or local turbulent Re_T values should be at least 100 as reported by Guo [40] and Guo and Maxwell [41]. Predictions of Polat et al. at the stagnation point differed from those of Gardon and Akfirat [5] by 10% at most. Their predictions showed that the effect of inlet turbulence level was greatest at the stagnation point which agreed with the available experimental evidence.

One of the earliest numerical simulations of a plane unconfined turbulent impinging jet is reported by Wolfshtein [6] for $H/w = 8$. He used the one-equation model described in Section 4.1. Good agreement with experiments for the stagnation point Nu for $H/w = 8$ and $Re_j < 11,000$ was obtained. However, as Re_j

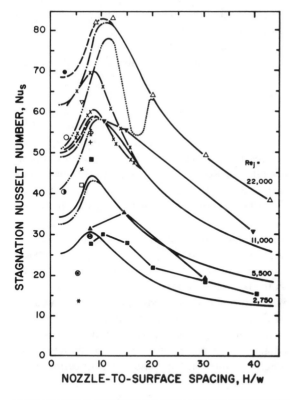

Figure 6 Stagnation Nusselt number as a function of nozzle-to-surface spacing for turbulent slot jets.

Symbol		Ref.	Re_j	w, m
— x —	exp.	[2]	10,800	0.0133
— x x —			10,300	0.0033
— — —	exp.	[5]		0.0064
— · — · —				0.0032
— · · — · · —				0.0016 normal operation
· · · · · · · · · ·				0.0016 with noise
Θ		[6]	2,750	
Φ			11,000	
⊞			22,000	
■		[13]	6,000	
▲			11,000	
▼			14,000	
△			22,000	
*		[35]	5,200	
⊛			8,100	
◐			20,200	
□		[36]	5,000 ⎫	
x			11,000 ⎬ $i = 0.7\%$	
▽			20,000 ⎭	
○			10,000 ⎫ $i = 10\%$	
●			20,000 ⎭	
+		[42,43]	11,000	

values increased, the stagnation point prediction worsened. Indeed, a minimum instead of a maximum was predicted at this location for Re > 11,000. He attributed this behavior to the increase of coarseness of the grid distribution used as Re increased, since the same number of grid lines was employed for high as well as low Re cases.

The stagnation Nu predicted by Huang et al. [42,43] for the case of a slot jet with $H/w = 8$ and $Re_j = 11,000$ is within 10% of that measured by [5].

Looney and Walsh [13] made predictions for various combinations of H/w and Re_j values. They reported that for $H/w < 8$ convergence problems were encountered. However the location of the peak in the stagnation Nu profiles around $H/w = 8$ in Fig. 6 was predicted well. For $Re_j = 14,000$ and $22,000$, the agreement between predictions and Gardon and Akfirat's data was quite good, but the cases of $Re_j < 11,000$ and $Re_j = 50,000$ (not shown in Fig. 6) deviated very much from the experiments.

Amano and Sugiyama [26] investigated heat transfer under a turbulent axisymmetric unconfined jet impinging on a flat plate. They used the high Re version of $(k - \epsilon)$ model with the wall function method for near-wall treatment. Three near-wall models, which are explained in Section 5.1, were tested. As shown in Fig. 7, they reported that the stagnation point heat transfer predictions improved by 30% when Model 2 instead of 3 was used as the near wall model. Utilizing Model 1, another 10–12% improvement was claimed. The remaining 10–25% difference between the experimental and the predicted results was explained in terms of poor modelling of the generation term in the ϵ equation. Since impinge-

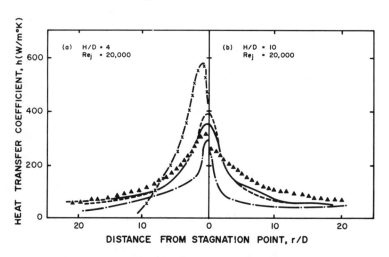

Figure 7 Heat transfer distribution under an axisymmetric jet. (▲) [26] Exp.; (— · — ·) [26] Model 1; (———) Model 2; (— — —) Model 3; (— x —) [39] Model 3. Model 1: Source terms for k and ϵ are computed according to the model described in Fig. 3. Model 2: Source term for k [subtract Eq. (36) from Eq. (34)] is according to the model described in Fig. 3 and ϵ is calculated using equilibrium relation (Eq. 33). Model 3: Eqs. (32) and (33).

ment point heat transfer is also affected by various parameters as mentioned in the next paragraph further investigation of this subject is needed before any definitive conclusions are drawn. It should also be noted that the prediction of the Nu distribution along the impingement surface seems worsened in expense of the better prediction claimed at the stagnation point Nu.

In Fig. 7 shows the recent simulation results of Ahmad [39] for a confined impinging turbulent axisymmetric jet for $H/D = 4$ and $Re_j = 20,000$. He also used the high Re version of $(k - \epsilon)$ model with Model 3 as the near-wall model. There is a substantial difference between these two studies. Although it is a known fact that the entrainment has a strong effect on heat transfer when the jet is not confined, this alone does not explain the large difference between the two curves.

At the stagnation point, in general, the agreement between experiments and the predictions is reported to be satisfactory with some exceptions. However it should be noted that the high Re versions of turbulence models were employed in all of these studies. Thus, Eqs. (32) and (37) were used as the so-called wall functions near the wall to bridge the low-Re part of the flow. It is obvious that these functions are not fully appropriate for use in the impingement region since they are derived from Couette flow assumptions. Thus, the general agreement between the numerical and experimental results at the stagnation point is surprising. If we take a closer look at Eqs. (32) and (37), it is obvious that the stagnation point heat transfer is dependent on location of the first grid point next to the intersection of the symmetry line and the wall lines (i.e., actual stagnation point). It is at this grid point that the so-called stagnation point heat transfer is calculated in numerical simulations. At the true stagnation point the velocity components and the wall shear stress are zero; therefore the present wall functions cannot be used. As noted earlier the grid line next to the wall should be in the fully turbulent region since wall functions are valid only for this type of flow. The next question is about the criteria one must use for locating where the first grid line parallel to the line of symmetry should be. The location of this line can be adjusted to obtain good agreement with the chosen experimental results at the stagnation point. On the other hand, as we move away from the stagnation point, better agreement is expected since the assumptions employed in deriving the wall functions become more appropriate for the flow. Therefore, we propose that in this region one should look for a grid independent solution and then adjust the so-called stagnation point heat transfer value if necessary by adjusting the location of the first grid line parallel to the symmetry line. It should be noted that the stagnation Nu values are particularly susceptible to uncertainties derived from finite resolution of the measurement technique employed.

The above discussion of stagnation point heat transfer is valid for the case of normal impingement. For oblique impingement the stagnation point is difficult to define. When there is superimposed cross flow or when the impingement surface is in motion, under certain flow and geometric conditions no stagnation point or line may exist at all. In such cases the wall functions may still be suitable for the flow, however handling of the other boundaries such as boundaries V or VII in Fig. 2 may become a problem.

Huang et al. [42,43] predicted flow and heat transfer under a confined slot jet for cases of cross flow and cross flow with the impingement surface in motion. They found that when a cross flow is introduced into the channel, between the confinement and the impingement surfaces, the maximum Nu number decreased in amount and shifted downstream from the jet center line (see Fig. 11). The amount of the decrease and the shift depended on the cross flow parameter, $M = Re_c/Re_j$ at a particular Re_j. The effect of the surface motion could not be simulated separately because of the difficulty in defining upstream boundary conditions. Hence they studied the combined effect of surface motion and cross flow. At a particular level of the cross flow parameter and Re_j, the effect of surface motion was found to decrease the amount of peak Nu number without affecting its location appreciably (see Fig. 12).

Polat et al. [44] tested high Re version of $(k - \epsilon)$ model for the case of multiple jet impingement to predict heat transfer distribution along the impingement surface. However, for the small S/H values, the stagnation region and the separating flow region under the exhaust ports, which are located symmetrically between the jets, were close enough to affect the intermediate flow region. This rendered the near wall model, which is based on the Couette flow assumptions, not suitable for the case. Hence a grid independent Nu profile could not be obtained. However, as can be seen in Fig. 8, a better agreement at the stagnation point could still be obtained due to the reasons explained previously.

Nu Profiles along the Impingement Surface. Profiles of the local Nu numbers along the impingement surface also show different characteristics depending on

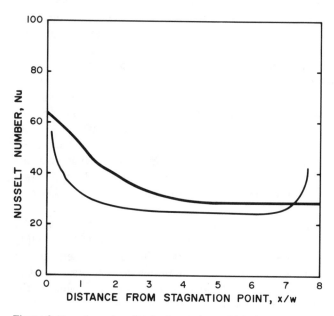

Figure 8 Nusselt number distribution under multiple jets. (——) Polat [44]; (——) Saad [2].

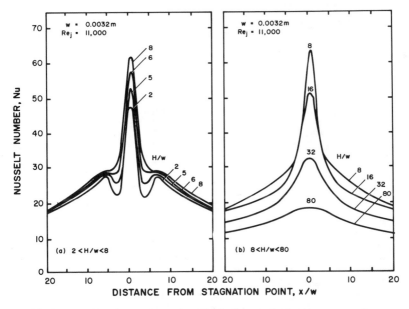

Figure 9 Experimental Nusselt number distribution on the impingement surface at various nozzle-to-surface distances [5].

H/w and Re_j. At smaller H/w values, the incoming jet turbulence level seems to affect the shape of the heat transfer distribution on the impingement surface. Figure 9 displays the experimental heat transfer data of Gardon and Akfirat [5] for an unconfined turbulent slot jet over a wide range of H/w values for $\mathrm{Re}_j = 11,000$. Three distinct patterns in the Nu distribution curves are evident:

1. $H/w > 14$, the distribution has the characteristic bell shape.
2. For $14 > H/w > 8$, the bell shape is modified slightly by a sudden change in slope around $x/w = 4$.
3. For $H/w < 8$, a hump begins to form at about $x/w = 7$. As H/w is reduced below 6, this hump becomes a well-defined secondary peak. As observed by [35] for confined impinging slot jets for $H/w = 2.6$ and 6, at much higher Re_j's the secondary peak becomes even more pronounced and its magnitude may eventually exceed the stagnation point Nu value.

Looney and Walsh [13] made flow and heat transfer predictions for an impinging jet for $8 < H/w < 43$ for various Re_j's. The confidence in their results for $H/w < 8$ suffered because of convergence problems they encountered. A partially or fully developed free jet, depending on H/w, was used as the inlet condition at approximately $0.55H$ above the plate. For the outflow condition, they specified the similarity velocity profile of a wall jet at about $0.55H$ downstream from the stagnation point. Figure 10 compares their results with those of [5]. The dependence of profile shape on Re_j is not evident in the predicted profiles. It

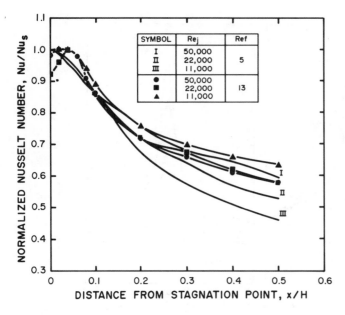

Figure 10 Comparison of the predicted Nu number profiles of [13] to the experimental results of [5] for $H/w > 8$.

seems that because of the low Nu_s found at $Re_j = 11,000$, a higher profile was found for this lowest Re_j case. However, the predicted profiles were reasonably in good agreement with data at higher Re_j's.

Looney and Walsh also tested the two different algebraic stress models proposed by Rodi [10] and Ljuboja and Rodi [12] for the case of plane free jets and impinging jets. Improved results were obtained using algebraic stress models in the case of the free jets. However no such improvement was noted in the case of impinging jets. Thus the standard $(k - \epsilon)$ model is preferred because of its relative simplicity. They detected that the main effect of these algebraic stress models used was to lower the peak in k near the stagnation point. This along with only small differences in the mean flow properties combined to give lower Nu number levels near the stagnation point.

Wolfshtein [6] reported the heat transfer predictions for $H/w = 8$ and $Re_j = 2750$, 11,000 and 22,000. He solved the impingement region flow and heat transfer using the relations given by [45] as the inlet conditions for flow. The inlet turbulence kinetic energy was deduced from data reported by [46]. The disagreement with the experiments of [5] for $Re_j = 2750$ and 11,000 away from the stagnation point was less than 20%. However at $Re_j = 22,000$, a minimum instead of a maximum at the stagnation point was found and the disagreement away from the stagnation point was more than 40%. Since a one-equation model of turbulence was used, the length scales were calculated using Eqs. (8) and (9). Later, Russell and Hatton [7] used the same model to simulate a plane impinging jet case and at the same time, they also made extensive turbulence flow measurements to verify the results of this model. However, the measurements themselves had an uncertainly level up to 30% while flow was not entirely two-dimensional due to the equipment design. When they used measured length scales in the computations instead of Eqs. (8) and (9), an improvement was noted.

The general features of the Nu distribution along the impingement surface are well predicted by Polat et al. [36] for $H/w = 2.6$ and 6 and for $5000 < \mathrm{Re}_j < 20,000$. The location of the experimentally observed secondary maximum of the Nu profile was predicted quite well to be at about $x/w = 7$. However a very steep decay of the Nu profile was predicted in the impingement region. The inflection point that marks the end of the impingement region was predicted to be located about 0.5–$1.5\ w$ which is experimentally found to be around 3–$4\ w$ from the stagnation line. The inflection point Nu numbers were always predicted higher than the experimental values at lower inlet turbulence levels. This was also observed by [47] who studied turbine blade cooling numerically using the low Re version of the $(k - \epsilon)$ model. The maximum deviation of the predicted Nu profiles of Polat et al. from experimental data was no more than 15% in the wall jet region but considerably higher in the zone of interest, i.e., the impingement and the transition zone.

Huang et al.'s [42,43] heat transfer prediction results for a confined slot jet impinging symmetrically as well as with superimposed cross flow are shown in Fig. 11 for $H/w = 8$ and for $\mathrm{Re}_j = 11,000$. Various inlet conditions were tested but the fully developed velocity, k and ϵ profiles for a two-dimensional channel flow were used for the results reported. In general the shape of the profiles were predicted quite well. The disagreement between the experimental and the predicted results are no more than 25% at any point for the no cross flow, stationary

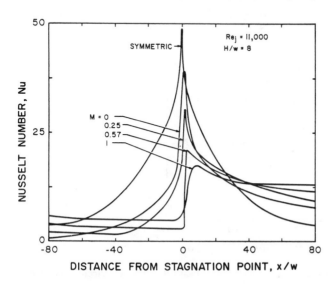

Figure 11 Nusselt number profiles for various levels of cross flow [42,43].

impingement surface case. No data are available for the other cases. Their results indicated that when a small amount of cross flow (M = Re_c/Re_j < 0.6) is introduced to the upstream of the jet some distance (x/w was about 80 in his case) away from the jet center line a small recirculation region developed near the impingement surface on the side of the jet where cross flow was introduced. At higher cross flows this recirculating region eventually disappears. The reattachment length of the main recirculation bubble on the side of the outflow boundary decreases with increasing cross flow. Figure 13 shows the shear stress distribution, along the impingement surface for various values of cross flow parameter, M, for Re = 11,000 and H/w = 8. As can easily be seen from this figure, for M < 0.6 the shear stress is positive for the flow upstream of the jet and the maximum value decreases and shifts downstream as M value increases. However at higher M values the shear stress changes sign and becomes negative indicating that the flow along the impingement surface is in the direction of cross flow. The shear stress downstream of the jet is always negative because the air exhausted in this direction.

Figure 11 shows the Nu distribution along the impingement surface computed by the above authors for the same cases as in Fig. 13. As expected the maximum Nu number decreases in magnitude and shifts downstream as M value increases. The downstream Nu number values increase with cross flow due to the fact that the total mass flow in this direction is increased as the cross flow is increased.

The effect of surface motion in the presence of cross flow was also studied by these authors. Figure 12 shows the effect of the surface motion on heat transfer at M = 0.57 for Re_j = 11,000 and H/w = 8. The upstream heat transfer rate

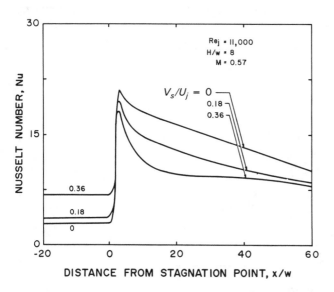

Figure 12 Nusselt number profiles at various levels of surface velocity to jet velocity ratio at M = 0.57 [42,43].

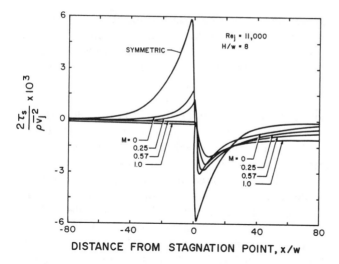

Figure 13 Shear stress distribution at various levels of cross flow [42,43].

increases while the downstream heat transfer decreases with increasing relative surface motion, V_s/U_j. The effect of surface motion on the average heat transfer was however very little due to its compensating effects on either side of the jets. This conclusion may not necessarily hold over other ranges of flow and geometric parameters.

In a recent study, Lampinen [8] applied a one equation model to an aerofoil dryer configuration to estimate the heat transfer coefficients on a moving (paper) surface. One of his typical results is shown in Fig. 14. In this specific impingement dryer for paper the jet impinges obliquely on the moving paper surface. To handle this problem economically, instead of using the true velocity profiles at the jet exit, he calculated the approximate projected velocity profiles due to the

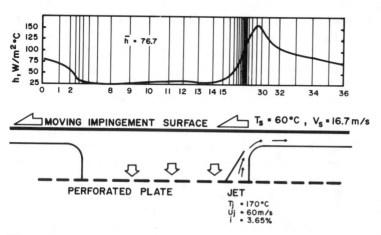

Figure 14 Heat transfer distribution on a moving surface under an inclined impinging jet [8].

jets at the inlet of the channel formed between the confinement and the paper surface. This simplified modeling appears to have worked quite well for this case, since the main concern was the estimation of the average heat transfer rates. A maximum deviation of 12% from the experimentally found heat transfer coefficients was observed.

Flow Predictions. Amano [16,24] and Amano and Brandt [25] made flow predictions for the case of an axisymmetric jet for $2 < H/D < 40$ and $10,000 < \text{Re}_j < 2,000,000$. They predicted a secondary peak in the skin friction profiles along the impingement surface for $H/w = 2$ and $\text{Re} > 100,000$. Secondary peaks of skin friction profiles were previously observed in a study of Baines and Keffer [48] experimentally. While Polat et al. [36] found that the secondary peak in the predicted Nu distribution disappeared with increasing inlet jet turbulence level, Amano predicted enhancement of the skin friction peak with its location shifting away from the center line. Amano and his co-workers used a modified near-wall treatment which is described in Section 5.1. They found that this near-wall treatment gave a prediction of the skin friction coefficient within 25% error of measured values. This indicates that the modifications made to the simpler near-wall model did not improve the predictions significantly.

Amano [16] and Amano and Brandt [25] also simulated an axisymmetric jet flowing into a cavity. They found that the center line velocity is insensitive (within 0.5%) to Re_j but changes with turbulence intensity at the nozzle exit.

Guo [40], Guo and Maxwell [41], and Agarwal and Bower [49] computed only the flow in impinging jet configurations. When we compare the streamline contours of the first two studies with those of the third an interesting picture emerges. These studies simulated a case where $H/w = 40$ and $\text{Re}_j = 25,000$ and $40,000$, respectively. The first two studies used the high Re version of $(k - \epsilon)$ model while the other used the low Re version of it. The flow domain numerically considered was $13w \times 18w$ in the first two studies and $1w \times 2.5w$ in the third. That means the third study simulated a region which was overlapped by wall functions in the case of the other two studies. Yet results of these studies indicated two recirculating regions, i.e., one that was found by Guo's studies at the outer edge of the flow and the other found deep inside the stagnation region by Agarwal and Bower. To the author's knowledge, this is the first time a small recirculation bubble in the stagnation region of a turbulent jet under these conditions has been reported in a numerical study. No experimental evidence exists to support such a prediction.

Guo [40] tried to simulate the flow domain of an inclined confined impinging jet numerically by selecting the computational domain in two ways. In one case the x axis coincided with the center line of the jet and, in the other case, the y axis coincided with the impingement surface and the inlet velocities were introduced accordingly. The first one was found to be economically unfavorable. But the results of the second were not even close to the experimental results. Several approaches were taken to improve these results but none of them worked. In this case, the location of the streamline at the jet center line was found by trial and error.

The impingement heat transfer and flow prediction results of various workers are summarized above. The commonly used high-Re version of the well-known $(k - \epsilon)$ model seems to produce quantitatively good results for prediction of the flow pattern. Predictions of pressure distribution and maximum velocity development along the impingement surface were reported to have differed at most 15% [42] from experiments. Although heat transfer predictions in the wall jet region agreed well with the experiments, results in the impingement region varied greatly between themselves and with the experiments. However, since near the impingement surface the so-called wall functions were used as the boundary condition, it is very hard to judge whether the failure of the turbulence model or of the near wall models contributed more to the disagreements between the predictions and the experiments.

The $(k - \epsilon)$ models used heretofore do not incorporate the redistribution effect of the pressure strain correlation terms on turbulence. This effect is believed to be important in the stagnation region due to the presence of highly favorable pressure gradients. Again, these models make the assumption of isotropy for the turbulent structure which is difficult to justify in the impingement region. It is in this region that the anisotropy between the longitudinal and lateral components of Reynolds stresses is amplified because of the rapid deceleration of the flow as measured by Hijikata et al. [50] among others. These authors also predicted the flow and heat transfer around the stagnation point of a cylinder by using a three equation model of turbulence. In this model the transport equations for k and ϵ were used in a form suitable for integration up to the wall with modifications to include the effect of anisotropy, $(v_r^2 - v_\theta^2)$. Another equation was derived for $(v_r^2 - v_\theta^2)$ by subtracting the transport equations of v_r^2 and v_θ^2 from each other. Although the contributions of these Reynolds stress components to the mean momentum equations are negligible away from a wall, apparently the $(v_r^2 - v_\theta^2)$ difference becomes appreciable as the stagnation point is approached since v_r^2 is amplified due to the severe velocity gradient while v_θ^2 is attenuated. However, it should be noted that just the reverse effect is produced due to the pressure-strain correlation term in this region, i.e., v_θ^2 is amplified and v_r^2 is suppressed.

Different aspects of the stagnation flow were considered in deriving the model equations which contained thirteen constants all together. This leads to the question of the universality of such empirical constants since this model is not tested for other types of flows. These authors reported good predictions of flow and heat transfer with this model. However, it is unfortunate that no effort was made to compare their results with the predictions made employing the conventional $(k - \epsilon)$ models. It is believed that such a comparison should be made before any judgment is made as to the superiority of this model to the $(k - \epsilon)$ models.

Extra strain effects due to streamline curvature are also not included in the turbulence models used in the aforementioned studies. Bradshaw [51] postulates that this effect is at least one order of magnitude more important than the normal pressure gradients and other extra strain effects. Turbulence models using the eddy viscosity concept have to be modified empirically in order to reproduce the characteristics of curved shear layers. In the stagnation region the effect of streamline curvature should be significant but not accounted for by the current forms of k

and ϵ equations. In a relatively recent paper Rodi and Scheuerer [52] compares the predictive capabilities of the two-equation models of Launder et al. [53] and Hanjalic et al. with the algebraic stress model of Gibson [54] in curved shear layers. In contrast with the $(k - \epsilon)$ models, the algebraic stress model does not need the extra empirical input to simulate the curvature effects and appears to have adequately predicted the features of the main flow in the test cases in the region with appreciable curvature. When the curvature effects were removed, the recovery of the turbulence quantities was predicted to be too fast by this model. On the other hand, the success of the two-equation models depended on the type of flow they were applied to. This is due to the empirical character of the modifications made to the models to account for the curvature effects. In conclusion none of the models investigated were found to be fully satisfactory to predict the details of the flow for the test cases. However, the algebraic stress model was recommended since it produced results only slightly inferior to those found by the parent Reynolds stress model.

Universality of the law of the wall is questionable under a number of influences in the stagnation region of an impinging jet. Hirata et al. [55] investigated the use of the scaling laws as the boundary condition near a wall in the numerical simulation of turbulent flows. They suggested that the integration of the equations up to the wall would result in better agreement with the experimental results. However, appreciable refinement in the modeling of near-wall flows would also be needed. In a more recent paper Launder [56] also suggested that the use of so-called wall functions for the near-wall flow may result in overlooking some of the important phenomena occurring in this region. He described another type of near wall treatment called the "parabolic sublayer (PSL)" method which gave better predictions with a similar economy of the wall-function method. The reader should refer to the original paper [57] for details of this procedure.

In summary, we might say that the high-Re version of $(k - \epsilon)$ model together with the previously described near-wall treatments is not adequate for reliable predictions of impingement heat transfer and turbulence quantities. No matter how refined the near-wall model is, it still carries the limitations of the log law of the wall. The effects of the turbulence properties are introduced by making an equilibrium assumption for the flow. Although this assumption holds better in the wall jet region, in the region of interest it is not valid at all. The low-Re version of the $(k - \epsilon)$ model and algebraic stress models have not been adequately tested for these types of flows and are recommended for future studies before any higher order models such as the Reynolds stress models or large eddy simulation models are considered.

9 CONCLUSIONS

1. The excellent agreement of laminar impinging jet predictions with experiments and among themselves shows that the numerical schemes are stable and able to produce accurate results.

2. In the case of turbulent impinging jets, the high Re version of the $(k - \epsilon)$ model has been used extensively among other eddy viscosity type models. The results show significant divergence between themselves and with experiments. Both the turbulence and the near-wall model may be responsible for the differences.

3. It is believed that, no matter how refined the near-wall models are, as long as the wall functions in the form of Eqs. (32) and (37) are used, it will fail to predict the stagnation region accurately. The validity of these wall-functions in this region is questionable.

4. The low-Re version of the $(k - \epsilon)$ model as well as the algebraic stress models have not been adequately tested for impingement flows. These relatively simple models need to be tested before more complex models are applied, e.g., large eddy simulation models.

5. Only two-dimensional impinging jet flows have been studied numerically in any depth although abundant experimental studies have mapped heat transfer under multiple, round and slot impinging jets. This is true for turbulent as well as laminar jets. Three-dimensional impingement-driven flows need to be examined numerically in the future.

6. A semi-confined, single turbulent impinging jet is recommended as a configuration that may be employed for future tests of newly developed turbulence models. While flow domain is relatively simple and well defined, the turbulent flow pattern in the impingement zone and the development of a recirculation zone provide stringent conditions for testing a turbulence model.

REFERENCES

1. E. Gutmark, M. Wolfshtein and I. Wygnanski, The Plane Turbulent Impinging Jet, *J. Fluid Mech.*, vol. 88, pt. 4, pp. 737–756, 1978.
2. N. R. Saad, Flow and Heat Transfer for Multiple Turbulent Impinging Slot Jets, Ph.D. thesis, McGill University, 1981.
3. J. M. F. Vickers, Heat Transfer Coefficients Between Fluid Jets and Normal Surfaces, *Ind. Eng. Chem.*, vol. 5i, no. 8, pp. 967–972, 1959.
4. K. J. McNaughton and C. G. Sinclair, Submerged Jets in Short Cylindrical Flow Vessels, *J. Fluid. Mech.*, vol. 25, no. 2, pp. 367–375, 1966.
5. R. Gardon and J. C. Akfirat, Heat Transfer Characteristics of Impinging Two-Dimensional Air Jets, *J. Heat Transfer*, vol. 88, pp. 101–108, 1966.
6. M. Wolfshtein, Convection Processes in Turbulent Impinging Jets, Ph.D. thesis, Imperial College, London, 1967.
7. P. J. Russell and A. P. Hatton, Turbulent Flow Characteristics of an Impinging Jet, *Proc. Instn. Mech. Engrs.*, vol. 186, no. 52, pp. 635–644, 1972.
8. M. J. Lampinen, Application of Turbulence Theory to the Determination of Heat Transfer Coefficients in an Aerofoil Dryer, *Drying Technology*, vol. 3, no. 2, pp. 171–219, 1985.
9. W. P. Jones and B. E. Launder, The Calculation of Low-Reynolds-Number Phenomena with a Two-Equation Model of Turbulence, *Int. J. Heat Mass Transfer*, vol. 16, pp. 1119–1130, 1973.
10. W. Rodi, Ph.D. thesis, University of London, 1972.
11. K. Hanjalic and B. E. Launder, A Reynolds Stress Model of Turbulence and its Application to Thin Shear Flows, *J. Fluid Mech.*, vol. 52, part 4, pp. 609–638, 1972.
12. M. Ljuboja and W. Rodi, Calculation of Turbulent Wall Jets with an Algebraic Stress Model,

Proc. Symp. on Turbulent Baundary Layers: Forced, Incompressible, Non-reacting, ed. H. E. Weber, pp. 131–138, 1979.

13. M. K. Looney and J. J. Walsh, Mean Flow and Turbulent Characteristics of Free and Impinging Jet Flows, *J. Fluid Mech.,* vol. 147, pp. 397–429, 1984.

14. J.-P. Jaussaud, W. J. M. Douglas and A. S. Mujumdar, Evaporation Under an Obliquely Impinging Laminar Ducted Slot Jet—A Numerical Study, *Int. Comm. Heat Mass Transfer,* vol. 11, pp. 335–344, 1984.

15. B. Huang, W. J. M. Douglas and A. S. Mujumdar, Heat Transfer Under a Laminar Swirling Impinging Jet—A Numerical Study, *Proc. 6th Int. Heat Tr. Conf.,* vol. 5, Toronto, pp. 311–316, 1978.

16. R. S. Amano, Numerical Study of a Turbulent Jet Impinging on a Flat Plate and Flowing into an Axisymmetric Cavity, Ph.D. thesis, University of California, Davis, 1980.

17. A. S. Mujumdar, Y.-K. Li and W. J. M. Douglas, Evaporation Under an Impinging Jet: A Numerical Study, *Can. J. Chem. Eng.,* vol. 58, pp. 448–453, 1980.

18. N. D. Deshpande and R. N. Vaishnav, Submerged Laminar Jet Impingement on a Plane, *J. Fluid Mech.,* vol. 114, pp. 213–236, 1982.

19. R. Ravuri and W. Tabakoff, A Numerical Solution for the Heat Transfer between an Axi-symmetric Air Jet and a Heated Plate, Report no. 73-38, Dept. of Aerospace Eng., University of Cincinnati, 1973.

20. S. Mikhail, S. M. Morcos, M. M. M. Abou-Ellail and W. S. Ghaly, Numerical Prediction of Flow Field and Heat Transfer from a Row of Laminar Slot Jets Impinging on a Flat Plate, *Proc. 7th Int. Heat Tr. Conf.,* vol. 3, Munchen, 1982.

21. H. S. Law, Mass Transfer Due to a Confined Laminar Impinging Two-Dimensional Jet, Ph.D. thesis, University of Alberta, 1982.

22. E. M. Sparrow, S. H. Lin and T. S. Lundgren, Flow Development in the Hydrodynamic Entrance Region of Tubes and Ducts, *Physics of Fluids,* vol. 7, no. 3, pp. 338–347, 1964.

23. C. C. Chieng and B. E. Launder, On the Calculation of Turbulent Heat Transport Downstream from an Abrupt Pipe Expansion, *Numerical Heat Transfer,* vol. 3, pp. 189–207, 1980.

24. R. S. Amano, Turbulence Effect on the Impinging Jet on a Flat Plate, *B. JSME,* vol. 26, no. 221, pp. 1891–1899, 1983.

25. R. S. Amano and H. Brandt, Numerical Study of Turbulent Axisymmetric Jets Impinging on Flat Plate and Flowing into an Axisymmetric Cavity, *Trans. ASME, J. Fluids Eng.,* vol. 106, pp. 410–417, 1984.

26. R. S. Amano and S. Sugiyama, An Investigation on Turbulent Heat Transfer of an Axisymmetric Jet Impinging on a Flat Plate, *B. JSME,* vol. 28, no. 235, pp. 74–79, 1985.

27. S. B. Pope and J. H. Whitelaw, The Calculation of Near-Wake Flows, *J. Fluid Mech.,* vol. 73, pp. 9–32, 1976.

28. D. B. Spalding, Heat Transfer from Turbulent Separated Flows, *J. Fluid Mech.,* vol. 27, p. 97, 1967.

29. C. L. Jayatillaka, The Influence of Prandtl Number and Surface Roughness on the Resistance of the Laminar Sub-layer to Momentum and Heat Transfer, *Prog. in Heat and Mass Transfer,* vol. 1, pp. 193–329, 1969.

30. A. D. Gosman, B. E. Launder and J. H. Whitelaw, Heat and Mass Transfer in Turbulent Recirculating Flows—Predictions and Measurements, *Short Course Notes,* McGill University, Chem. Eng. Dept., August, 1976.

31. N. R. Saad, A. S. Mujumdar and W. J. M. Douglas, Prediction of Heat Transfer Under an Axisymmetric Laminar Impinging Jet, *Ind. Eng. Chem. Fund.,* vol. 16, no. 1, pp. 148–154, 1977.

32. A. R. P. van Heiningen, A. S. Mujumdar and W. J. M. Douglas, Numerical Prediction of the Flow Field and Impingement Heat Transfer Due to a Laminar Slot Jet, *ASME Paper, #75-WA/ HT-99,* 1975.

33. E. M. Sparrow and L. Lee, Analysis of Flow Field and Impingement Heat and Mass Transfer due to a Non-uniform Slot Jet, *Trans. ASME, Series C, J. Heat Transfer,* pp. 191–197, 1975.

34. E. M. Sparrow and T. C. Wong, Impingement Transfer Coefficients Due to Initially Laminar Slot Jets, *Int. J. Heat Mass Transfer,* vol. 18, pp. 597–605, 1975.

35. A. R. P. van Heiningen, Heat Transfer under an Impinging Slot Jet, Ph.D. thesis, McGill University, Montreal, 1982.
36. S. Polat, A. S. Mujumdar and W. J. M. Douglas, Heat Transfer Distribution under a Turbulent Impinging Jet—A Numerical Study, *Drying Technology,* vol. 3, no. 1, pp. 15–38, 1985.
37. B. Huang, Heat Transfer under Inclined Slot Jet Impinging on a Moving Impingement Surface, Ph.D. thesis, in preparation, Chem. Eng. Dept., McGill University, 1988.
38. C. O. Folayan, J. H. Whitelaw, Impingement Cooling and its Application to Combustor Design, *Tokyo Joint Gas Turbine Congress,* May 22–27, Tokyo, pp. 69–76, 1977.
39. I. Ahmad, Chem. Eng. Dept., McGill University, Personal Communications [1986].
40. C. Y. Guo, Numerical Modeling of a Turbulent Plane Jet Impingement on a Solid Wall, Ph.D. thesis, University of Illinois, Urbana, 1982.
41. C. Y. Guo and W. H. C. Maxwell, Numerical Modeling of Normal Turbulent Plane Jet Impingement on a Solid Wall, *J. Eng. Mech.,* vol. 110, no. 10, pp. 1498–1509, 1984.
42. G. P. Huang, A. S. Mujumdar and W. J. M. Douglas, Prediction of Heat Transfer under a Plane Turbulent Impinging Jet Including Effects of Crossflow and Wall Motion, *Drying '82,* ed. A. S. Mujumdar, Hemisphere, 1982.
43. G. P. Huang, A. S. Mujumdar and W. J. M. Douglas, Numerical Prediction of Fluid Flow and Heat Transfer under a Turbulent Impinging Slot Jet With Surface Motion and Crossflow, *ASME paper, #84-WA/HT-33,* 1984.
44. S. Polat, A. S. Mujumdar and W. J. M. Douglas, Numerical Prediction of Multiple Impinging Turbulent Slot Jets, *Drying '86, Proc. of Fifth Int. Drying Conf.,* August 13–15, Cambridge, pp. 868–879, 1986.
45. J. J. Schauerer and R. H. Eustis, The Flow Development and Heat Transfer Characteristics of Plane Turbulent Impinging Jets, *Tech. Report #3,* Dept. of Mech. Eng., Stanford University, 1963.
46. G. Heskestad, Hot Wire Measurements in a Plane Turbulent Jets, *ASME Trans., J. App. Mech., paper #65-APM-H,* 1956.
47. T. R. Shembharkar and B. R. Pai, Prediction of Heat Transfer in Turbine Blades with 2-Equation Model of Turbulence, *Proc. 7th Int. Heat Tr. Conf.,* vol. 3, Munchen, pp. 289–294, 1982.
48. W. D. Baines and J. F. Keffer, Shear Stress and Heat Transfer at a Stagnation Point, *J. Heat and Mass Transfer,* vol. 19, pp. 21–26, 1976.
49. R. K. Agarwal and W. W. Bower, Navier-Stokes Computations of Turbulent Compressible Two-Dimensional Impinging Jet Flowfields, *AIAA J.,* vol. 20, no. 5, pp. 577–584, 1982.
50. K. Hijikata, H. Yoshida and Y. Mori, Theoretical and Experimental Study of Turbulence Effects on Heat Transfer around the Stagnation Point of a Cylinder, *Proc. 7th Int. Heat Tr. Conf.,* vol. 3, Munchen, pp. 165–170, 1982.
51. P. Bradshaw, Effect of Streamwise Curvature on Turbulent Flows, *AGARDograph #169,* 1973.
52. W. Rodi and G. Scheuerer, Calculation of Curved Shear Layers with Two-equation Turbulence Models, *Physics of Fluids,* vol. 26, no. 6, pp. 1422–1436, 1983.
53. B. E. Launder, C. H. Priddin and B. I. Sharma, The Calculation of Turbulent Boundary Layers on Spinning and Curved Surfaces, *J. Fluids Eng.,* vol. 99, pp. 237–242, 1977.
54. M. M. Gibson, An Algebraic Stress and Heat-Flux Model for Turbulent Shear Flow with Streamline Curvature, *Int. J. Heat Mass Transfer,* vol. 21, pp. 1609–1617, 1981.
55. M. Hirata, H. Tanaka, H. Kawamura and N. Kasogi, Heat Transfer in Turbulent Flows, *Proc. 7th Int. Heat Tr. Conf.,* vol. 1, Munchen, pp. 31–57, 1982.
56. B. E. Launder, Numerical Computation of Convection Heat Transfer in Complex Turbulent Flows: Time to Abandon Wall Functions?, *Int. J. Heat Mass Transfer,* vol. 29, no. 9, pp. 1485–1491, 1984.
57. H. Iacovides and B. E. Launder, PSL—An Economical Approach to the Numerical Analysis of Near-Wall, Elliptic Flow, *J. Fluids Eng.,* vol. 106, pp. 241–242, 1984.

SOME NUMERICAL ASPECTS OF CAVITATION BUBBLE COLLAPSE

A. Shima and Y. Tomita

ABSTRACT

Cavitation damage is known to be predominantly caused by impulsive pressures produced from collapsing bubbles in liquid. In the final stage of the bubble collapse, some important physical phenomena are induced. Mathmetically speaking the nonlinear effect is important in this stage. Until now, many numerical attempts have been made to solve the problems associated with bubble collapse, including nonlinear effect. This chapter presents some numerical methods that have been applied to the problems of cavitation bubble collapse. First, spherical bubble collapse and then the dynamics of nonspherical bubble collapse are discussed.

1 INTRODUCTION

When liquid pressure drops locally below a certain level, an interesting physical phenomenon occurs called cavitation. Cavitation is often compared with boiling resulting from the differences in liquid temperatures. The simplest appearance of cavitation is a single bubble which possesses a closed surface dividing the region of concern into two parts. The bubble behavior is closely related to cavitation-induced phenomena, such as vibration, noise, material damage, and sonoluminescence which are of frequent occurrence in various kinds of hydraulic machineries. Much research on cavitation bubble dynamics has been done vigorously, and systematic reviews have been reported by Hsieh [1], Flynn [2], Plesset and Prosperetti [3], and Neppiras [4]. Bubble dynamics is also summarized in the books on cavitation (e.g., Knapp et al. [5], Hammitt [6], Lauterborn [7], Isay [8], and Anton [9]).

In connection with cavitation damage, the study of the impulsive pressure generation is considered as the most important research task. Rayleigh [10] first discussed the impulsive pressure produced from a collapsing cavity in relation to cavitation damage. He showed that the energy of a cavity concentrates within a narrow region of liquid in the neighborhood of the cavity wall in the later stage of its collapse. He also derived the following collapse time necessary for filling up the cavity with liquid, which is generally referred to as Rayleigh's collapse time:

$$t_c = 0.915R_0 \sqrt{\frac{\rho_\infty}{P_\infty}} \tag{1}$$

where R_0 is the initial radius of a cavity, ρ_∞ and P_∞ are the density and the pressure of liquid at infinity, respectively. For instance, a bubble with the initial radius of 1 mm collapses within 10^{-4} s under the pressure difference of 0.1 MPa. A bubble shrinks slowly in the initial stage of collapsing process and gradually accelerates its motion. In the final stage of collapse, the bubble motion becomes so rapid that it needs the temporal resolution of about a hundredth of the bubble collapse time, for example 1 μs in the case of the above-mentioned conditions, to observe the phenomena in detail. The bubble collapse time is proportional to the initial bubble radius, which is apparent from Eq. (1). The difficulty for measuring impulsive pressure caused by the bubble collapse rapidly increases when a bubble becomes small in size.

The numerical method is one of the effective means for understanding the aspects of a collapsing bubble. The problem of the single bubble motion belongs to the type of free boundary-value problems and is divided into two categories: (1) spherical and (2) nonspherical bubble collapse. This chapter presents some numerical techniques applied to bubble dynamics and discusses the aspects of cavitation bubble collapse.

2 SPHERICAL BUBBLE COLLAPSE IN INFINITE VOLUME OF LIQUID

Cavitation damage results from collapsing bubbles near a solid wall. From this point of view, it is not always adequate to discuss only the mechanism of impulsive pressure caused by the spherical bubble collapse. However, solving problems for nonspherical collapse of a bubble is very difficult if one takes into consideration all existing real effects. If the assumption of spherical symmetry is introduced, the complexity of the problem is greatly reduced. In that case, however, it still remains formidable because in the later process of bubble collapse, the motion of the bubble wall exhibits a high-speed two-phase flow with accompanying heat and mass transfers. In this chapter, several numerical methods applying to the spherical bubble collapse are presented and effects of both liquid compressibility and heat transfer on bubble behavior are considered.

2.1 Collapse of a Spherical Gas Bubble in Compressible Liquid

Liquid compressibility is one of the most important factors in bubble collapse. Herring [11] studied the motion of a gas globe produced by underwater explosion, taking into consideration a first-order correction for liquid compressibility. Trilling [12] also followed this problem by using the acoustic assumption but extended its solution to include the velocity and pressure fields in the liquid. Gilmore [13] obtained a solution with higher-order compressibility terms on the basis of the Kirkwood-Bethe hypothesis [14], it was proved that this hypothesis loses its validity when the bubble wall velocity approaches the sonic speed of liquid and the nonlinear effect of a wave becomes dominant [15].

Hickling and Plesset [16] obtained numerical solutions for the collapse of a gas-filled bubble in a compressible liquid as well as the velocity and pressure fields in the liquid. The equations of motion expressing the conservation of mass and momentum for a spherically symmetric system can be written as follows in the Eulerian form:

$$\frac{D\rho}{Dt} = -\rho\left(\frac{\partial v}{\partial r} + \frac{2v}{r}\right) \tag{2}$$

$$\rho\frac{Dv}{Dt} = -\frac{\partial P}{\partial r} \tag{3}$$

where P and ρ are the pressure and density in the liquid and v is the radial particle velocity. The operator appearing in above equations means the derivative with respect to time following the motion of the liquid and expressed by

$$\frac{D}{Dt} = \frac{\partial}{\partial t} + v\frac{\partial}{\partial r} \tag{4}$$

The empirical pressure-density relation for isentropic compression is given by

$$\frac{P + B}{P_\infty + B} = \left(\frac{\rho}{\rho_\infty}\right)^n \tag{5}$$

This is the equation of state for liquid which is called the modified Tait equation [17]. Using this equation, the velocity of sound in liquid is defined by the expression

$$C^2 = \frac{dP}{d\rho} = \frac{n(P + B)}{\rho} \tag{6}$$

A method of characteristics is applied to solve the problem. The characteristic equations for spherically symmetric flow are

$$\frac{\partial r}{\partial \eta} = (v + C) \frac{\partial t}{\partial \eta} \tag{7}$$

$$\frac{\partial r}{\partial \zeta} = (v - C) \frac{\partial t}{\partial \zeta} \tag{8}$$

where η and ζ are the outward-going and inward-going characteristics. Since ρ changes quite slowly in liquid, we can get $\delta\rho = \delta P/C^2$. So, Eqs. (2) and (3) become

$$\frac{\partial v}{\partial \eta} + \frac{1}{\rho C} \cdot \frac{\partial P}{\partial \eta} + \frac{2Cv}{r} \cdot \frac{\partial t}{\partial \eta} = 0 \tag{9}$$

$$\frac{\partial v}{\partial \zeta} - \frac{1}{\rho C} \cdot \frac{\partial P}{\partial \zeta} - \frac{2Cv}{r} \cdot \frac{\partial t}{\partial \zeta} = 0 \tag{10}$$

The numerical integration of Eqs. (9) and (10) can be carried out by using the standard procedure with the imposed initial and boundary conditions. In calculating the region in liquid after the bubble rebounds, it is convenient to use the Lagrangian formulation where the properties of the liquid are obtained by following the particle motion. The Lagrangian coordinates (y, t) are related to the Eulerian coordinates (r, t) by the following relation:

$$y = \int_{r(o,t)}^{r(y,t)} \rho r^2 dr \tag{11}$$

Figure 1 shows the pressure distribution in liquid after bubble collapse. A compression wave generates and gradually steepens while traveling outward in the rebounding process. The numerical solution was terminated where the compression wave had steepened into a vertical front. Of course, the computational difficulties due to the appearance of a shock wave may be overcome by introducing an artificial frictional term [18,19]. The peak pressure in the wave decreases almost inversely with increasing the propagating distance from the collapse center. In the case where the initial gas pressure is 10^2 Pa, for example at $r/R_0 \sim 2$, the peak intensity still remains about 20 MPa. This shock wave pressure is considered to be an important factor contributing to material damage. Ivany and Hammitt [20] made numerical calculations including both effects of the viscosity and surface tension of the liquid.

Later, an available method was applied to the problem of compressible flow. It is the Poincaré-Lighthill-Kuo (PLK) coordinate perturbation technique which was first introduced by Poincaré and developed by Lighthill and Kuo [21]. Benjamin [15] first applied this method to the problem of bubble collapse, and Jahsman [22] compared his results with earlier work. Using this method, Tomita and Shima [23] obtained the equation of motion of the bubble as well as the equation

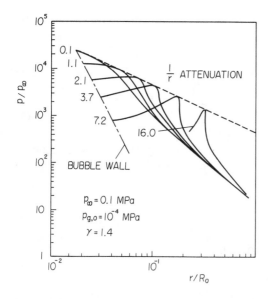

Figure 1 Pressure distribution in liquid immediately after the bubble collapse (Hickling and Plesset [16]).

describing the pressure field in liquid with the second-order correction of liquid compressibility.

In the spherically symmetric and isentropic flow, the velocity potential Φ exists

$$\mathbf{V} = \nabla\Phi \tag{12}$$

where $\mathbf{V}$ is the vector velocity. From Eqs. (2), (3), and (5), we have a hyperbolic partial differential equation with respect to Φ.

$$\Phi_{rr} + \frac{2}{r}\Phi_r - \frac{1}{C_\infty^2}\Phi_{tt} = \frac{1}{C_\infty^2}\left[2\Phi_r\Phi_{rt} + \frac{2(n-1)}{r}\Phi_r\Phi_t + (n-1)\Phi_{rr}\Phi_t \right.$$
$$\left. + \frac{(n+1)}{2}\Phi_r^2\Phi_{rr} + \frac{(n-1)}{r}\Phi_r^3 \right] \tag{13}$$

This equation can be solved to get an approximate solution by using the PLK method. In this case, the equation should be solved only along the outward-going characteristic because the problem under consideration is of the collapse of a bubble in an infinite volume of liquid. The equation concerned with the propagating velocity along the outward-going characteristic is given by

$$\left(\frac{dt}{dr}\right)_{\eta=\text{const}} = \frac{1}{v+C} \tag{14}$$

According to the PLK method, Φ, r, and t may be expanded as follows:

$$\Phi(r, \eta) = \phi_0(r, \eta) + \frac{1}{C_\infty} \phi_1(r, \eta) + \frac{1}{C_\infty^2} \phi_2(r, \eta) + \cdots$$

$$r = r$$

$$t = \eta + \frac{1}{C_\infty} t_1(r, \eta) + \frac{1}{C_\infty^2} t_2(r, \eta) + \cdots \tag{15}$$

where η is the initial time on the outward-going characteristic and satisfies the condition $\eta = t$ on $r = R$. When a compression wave steepens into a shock front, the numerical solution given by using the normal perturbation technique tends to be divergent near the front, which is a mathematically singular point. However, if the PLK method is applied to the problem in which not only the dependent variable Φ but also the independent variable t is expanded in power series of $1/C_\infty$, an approximate solution with uniform accuracy can be given over the whole domain of solution. Substituting Eq. (15) into Eq. (13) and taking into consideration the boundary conditions, we finally obtain the equation of motion of the bubble:

$$R\ddot{R}\left(1 - \frac{2\dot{R}}{C_\infty} + \frac{23\dot{R}^2}{10C_\infty^2}\right) + \frac{3}{2}\dot{R}^2\left(1 - \frac{4\dot{R}}{3C_\infty} + \frac{7\dot{R}^2}{5C_\infty^2}\right)$$

$$+ \frac{1}{\rho_\infty}\left[P_\infty - P_{2,r=R} - \frac{R}{C_\infty}\dot{P}_{1,r=R} + \frac{1}{C_\infty^2}\left\{2R\dot{R}\dot{P}_{1,r=R}\right.\right.$$

$$\left.\left. + \frac{1}{2}(P_\infty - P_{1,r=R})\left[\dot{R}^2 + \frac{3}{\rho_\infty}(P_\infty - P_{1,r=R})\right]\right\}\right] = 0 \tag{16}$$

where $P_{i,r=R}$ is the pressure in liquid at the bubble wall and given by

$$P_{1,r=R} = P_{g,0}\left(\frac{R_0}{R}\right)^{3\gamma} - \frac{2\sigma}{R} - \frac{4\mu}{R}\dot{R} \tag{17}$$

$$P_{2,r=R} = P_{1,r=R} - \left(\frac{4\mu}{3} + \zeta\right)\frac{\dot{P}_{1,r=R}}{\rho_\infty C_\infty^2} \tag{18}$$

In the same way, Eq. (14) becomes

$$t = \eta + \frac{1}{C_\infty}[r - R(\eta)] + \frac{R^2\dot{R}}{C_\infty^2}\left[\frac{1}{r} - \frac{1}{R(\eta)}\right] \tag{19}$$

The third term on the right-hand side of Eq. (19) comes from the nonlinear effect of a wave motion and represents the distortion from the Mach cone. Tomita and Shima [23] also obtained the ordinary differential equation describing the pressure distribution in liquid and found again that the maximum impulsive pressure attenuates inversely with increasing propagating distance. Fujikawa and Akamatsu [24] further developed this method and discussed the pressure wave produced by the collapse of a bubble in liquid taking account of the effect of nonequilibrium condensation of vapor within the bubble. Lastman and Wentzell [25,26] compared some existing theoretical models associated with bubble collapse in a compressible liquid.

2.2 Effect of Heat Transfer on Bubble Collapse

In the later stage of collapse of a bubble containing both vapor and noncondensable gas, the effect of heat transfer plays an important role on bubble motion. When bubble collapse proceeds, the temperature inside the bubble gradually increases and heat is conducted because of the temperature difference between the bubble wall and ambient liquid. As a result, thermal boundary layers develop both inside and outside the bubble. At the bubble wall, phase change of the vapor takes place resulting in heat and mass transfer. These phenomena essentially depend on the rate of the transport process. If the bubble wall velocity is relatively smaller than the transport rate, the thermodynamic equilibrium at the bubble wall will remain during the whole process of the bubble motion. In the opposite condition, on the other hand, there would not be an equilibrium and part of the vapor will behave like the noncondensable gas. Consequently there is a thin but finite nonequilibrium region at the bubble wall.

The collapsing process of a bubble, including the effect of heat transfer, should be determined by solving simultaneously the equation of motion of the bubble with the energy equation. Equilibrium bubble collapse was studied by Zwick and Plesset [27], Florschuetz and Chao [28], and Cho and Seban [29]. Theofanous et al. [30] and Mitchell and Hammitt [31] studied the nonequilibrium collapse of a bubble in an incompressible liquid and, later, Tomita and Shima [32], Fujikawa and Akamatsu [24] and Matsumoto [33] discussed it in a compressible liquid. The rate of mass transfer of the vapor at the phase interface is expressed by [34]

$$\dot{m}_v = \frac{\alpha_M}{\sqrt{2\pi R_v}} \left[\frac{P_L^*(T_L)}{\sqrt{T_L}} - \Gamma \frac{P_v}{\sqrt{T_m}} \right] \tag{20}$$

where T_L and T_m are the temperatures of the liquid and the vapor at the phase interface; P_L^* is the equilibrium vapor pressure; R_v is the gas constant of tne vapor; Γ is the correction factor expressing the deviation from the Maxwell velocity distribution; and α_M is the accommodation coefficient for evaporation or condensation. Although the value between zero and unity is available for α_M, the correct value is still unknown [35].

Fujikawa and Akamatsu [24] have discussed in detail the equation of bubble motion taking into account the movement of the bubble wall due to phase change. For example, the equation of bubble motion with first-order correction of the liquid compressibility is written as

$$R\ddot{R}\left(1 - \frac{2\dot{R}}{C_\infty} + \frac{\dot{m}_v}{\rho_\infty C_\infty}\right) + \frac{3}{2}\dot{R}^2\left(1 - \frac{4\dot{R}}{3C_\infty} + \frac{4\dot{m}_v}{3\rho_\infty C_\infty}\right)$$

$$-\frac{\dot{m}_v R}{\rho_\infty}\left(1 - \frac{2\dot{R}}{C_\infty} + \frac{\dot{m}_v}{\rho_\infty C_\infty}\right)$$

$$-\frac{\dot{m}_v}{\rho_\infty}\left(\dot{R} + \frac{\dot{m}_v}{2\rho_\infty}\right) + \frac{1}{\rho_\infty}\left(P_\infty - P_{r=R} - \frac{R}{C_\infty}\dot{P}_{r=R}\right) = 0 \qquad (21)$$

As to the gas temperature inside the bubble, Flynn [2] analyzed in detail the simple case without phase change, and Fujikawa and Akamatsu [24] applied his method to the case where phase change exists.

Now we introduce various methods for solving the energy equation of the liquid. The equation to be solved is that of a nonsteady heat conduction with spherical symmetry:

$$\frac{\partial T}{\partial t} + v\frac{\partial T}{\partial r} = D_T\left(\frac{\partial^2 T}{\partial r^2} + \frac{2}{r}\frac{\partial T}{\partial r}\right) \qquad (22)$$

where T is the liquid temperature and D_T is the thermal diffusivity of the liquid.

2.2.1 Successive approximation. The solution of Eq. (22) was first analyzed by using the method of successive approximation. Plesset and Zwick [36] obtained the interfacial liquid temperature by assuming that appreciable temperature variations occur only in a thin layer adjacent to the bubble wall. The zero-order solution without heat generation may be expressed by

$$T_L(t) = T_\infty - \sqrt{\frac{D_T}{\pi}}\int_0^t \frac{R^2(x)(\partial T/\partial r)_{r=R(x)}}{[\int_x^t R^4(y)\,dy]^{1/2}}\,dx \qquad (23)$$

Plesset and Zwick also derived the first-order correction for the temperature field making use of the procedure of successive approximation. These approximate solutions rapidly converge as far as the assumption of the thin "thermal boundary layer" is valid. However, Cho and Seban [29] showed that the Plesset and Zwick approximation tends to give a greater heat flux into the liquid when a bubble oscillates. It is also doubtful that the thickness of the thermal boundary layer is still small compared with the bubble radius in the final process of bubble collapse, because in this stage the temperature field in the liquid is fully developed.

2.2.2 Integral method. The integral method, which needs no information about the thickness of the thermal boundary layer but needs only a temperature profile, is another useful approximate solution for the energy equation. This is available as far as the bubble motion monotonously changes. Let us assume a parabolic temperature profile as follows:

$$\frac{T - T_L}{T_\infty - T_L} = 2\left(\frac{r - R}{\delta}\right) - \left(\frac{r - R}{\delta}\right)^2 \tag{24}$$

where δ is the thickness of the thermal boundary layer. The above expression satisfies the following boundary conditions.

$$T = T_L \quad \text{at} \quad r = R \tag{25}$$

$$T = T_\infty \quad \text{and} \quad \frac{\partial T}{\partial r} = 0 \quad \text{at} \quad r = R + \delta$$

Substituting Eq. (24) into Eq. (22) and integrating from R to $R + \delta$ with respect to r, we can obtain an ordinary differential equation. If the liquid is incompressible and $\rho_v \ll \rho_\infty$, the equation may be reduced to the following form:

$$\frac{\dot{T}_L}{T_L - T_\infty}\left[1 + \frac{1}{2}\left(\frac{\delta}{R}\right) + \frac{1}{10}\left(\frac{\delta}{R}\right)^2\right] + \frac{\dot{\delta}}{\delta}\left[1 + \frac{\delta}{R} + \frac{3}{10}\left(\frac{\delta}{R}\right)^2\right]$$

$$+ \frac{2\dot{R}}{R}\left[1 + \frac{1}{4}\left(\frac{\delta}{R}\right)\right] = \frac{6D_T}{\delta^2} \tag{26}$$

Supposing that the thermal boundary layer is very small compared with the bubble radius (i.e., $\delta \ll R$), the above expression comes to the known result obtained by Theofanous et al. [30] and Mitchell and Hammitt [31]. Tomita and Shima [32] further studied the problem where the liquid is considered to be compressible. This method is applicable up to the collapse point of a bubble, because the complicated temperature field may be formed in the rebounding process.

2.2.3 Finite difference method. Bubble motion including the rebounding process has been numerically solved by several investigators [29,32,33,37]. The solution of the finite difference method by Tomita and Shima [32] is outlined in the following. First a coordinate transformation is made:

$$\xi(r, \eta) = 2\left(\frac{r - R}{\delta^*}\right) - \left(\frac{r - R}{\delta^*}\right)^2 \tag{27}$$

where δ^* is defined as the thickness of the calculated region measured from the bubble wall. A new variable ξ is introduced to normalize the calculated region, so that ξ gives zero at $r = R$ and unity at $r = R + \delta^*$. This transformation divides the region near the bubble wall into small intervals in order to obtain sufficient accuracy of solution. The energy equation (22) may be transformed by virtue of the variable ξ as follows:

$$A(\xi, \eta)\frac{\partial T}{\partial \eta} = B(\xi, \eta)\frac{\partial^2 T}{\partial \xi^2} + C(\xi, \eta)\frac{\partial T}{\partial \xi} + D(\xi, \eta)\frac{\partial^2 T}{\partial \xi \partial \eta} \tag{28}$$

A multipoint implicit-type finite difference method by Saito and Shima [38] was applied to Eq. (28). Figure 2 shows grids for numerical computation which consist

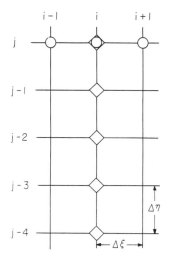

Figure 2 Grids for numerical computation.

of three nodal points in space and of five nodal points in time, where $\Delta\xi$ and $\Delta\eta$ denote the space and time intervals, respectively. Using this method, the temperature variation with time can be obtained with good accuracy because of taking five nodal points in time. Finally the solutions are determined by solving the following simultaneous equations:

$$-e_i T_{i-1} + f_i T_i - g_i T_{i+1} = h_i \qquad (i = 1 \sim N) \tag{29}$$

where the temperature T_i in the left-hand is the short form for $T_{i,j}$, which is the liquid temperature at the point $(i\Delta\xi, j\Delta\eta)$, and so on, and h_i in the right-hand side includes some variables prior to time j. Noting that the latent heat term usually dominates over the others and applying a three-point method to the temperature gradient, the equation of energy balance at the interface may be expressed by

$$\frac{-3T_{1,j} + 4T_{2,j} - T_{3,j}}{2\Delta\xi} = \frac{\dot{m}_v(Ti_j^{(n)})L\delta^*}{2\lambda} \tag{30}$$

where $T_{1,j}$ is the liquid temperature at the interface, and $T_{1,j}^{(n)}$ is the nth approximate value of $T_{1,j}$ in which the first approximate value, $T_{1,j}^{(1)}$, can be estimated by using the extrapolation formula with four approximate values at the previous time level. L and λ denote the latent heat and the thermal conductivity, respectively. The $(N - 1)$th simultaneous equations obtained from Eqs. (29) and (30) can be solved by means of the Gauss elimination method. In order to keep the accuracy of the numerical solution, δ^* was suitably controlled in course of calculation. When δ^* varies, the following relation must be geometrically satisfied between the old coordinate ξ and the new one ξ':

$$\xi = \frac{\delta^{*\prime}}{\delta^*}(1 - \sqrt{1 - \xi'})\left[2 - \frac{\delta^{*\prime}}{\delta^*}(1 - \sqrt{1 - \xi'})\right] \tag{31}$$

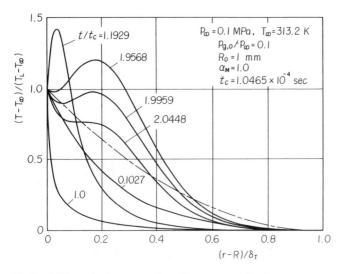

Figure 3 Dimensionless expression of temperature distribution within thermal boundary layer (Tomita and Shima [32]).

where $\delta*'$ is the thickness of the new calculated region. Figure 3 shows the dimensionless expression of the temperature distribution in the thermal boundary layer, where the solid lines denote solutions by the finite difference method and the dashed line shows the solution by the integral method. In Fig. 3, t_c refers to the collapse time of a bubble and δ_T denotes the thickness of the thermal boundary layer defined as the position of $T/T_\infty = 1.005$ as its outer edge. δ_T increases up to the size of the bubble radius in the final stage of bubble collapse.

3 SOME NUMERICAL METHODS FOR NONSPHERICAL BUBBLE COLLAPSE

Kornfeld and Suvorov [39] found important evidence from the observations of bubbles produced by a magnetostriction oscillator that the bubble surface tends to be deformed from the spherical shape because of instability, and they suggested that the impulsive pressure caused by the impact of a liquid jet formed during bubble collapse is a dominant factor contributing to cavitation damage. Since then, much study has been vigorously pursued about the collapse of a bubble near a solid wall. As already mentioned, however, the variation of the bubble shape with time is so rapid that it is difficult to follow the bubble motion experimentally in detail. The theoretical approach concerning this problem was first made by Rattray [40]. With advances in computer technology, various kinds of techniques in numerical calculation have been developed and some of them have been applied to the problems of the nonspherical bubble collapse. The following discussion involves typical numerical methods.

3.1 Perturbation Method

The great part of earlier studies dealt with the problem by means of a perturbation method in which the deviation from spherical shape is assumed to be small and all quantities are expanded by the series of a spherical harmonic. For instance, Rattray [40] solved the problem of collapse of an initially spherical bubble near a rigid wall in a nonviscous incompressible liquid. Later Shima [41] studied this problem by a similar way, but including gas inside the bubble. As a result, it was suggested that the shape of a bubble elongates in the normal direction to the wall in the relatively earlier stage of collapse, and a liquid jet directed toward the wall may be formed in the later stage of collapse. Plesset and Mitchell [42] pointed out the instability of the spherical shape of a vapor cavity in the final stage of collapse on the basis of a linearized perturbation analysis. Naudé and Ellis [43] analyzed theoretically the collapse process of a nonhemispherical cavity in contact with a solid boundary taking the second-order effects into account to improve the solution, and demonstrated that the liquid jet enters the cavity and strikes the solid boundary. Figure 4 shows a spherical coordinate system utilizing the formulation of the problem. With some imposed assumptions the problem is reduced to solving Laplace's equation with respect to the velocity potential Φ.

$$\frac{1}{r^2}\frac{\partial}{\partial r}(r^2\,\Phi_r) + \frac{1}{r^2\sin\theta}\cdot\frac{\partial}{\partial\theta}(\sin\theta\cdot\Phi_\theta) = 0 \tag{32}$$

The following boundary conditions must be satisfied:

$$\Phi = 0 \qquad \text{as } r \to \infty \tag{33}$$

$$\frac{1}{r}\,\Phi_\theta = 0 \qquad \text{on } \theta = \frac{\pi}{2} \tag{34}$$

$$\Phi_r - \frac{1}{R^2}\Phi_\theta R_\theta = R_t \qquad \text{on } r = R\,(\theta,\, t) \tag{35}$$

$$\Phi_t + \frac{1}{2}\,\Phi_r^2 + \frac{1}{2R^2}\,\Phi_\theta^2 = \frac{1}{\rho_\infty}(P_\infty - P_c) \qquad \text{on } r = R(\theta,\, t) \tag{36}$$

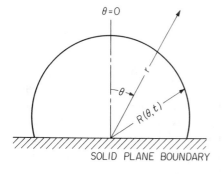

$\theta = 0$

$R(\theta, t)$

SOLID PLANE BOUNDARY   **Figure 4** Spherical coordinate system.

A solution to Eq. (32) which satisfies Eqs. (33) and (34) can be written as follows:

$$\Phi (r, \theta, t) = \frac{\phi_0 (t)}{r} + \sum_{n=1}^{\infty} \frac{\phi_{2n}(t)}{r^{2n+1}} P_{2n} (\cos \theta) \tag{37}$$

where $P_{2n} (\cos \theta)$ are Legendre polynomials, and ϕ_0 and ϕ_{2n} are time-dependent coefficients in the expansion. Furthermore, the bubble shape is expressed by

$$R (\theta, t) = R_0 (t) + \sum_{n=1}^{\infty} R_{2n} (t) P_{2n} (\cos \theta) \tag{38}$$

Substituting Eqs. (37) and (38) into Eqs. (35) and (36), we finally get the family of ordinary differential equations. The equations obtained are solved with the initial conditions of given $R (\theta, 0)$ and $dR (\theta, 0)/dt$.

3.1.1 First perturbation procedure. Neglecting all interactions in Eqs. (37) and (38) and using the orthogonality of Legendre polynomials, we have the following equation for R_{2n}:

$$(1 - x) x \frac{d^2 y_{2n}}{dx^2} + \left(\frac{1}{3} - \frac{5}{6} x \right) \frac{dy_{2n}}{dx} - \frac{1}{6} (2n - 1) y_{2n} = 0 \tag{39}$$

where $x = [R_0(0)/R_0]^3$ and $y_{2n} = R_{2n}/R_{2n} (0)$. The above equation is of hypergeometric form and the general solution can be easily obtained.

3.1.2 Second perturbation procedure. Considering a more realistic cavity for which $R_0 > R_2 > R_4 > R_6$, and R_{2n} is negligibly small for $n \geq 4$ in Eq. (38), we obtain the following differential equations:

$$(1 - x) x \frac{d^2 y'_{2n}}{dx^2} + \left(\frac{1}{3} - \frac{5}{6} x \right) \frac{dy'_{2n}}{dx} - \frac{1}{6} (2n - 1) y'_{2n} = F_{2n} (x) \tag{40}$$

where $y'_{2n} = R_{2n}/R_{2n} (0)$ and F_{2n} is a forcing function. Figure 5 shows theoretical cavity shapes for different values of $R_0/R_0(0)$. We can notice the formation of a liquid jet which strikes a solid boundary. However, the present method lacks accuracy for getting reasonable estimates concerning the configuration of its tip as well as the velocity impacting on the boundary because of the excitation of the higher-order harmonics.

3.2 Variational Method

In order to understand the essential features of nonspherical collapse of a bubble due to the presence of a solid boundary, many studies have neglected all other real effects. In real situations, however, some of these other effects significantly influence bubble collapse. For approximate analysis of nonspherical bubble collapse, Hsieh [44] proposed a variational method, which is relatively easy to treat various boundary conditions. In this case, both motions of a bubble and a sur-

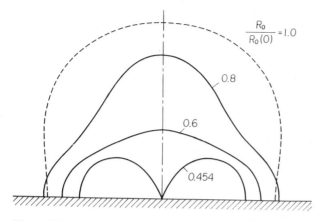

Figure 5 Theoretical cavity shapes for different values of $R_0/R_0(0)$ (Naudé and Ellis [43]).

rounding liquid are given by the extreme conditions of the functional. The variational principle is similar to Hamilton's principle which expresses the motion of a particle in a conservative system.

Now let us consider the collapse of a bubble near a solid wall which is the same problem as that presented by Rattray [40]. Sato and Shima [45] used a spherical coordinate system (shown in Fig. 6) to formulate the problem. The origin of the coordinate system, 0, is initially taken at a certain location on the symmetric axis different from the initial center of a bubble, $0'$, in order to improve the convergence of the solution in the final part of collapse. The functional J can be written as

$$
J = 2\pi \int_{t_1}^{t_2} dt \left[\int_0^{\pi/2} \sin\theta \, d\theta \int_R^{L'/\cos\theta} \rho_\infty \left(\Phi_t + \frac{1}{2} \Phi_r^2 + \frac{1}{2r^2} \Phi_\theta^2 \right) r^2 \, dr \right.
$$
$$
+ \int_{\pi/2}^\pi \sin\theta \, d\theta \int_R^\infty \rho_\infty \left(\Phi_t + \frac{1}{2} \Phi_r^2 + \frac{1}{2r^2} \Phi_\theta^2 \right) r^2 \, dr
$$
$$
\left. + \int_0^\pi \sin\theta \, d\theta \left(\frac{P_\infty - P_g - P_v}{3} R^3 + \sigma R \sqrt{R^2 + R_\theta^2} \right) \right] \tag{41}
$$

Taking an extremum of the functional J, that is $\delta J = 0$, the free boundary-value problem can be regarded to be equivalent to the variational problem. We have Laplace's equation (32) as Euler's equation and the following equations as natural boundary conditions:

$$
\Phi_r \cos\theta - \frac{1}{r} \Phi_\theta \sin\theta = 0 \qquad \text{at } r = \frac{L'}{\cos\theta}, \quad 0 < \theta < \frac{\pi}{2} \tag{42}
$$

$$
\Phi_r - \frac{1}{R^2} \Phi_\theta R_\theta = R_t \qquad \text{on } r = R(\theta, t) \tag{43}
$$

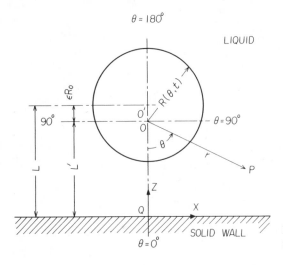

Figure 6 Spherical coordinate system for variational method.

$$\Phi_t + \frac{1}{2}\Phi_r^2 + \frac{1}{2R^2}\Phi_\theta^2 = \frac{1}{\rho_\infty}\left[P_\infty - P_g - P_v + \sigma\left(\frac{1}{R_1} + \frac{1}{R_2}\right)\right]$$

$$\text{on } r = R(\theta, t) \tag{44}$$

where R_1 and R_2 are the principal radii of curvature of the bubble surface, and the mean curvature is given by

$$\frac{1}{2}\left(\frac{1}{R_1} + \frac{1}{R_2}\right) = \frac{1}{2}\left[\frac{R - R_\theta \cot\theta}{R(R^2 + R_\theta^2)^{1/2}} + \frac{R^2 + 2R_\theta^2 - RR_{\theta\theta}}{(R^2 + R_\theta^2)^{3/2}}\right] \tag{45}$$

Assuming that the noncondensable gas inside the bubble undergoes an adiabatic process during the whole motion of the bubble, the gas pressure may be expressed by

$$P_g = P_{g,0}\left[\frac{V(0)}{V(t)}\right]^\gamma \tag{46}$$

where γ is the polytropic index and $V(t)$ is the bubble volume at time t given by

$$V(t) = \frac{2}{3}\pi\int_0^\pi R^3(\theta, t)\sin\theta\, d\theta \tag{47}$$

Taking the boundary conditions, $\Phi = 0$ as $r \to \infty$ and Eq. (41), we choose the trial functions for Φ and R as follows:

$$\Phi(r, \theta, t) = \sum_{n=0}^{N} \phi_n(t)\left\{\frac{P_n(\cos\theta)}{r^{n+1}} + \frac{P_n(\cos\Theta_s)}{r_s^{n+1}}\right\} \tag{48}$$

$$R(\theta, t) = R_0(t) + \sum_{n=1}^{N} R_n(t)P_n(\cos\theta) \tag{49}$$

where

$$r_s = [r^2 + (2L')^2 - 4L'r \cos \theta]^{1/2} \qquad (50)$$

$$\cos \Theta_s = \frac{2L' - r \cos \theta}{r_s} \qquad (51)$$

The variations of Φ and R are

$$\delta\Phi = \sum_{n=0}^{N} \delta\phi_n \left[\frac{P_n(\cos \theta)}{r^{n+1}} + \frac{P_n(\cos \Theta_s)}{r_s^{n+1}} \right] \qquad (52)$$

$$\delta R = \sum_{n=0}^{N} \delta R_n P_n(\cos \theta) \qquad (53)$$

Substituting Eqs. (48), (49), (52), and (53) into the first variation δJ and taking into consideration of such conditions that the trial function (48) satisfies the Laplace's equation (32) and that $\delta\phi_n$ and δR_n are arbitrary, we have the following Euler's equations:

$$\int_0^\pi \left(R_t - \Phi_r + \frac{1}{R^2} R_\theta \Phi_\theta \right) \left\{ \frac{P_k(\cos \theta)}{R^{k+1}} + \frac{P_k(\cos \Theta_s)}{R_s^{k+1}} \right\} R^2 \sin \theta \, d\theta = 0 \qquad (54)$$

$$\int_0^\pi \left[\Phi_t + \frac{1}{2} \Phi_r^2 + \frac{1}{2R^2} \Phi_\theta^2 + \frac{1}{\rho_\infty} (P_g + P_v - P_\infty) \right.$$

$$\left. - \frac{\sigma}{\rho_\infty} \left(\frac{1}{R_1} + \frac{1}{R_2} \right) \right] R^2 P_k(\cos \theta) \sin \theta \, d\theta = 0 \qquad (55)$$

where $k = 0, 1, \ldots, N$. Finally the equation of bubble motion can be obtained by solving the simultaneous ordinary differential equation. It is written in matrix form as

$$\ddot{\mathbf{R}} = \mathbf{F}^{-1}[(\dot{\mathbf{M}}\mathbf{M}^{-1}\mathbf{F} - \dot{\mathbf{F}})\dot{\mathbf{R}} + \mathbf{M}(\mathbf{F}^{-1})^T \mathbf{S}] \qquad (56)$$

where $\mathbf{R}$ and $\mathbf{S}$ are the column vectors, $\mathbf{M}$ and $\mathbf{F}$ the matrices, and the superscripts -1 and T mean the inverse and transposed matrices, respectively. On the other hand, the pressure distribution in the liquid can be obtained by Bernoulli's equation

$$P = P_\infty - \rho_\infty \left(\Phi_t + \frac{1}{2} \Phi_r^2 + \frac{1}{2r^2} \Phi_\theta^2 \right) \qquad (57)$$

Equation (56) can be numerically integrated by means of the Runge-Kutta-Gill method. Since the variational method can be easily used to treat the boundary conditions, the problems concerning the bubble collapse under various situations have been solved by the authors [45–47]. Figure 7 shows velocity distributions on the bubble surfaces at two points of the collapsing process of an attached bub-

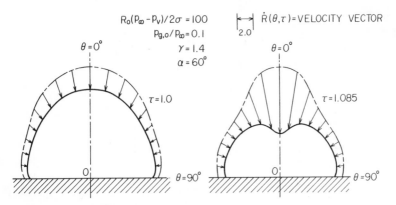

Figure 7. Velocity distributions on bubble surfaces at two points of the collapsing process of an attached bubble with a contact angle of 60° (Shima and Sato [47]).

ble with the contact angle kept at 60°, where $\tau = t\sqrt{(P_\infty - P_v)/\rho_\infty}/R_0$ is dimensionless time [47]. From these results it is readily seen that the velocity on the symmetric axis, that is on $\theta = 0°$, is well developed compared with those on the other portions of the bubble surface as the bubble collapses. However, it is difficult to continue the numerical calculation up to the stage where a liquid microjet will be formed because the instability of the solution may be amplified.

3.3 Finite Difference Method

To clarify the detailed aspects about a liquid jet formed within a bubble, the nonlinear effects, which play an important role in the final stage of collapse, should be precisely simulated in numerical calculation. Plesset and Chapman [48] first succeeded in obtaining the liquid jet velocity impacting a solid wall and discussed the induced impulsive pressure in connection with cavitation damage. The method used was the finite difference one in which Laplace's equation is numerically formulated. It was solved by using the Liebmann iterative method. The following paragraphs describe the numerical procedure for the problem.

The first problem concerned is the collapse of a vapor bubble near a solid wall in a nonviscous incompressible liquid which is the nonspherical version of the classical Rayleigh problem. Since the flow is irrotational, the velocity potential ϕ which satisfies Laplace's equation exists. The method of flow simulation is based on a series of small time steps. The shape and the potential distribution of the free boundary forming the bubble is known at the beginning of each time step. The boundary conditions determine the potential throughout the liquid. Then the velocities on the free surface can be calculated. If the time step Δt is small enough, the velocities will remain constant during the time step. So the displacement of a certain point on the free surface is approximately expressed by

$$\Delta \mathbf{X} = \mathbf{V}\Delta t \tag{58}$$

In the same manner, the change in the potential of a displaced point on the free surface is approximately given by

$$\Delta\phi = \left[\frac{P_\infty - P_v}{\rho_\infty} + \frac{1}{2}v^2 \right]\Delta t \tag{59}$$

The velocities are, of course, calculated at the beginning of the time step. After the free boundary has been displaced and the potentials on it changed accordingly, the new bubble shape with the new potential distribution on the free surface can be used for another time step. A finite difference scheme based on cylindrical coordinates can be applied to this problem which includes the higher deformation of the bubble shape. Lines of both horizontal ($z =$ constant) and vertical ($r =$ constant) families are separated by a constant distance h which is called the mesh length. A typical nodal point and its four neighboring nodal points, each a distance h from the center point, form a regular star. If a star is near the free surface, some of its outer nodal points may fall inside the bubble. Such stars are called irregular stars because the nodal point inside the bubble must be replaced by a free surface point of known potential, producing a leg shorter than the mesh length h. Figure 8 shows the numbering system which is applied to both regular and irregular stars. The finite difference equation at a star is derived by expanding the potential about the central point and neglecting the higher derivatives. The equation for regular stars, except the symmetric axis, may be written as follows:

$$\phi_0 = \frac{1}{4}\left[\phi_4 + \phi_2 + \phi_1\left(1 + \frac{h}{2r_0}\right) + \phi_3\left(1 - \frac{h}{2r_0}\right) \right] \tag{60}$$

For a regular star centered on the symmetric axis, the finite difference approximation becomes

$$\phi_0 = \frac{1}{6}(\phi_2 + \phi_4 + 4\phi_1) \tag{61}$$

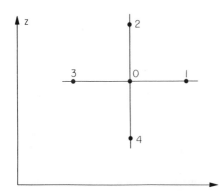

Figure 8 Numbering system for stars.

For the resulting regular stars at $r = h$, the equation of potential becomes

$$\phi_0 = \frac{1}{2}(\phi_2 + \phi_4 + \phi_1 - \phi_3) \tag{62}$$

Stars centered on the solid wall must satisfy the following relation because the wall forms a plane of symmetry

$$\phi_2 = \phi_4 \tag{63}$$

The Liebmann iterative method is used with overrelaxation to find the potential distribution that solves all star equations simultaneously. The condition at infinity is applied to the outer boundary. The outer boundary relates to the net excluding the free boundary, the axis of symmetry, and the solid wall (see Fig. 9). The net used to establish the outer boundary potentials had a radius of 40 mesh lengths and the initial bubble radius consisted of 5 mesh lengths. According to the bubble collapse, the scale of this net is divided into halves several times. Three or four progressively finer nets are applied successively to obtain more detailed expression of liquid near the bubble. Figure 10 shows the linear interpolation between the end points to determine new boundary points. A detailed description of this numerical method can be found in a reference by Plesset and Chapman [49].

Numerical calculations were carried out for $L/R_0 = 1.5$ and $L/R_0 = 1.0$. In the former case, it is found that the bubble shape with time is well coincident with the results experimentally obtained by Lauterborn and Bolle [50] from laser-induced bubbles. The numerical result for the case where $L/R_0 = 1.0$ is shown in Fig. 11. In this case, the impacting velocity of a liquid microjet on the solid wall, v_j, can be calculated as about 130 m/s. Instantaneously, after the jet impacts on the wall, an impulsive pressure occurs. This is the so-called water-hammer pressure, which may be written as

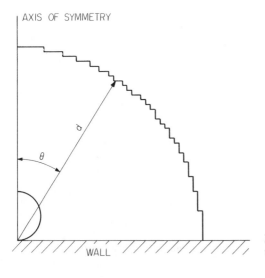

AXIS OF SYMMETRY

WALL

Figure 9 Net used for applying the condition at infinity.

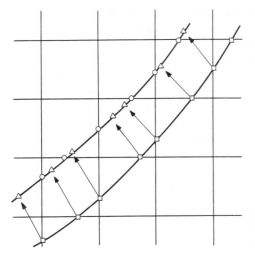

Figure 10 Linear interpolation to obtain new boundary points; □, original points; △, displaced points; ○, new points.

$$P_{wh} = \rho_\infty C_\infty v_j \tag{64}$$

This equation gives the water-hammer pressure of 200 MPa when v_j is 130 m/s.

Such high pressure may cause damage in some kinds of materials if it is of sufficient duration. This can be easily estimated from the experimental evidence associated with the material fracture resulting from a liquid jet impact (e.g., Bowden and Brunton [51], Lush [52] and Field et al. [53]). In contrast, for the case of a liquid microjet formed inside a bubble, the duration of an induced impulsive pressure is very short. For instance, the time is estimated to be 10^{-7} s for a bubble with the radius of 1 mm. This is too short a time for plastic deformation to occur. On the other hand, the stagnation pressure, which is followed by the water-hammer pressure, is about 9 MPa, and its duration can be roughly estimated to be 4

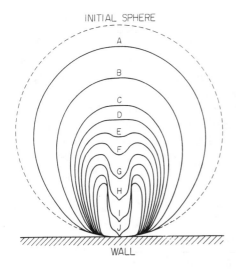

Figure 11 Variation of bubble shape with time; time is marching from A to J (Plesset and Chapman [48]).

μs. It is readily found from Fig. 11 that the diameter of a liquid microjet is about one-tenth of the initial bubble radius. This value agrees reasonably with the results obtained by Lenoir [54] and Tomita and Shima [55]. Further, Hwang and Hammitt [56] numerically described the pressure distribution on a solid surface immediately following the impact of a liquid jet. They showed that the maximum pressure exceeds the water-hammer pressure due to the wave interaction resulting from the shape change of the tip of the liquid jet. In order to simulate the impulsive pressure caused by the liquid jet impact, therefore, the impacting velocity of a liquid microjet formed inside a bubble should be calculated precisely, and the existing real effects such as the viscosity and compressibility of liquid, which are omitted in the above, must be taken into consideration. Chapman and Plesset [57] discussed the nonlinear effect during nonspherical bubble collapse by means of the finite difference method and found that the linear solution tends to give a higher impacting velocity of the induced liquid microjet compared with one obtained by the nonlinear solution.

3.4 Finite Element Method

It is well understood that existing real effects, such as surface tension and gas pressure inside a bubble, which show up as dynamic conditions at the bubble wall, can be easily simulated by means of the variational method. However, very few studies on nonspherical bubble collapse have considered other realistic effects, such as the viscosity and compressibility of liquid, which are the important and essential attributes of liquid. As far as we know, only the viscous effect was discussed, with the exception of problems for nearly spherical bubbles. Mitchell and Hammitt [58] simulated various nonspherical bubble collapses in an incompressible viscous liquid using a modified marker-and-cell technique and demonstrated the viscous effect to be weak on the bubble motion. In this case, however, since all calculations were carried out prior to the initial stage of the liquid jet formation, the effect of viscosity on the impacting jet velocity is not always clear.

The finite element method, however, which has been used in solid and structural mechanics and which has recently been applied to a wide range of problems in continuum mechanics, is used for the analysis of the behavior of a nonspherical bubble in a viscous incompressible liquid by Nakajima and Shima [59]. The finite element method for viscous flow problems [60,61] described by the Navier-Stokes equations can be formulated by means of the method of weighted residuals [62].

The flow field is asssumed to be axially symmetric, so that the cylindrical coordinates are used to describe the problem. For a viscous incompressible liquid, the equations in dimensionless form of continuity and momentum can be written as follows:

$$\frac{\partial v_r}{\partial r} + \frac{v_r}{r} + \frac{\partial v_z}{\partial z} = 0 \tag{65}$$

$$\frac{\partial v_r}{\partial t} + v_r \frac{\partial v_r}{\partial r} + v_z \frac{\partial v_r}{\partial z} = \frac{\partial \tau_{rr}}{\partial r} + \frac{\partial \tau_{zr}}{\partial z} + \frac{\tau_{rr} - \tau_{\varphi\varphi}}{r} \tag{66}$$

$$\frac{\partial v_z}{\partial t} + v_r \frac{\partial v_z}{\partial r} + v_z \frac{\partial v_z}{\partial z} = \frac{\partial \tau_{zr}}{\partial r} + \frac{\partial \tau_{zz}}{\partial z} + \frac{\tau_{zr}}{r} \tag{67}$$

where r, z, and φ are the cylindrical coordinates; v_r and v_z are the velocity components in r and z directions, respectively; and $\tau_{ij} = (i, j = r, z, \varphi)$ is the stress tensor. The flow domain is divided into a finite number of connected subdomains V^e called finite elements. The approximations of the velocity and pressure fields within each finite element can be expressed by

$$v_i(r, z, t) = N^L(r, z)v_i^L(t) \qquad i = r, z \tag{68}$$

$$P(r, z, t) = N^L(r, z)P^L(t) \tag{69}$$

where $v_i^L(t)$ and $P^L(t)$ are the values of the velocity components and pressure at node at the time t, and $N^L(r, z)$ are the shape functions. The repeated nodal indices L in Eqs. (68) and (69) are summed from unity to L_e, where L_e is the total number of nodes of finite elements. The shape functions are defined so as to have the properties [59] given in Eq. (70), and it is required that v_i and P are continuous across interelement boundaries [63].

$$N^L(r, z) = 0 \qquad r, z \notin V^e, \qquad N^L(r_K, z_K) = \delta_{LK} \qquad \sum_{L=1}^{L_e} N^L(r, z) = 1 \tag{70}$$

Substituting Eqs. (68) and (69) into Eqs. (65)–(67), multiplying Eqs. (65)–(67) by the weight function $N^L(r, z)$ and integrating over the domain (i.e., applying the Galerkin method to the present problem) we obtain

$$\sum_{e=1}^{M} \iint_{V^e} \left(N^K N_{,r}^L v_r^L + \frac{N^K N^L}{r} v_r^L + N^K N_{,z}^L v_z^L \right) r \, dr \, dz = 0 \tag{71}$$

$$\sum_{e=1}^{M} \iint_{V^e} \left[\left(N^K N^L \dot{v}_r^L + N^K N^i v_r^i N_{,r}^L v_r^L + N^K N^i v_z^i N_{,z}^L v_r^L \right) \right.$$
$$- \left(\frac{N^K}{r} + N_{,r}^K \right) N^L P^L + \frac{1}{Re} \left(2N_{,r}^K N_{,r}^L v_r^L + N_{,z}^K N_{,z}^L v_r^L + N_{,z}^K N_{,z}^L v_r^L \right.$$
$$+ \left. \left. \frac{2N^K N^L}{r^2} v_r^L \right) \right] r \, dr \, dz = \sum_{e=1}^{M} \int_{A^e} N^K S_r \, dA^e \tag{72}$$

$$\sum_{e=1}^{M} \iint_{V^e} \left[(N^K N^L \dot{v}_z^L + N^K N^i v_r^i N_{,r}^L v_z^L + N^K N^i v_z^i N_{,z}^L v_z^L) - N_{,z}^K N^L P^L \right.$$
$$+ \left. \frac{1}{Re} (N_{,r}^K N_{,r}^L v_z^L + N_{,z}^K N_{,r}^L v_r^L + 2N_{,z}^K N_{,z}^L v_z^L) \right] r \, dr \, dz$$
$$= \sum_{e=1}^{M} \int_{A^e} N^K S_z \, dA^e \tag{73}$$

where S_r and S_z are the stress components, $\dot{v}_r^L = dv_r^L/dt$, Re is the Reynolds number, defined as $Re = R_0 \sqrt{p_\infty/\rho_\infty}/\nu$ and ν is the kinematic viscosity. Equations

(71)–(73) consist of N simultaneous equations and can be solved by the step-by-step numerical integration method. If the forward difference time scheme is introduced for dealing with the transient terms, these equations can be written in the following matrix form:

$$
\begin{bmatrix}
\dfrac{1}{\Delta t}\mathbf{A} + \mathbf{E} & -\mathbf{B}^T & \mathbf{D} \\[2mm]
-\mathbf{B} & 0 & \mathbf{C} \\[2mm]
\mathbf{D}^T & -\mathbf{C}^T & \dfrac{1}{\Delta t}\mathbf{A} + \mathbf{F}
\end{bmatrix}
\begin{Bmatrix}
\mathbf{v}_{r,t+\Delta t} \\[2mm]
\mathbf{P}_{t+\Delta t} \\[2mm]
\mathbf{v}_{z,t+\Delta t}
\end{Bmatrix}
=
\begin{Bmatrix}
\mathbf{f}_{r,t} + \dfrac{1}{\Delta t}\mathbf{A}\mathbf{v}_{r,t} - \mathbf{G}\mathbf{v}_{r,t} \\[2mm]
0 \\[2mm]
\mathbf{f}_{z,t} + \dfrac{1}{\Delta t}\mathbf{A}\mathbf{v}_{z,t} - \mathbf{G}\mathbf{v}_{z,t}
\end{Bmatrix}
\quad (74)
$$

In numerical calculation, six-node triangular elements, shown in Fig. 12, are employed. The shape function for the element in Fig. 12 is expressed by

$$\{N^L\} = \{2\zeta_1^2 - \zeta_1,\ 4\zeta_1\zeta_2,\ 2\zeta_2^2 - \zeta_2,\ 4\zeta_2\zeta_3,\ 2\zeta_3^2 - \zeta_3,\ 4\zeta_3\zeta_1\} \quad (75)$$

where ζ_1, ζ_2, and ζ_3 are the area coordinates and are satisfied by the following relation

$$\zeta_1 + \zeta_2 + \zeta_3 = 1 \quad (76)$$

On the other hand, the relationship between the $r - z$ coordinates and the area coordinates of the triangle is given as follows:

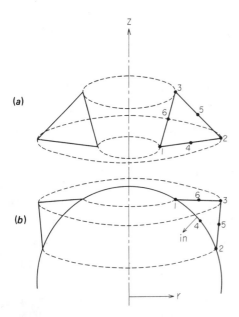

Figure 12 Cylindrical coordinate system and typical finite elements.

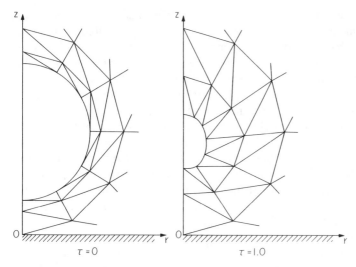

Figure 13 Deformation of elements adjacent to the bubble surface.

$$\begin{Bmatrix} \zeta_1 \\ \zeta_2 \\ \zeta_3 \end{Bmatrix} = \frac{1}{2A} \begin{bmatrix} a_1 & b_1 & c_1 \\ a_2 & b_2 & c_2 \\ a_3 & b_3 & c_3 \end{bmatrix} \begin{Bmatrix} 1 \\ r \\ z \end{Bmatrix} \tag{77}$$

where A is the area of triangle element. Figure 12b shows an isoparametric element which is available to represent a close approximation of the bubble surface geometry. The definition of the volume element for integrating over the finite element is given by

$$dV = r \, dr \, dz = r \, |J| \, d\zeta_i \, d\zeta_j \tag{78}$$

where J is the Jacobian. The area element can be written as

$$dA^e = R \sqrt{R_{,\zeta}^2 + Z_{,\zeta}^2} \, d\zeta_1 \tag{79}$$

The bubble volume is obtained from the following integration:

$$V = \sum_{e=1}^{M} \pi \int_0^1 R^2 Z_{,\zeta} \, d\zeta_1 \tag{80}$$

The simultaneous equations obtained from Eq. (74) can be solved for the velocity and pressure variables by means of the Gauss elimination method. The nodal coordinates near the bubble surface at time $t + \Delta t$ are simply determined by the following relations

$$r_{L,t+\Delta t} = r_{L,t} + \Delta t \cdot v_{r,t+\Delta t}^L \tag{81}$$

$$Z_{L,t+\Delta t} = Z_{L,t} + \Delta t \cdot v_{z,t+\Delta t}^L \tag{82}$$

Figure 13 shows the deformation of elements adjacent to the bubble surface collapsing near a solid boundary. All elements are divided into two groups. They are a fixed and movable element represented by triangular and isoparametric elements, respectively. For fixed elements, the coefficient matrix on the left-hand side of Eq. (74) remains constant throughout the calculation as long as the time increment Δt is not changed, so that its matrix is formed initially and eliminated over fixed nodes. The results of eliminations are placed in storage. For movable elements, the matrix formation is carried out at each time step. The vectors on the right-hand side of Eq. (74) are also formed by using the results obtained at the previous step. Figure 14 shows a numerical example in the case of $P_{g,0} = 0$, $\sigma = 0$ and Re $= 1000$. An initially spherical bubble is located at the distance of $1.5\ R_0$ from a solid boundary. The flow domain is ended at the distance of about $5.5\ R_0$ from the initial bubble center. For the discretization, 50 elements are employed and there are 121 nodes. In the final stages of collapse the bubble becomes an irregular shape because of the small number of elements forming the bubble surface. Figure 15 examines the viscous effect on the downward velocity on the upper portion of the bubble at the symmetric axis. The results suggest that the impacting jet velocity on a solid boundary may significantly decrease owing to the liquid viscosity. To discuss the viscous effect in detail, however, more accurate numerical calculation should be carried out in the future.

4 CONCLUDING REMARKS

Some numerical techniques applying to the problems of cavitation bubble collapse have been presented. The formation of a liquid microjet comes from the non-spherical bubble collapse, in contrast to the occurrence of a spherical shock wave

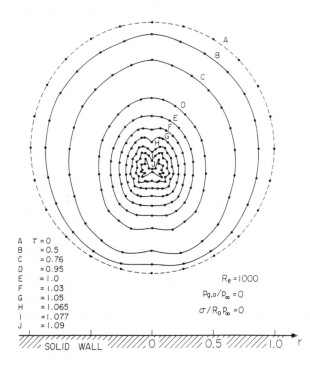

A	$\tau = 0$
B	$= 0.5$
C	$= 0.76$
D	$= 0.95$
E	$= 1.0$
F	$= 1.03$
G	$= 1.05$
H	$= 1.065$
I	$= 1.077$
J	$= 1.09$

$R_e = 1000$

$P_{g,0}/P_\infty = 0$

$\sigma/R_0 P_\infty = 0$

SOLID WALL   0   0.5   1.0 r

Figure 14 Variation of bubble shape with time in the case of Re = 1000. A spherical bubble is initially located at the distance of $1.5\ R_0$ from a solid boundary (Nakajima and Shima [58]).

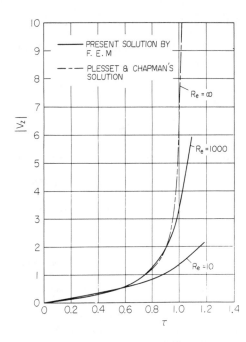

Figure 15 Viscous effect on jet velocity; V_z is the downward velocity on the upper portion of the bubble at the symmetric axis (Nakajima and Shima [58]).

resulting from spherical bubble collapse. The impulsive pressures caused by these two important factors should be precisely obtained in order to clarify the mechanism of cavitation damage. What yet needs to be done is to solve the problem of nonspherical bubble collapse in a viscous compressible liquid. However, it will be quite difficult, even if computer technology does make rapid progress. In the case of the bubble collapse, in general, the numerical solution breaks down when a liquid microjet penetrates the opposite bubble wall. Nevertheless, the phenomenon occurs. Recently, Tomita and Shima [53] found important experimental evidence that the damage pattern resulting from spark-induced bubble collapse may be caused directly by local high pressure generated at the collision between the contracting bubble surface and the radial flow following liquid jet impact on a solid boundary. The problem of high-speed liquid-liquid impact is considered to be an important research task to make systematically clear the impulsive pressure generation. Therefore, a suitable theoretical model is desired.

This chapter has discussed the problem of collapse of a single bubble in which a rigid wall is assumed to be a boundary. The reason results from obtaining some essential information associated with cavitation damage. In connection with the suppression of cavitation damage, Gibson [64] and Gibson and Blake [65,66] studied bubble collapse near various kinds of boundaries. The problems of bubble collapse near a free surface [52,67,68] and the nonspherical collapse of two bubbles [69,70] have been numerically solved. In addition, the theoretical model concerning a cluster of bubbles was proposed and numerically calculated by Hansson and Mørch [71] and Chahine [72]. Recently, the boundary-integral method was applied to the problems of the growth and collapse of transient cavities near boundaries by Blake et al. [73,74].

ACKNOWLEDGMENTS

We thank Mr. N. Miura for his drawings and Miss N. Inomata for typing this manuscript.

REFERENCES

1. D. Y. Hsieh, Some Analytical Aspects of Bubble Dynamics, *J. Basic Eng., Trans. ASME, Ser. D*, Vol. 87, pp. 991–1005, 1965.
2. H. G. Flynn, Cavitation Dynamics. I. A Mathematical Formulation, *J. Acous. Soc. Am.*, Vol. 57, pp. 1379–1396, 1975.
3. M. S. Plesset and A. Prosperetti, Bubble Dynamics and Cavitation, *Ann. Rev. Fluid Mech.*, Vol. 9, pp. 145–185, 1977.
4. E. A. Neppiras, Acoustic Cavitation, *Physics Reports* (Review Section of Physics Letters), Vol. 61, pp. 159–251, 1980.
5. R. T. Knapp, J. W. Daily and F. G. Hammitt, *Cavitation*, p. 94, McGraw-Hill, New York, 1970.
6. F. G. Hammitt, *Cavitation and Multiphase Flow Phenomena*, p. 136, McGraw-Hill, New York, 1980.
7. W. Lauterborn, *Cavitation and Inhomogeneities in Underwater Acoustics*, p. 1, Springer-Verlag, Berlin, Heidelberg, New York, 1980.
8. W. H. Isay, *Kavitation*, p. 1, Schiffahrts-Verlag, "Hansa" C. Schroedter & Co., 1981.
9. I. Anton, *Cavitatia*, Vol. 1, p. 79, Editura Academiei Republicii Socialiste România, 1984.
10. Lord Rayleigh, On the Pressure Developed in a Liquid During the Collapse of a Spherical Cavity, *Phil. Mag.*, Vol. 34, pp. 94–98, 1917.
11. C. Herring, Theory of Pulsations of a Gas Bubble Produced by an Underwater Explosion, Columbia University. NDRC Rept. C-4-sr10-010, 1941.
12. L. Trilling, The Collapse and Rebound of a Gas Bubble, *J. Appl. Phys.*, Vol. 23, pp. 14–17, 1952.
13. F. R. Gilmore, The Collapse and Growth of a Spherical Bubble in a Viscous Compressible Liquid, *Hydro. Lab., Cal. Inst. Technol.*, Rept. 26–4, 1952.
14. J. G. Kirkwood and H. A. Bethe, The Pressure Wave Produced by an Underwater Explosion, OSRD Rept. 588, 1942.
15. T. B. Benjamin, Pressure Waves from Collapsing Cavities, Proc. 2nd Symp. on Naval Hydrodynamics, Washington, pp. 207–233, 1958.
16. R. Hickling and M. S. Plesset, Collapse and Rebound of a Spherical Bubble in Water, *Phys. Fluids*, Vol. 7, pp. 7–14, 1964.
17. J. O. Hirschfelder, C. F. Curtiss and R. B. Bird, *Molecular Theory of Gases and Liquids*, p. 261, John Wiley & Sons, Inc., New York, 1967.
18. J. von Neumann and R. D. Richtmeyer, A Method for the Numerical Calculation of Hydrodynamic Shocks, *J. Appl. Phys.*, Vol. 21, pp. 232–237, 1950.
19. P. D. Lax, Weak Solutions of Nonlinear Hyperbolic Equations and their Numerical Computation, *Commun. on Pure and Appl. Math.*, Vol. 7, pp. 159–193, 1954.
20. R. D. Ivany and F. G. Hammitt, Cavitation Bubble Collapse in Viscous Compressible Liquid-Numerical Analysis, *J. Basic Eng., Trans. ASME*, Ser. D, pp. 977–985, 1965.
21. H. S. Tsien, The Poincaré-Lighthill-Kuo Method, *Advances in Appl. Mech.*, Vol. 4, pp. 281–349, 1956.
22. W. E. Jahsman, Collapse of a Gas-Filled Spherical Cavity, *J. Appl. Mech., Trans. ASME, Ser. E*, Vol. 35, pp. 579–587, 1968.
23. Y. Tomita and A. Shima, On the Behavior of a Spherical Bubble and the Impulse Pressure in a Viscous Compressible Liquid, *Bull. JSME*, Vol. 20, pp. 1453–1460, 1977.
24. S. Fujikawa and T. Akamatsu, Effect of the Non-Equilibrium Condensation of Vapour on the Pressure Wave Produced by the Collapse of a Bubble in a Liquid, *J. Fluid Mech.*, Vol. 97, pp. 481–512, 1980.

25. G. L. Lastman and R. A. Wentzell, Cavitation of a Bubble in an Inviscid Compressible Liquid, with Comparisons to a Viscous Incompressible Liquid, *Phys. Fluids,* Vol. 22, pp. 2259–2266, 1979.

26. G. L. Lastman and R. A. Wentzell, Comparison of Five Models of Spherical Bubble Response in a Inviscid Compressible Liquid, *J. Acoust. Soc. Am.,* Vol. 69, pp. 638–642, 1981.

27. S. A. Zwick and M. S. Plesset, On the Dynamics of Small Vapor Bubbles in Liquids, *J. Math. Phys.,* Vol. 33, pp. 308–330, 1954.

28. L. W. Florschuetz and B. T. Chao, On the Mechanics of Vapor Bubble Collapse, *J. Heat Transfer, Trans. ASME, Ser. C,* Vol. 87, pp. 209–220, 1965.

29. S. M. Cho and R. A. Seban, On Some Aspects of Steam Bubble Collapse, *J. Heat Transfer, Trans. ASME, Ser. C,* Vol. 91, pp. 537–542, 1969.

30. T. G. Theofanous, L. Biasi, H. S. Isbin and H. K. Fauske, Nonequilibrium Bubble Collapse: A Theoretical Study, *Chem. Eng. Progr. Symp. Ser.,* Vol. 66, pp. 37–47, 1970.

31. T. M. Mitchell and F. G. Hammitt, On the Effect of Heat Transfer upon Collapsing Bubbles, *Nucl. Sci. Eng.,* Vol. 53, pp. 263–276, 1974.

32. Y. Tomita and A. Shima, The Effect of Heat Transfer on the Behavior of a Bubble and the Impulse Pressure in a Viscous Compressible Liquid, *Z. Angew. Math. Mech.,* Bd. 59, S. 297–306, 1979.

33. Y. Matsumoto, Contribution of Homogeneous Condensation inside Cavitation Nuclei to Cavitation Inception, *Int. Symp. on Cavitation Inception,* ASME, pp. 27–32, 1984.

34. R. W. Schrage, *A Theoretical Study of Interphase Mass Transfer,* p. 36, Columbia Univ. Press, New York, 1953.

35. S. Fujikawa, T. Akamatsu, J. Yahara and H. Fujioka, Studies of Liquid-Vapour Phase Change by a Shock Tube, *Appl. Sci. Res.,* Vol. 38, pp. 363–372, 1982.

36. M. S. Plesset and S. A. Zwick, A Nonsteady Heat Diffusion Problem with Spherical Symmetry, *J. Appl. Phys.,* Vol. 23, pp. 95–98, 1952.

37. Y. Matsumoto and A. E. Beylich, Influence of Homogeneous Condensation inside a Small Gas Bubble on Its Pressure Response, *J. Fluids Eng., Trans. ASME,* Vol. 107, pp. 281–285, 1985.

38. T. Saitoh and A. Shima, Numerical Solution for the Spherical Bubble Growth Problem in an Uniformly Ultraheated Liquid, *J. Mech. Eng. Sci.,* Vol. 19, pp. 101–107, 1977.

39. M. Kornfeld and L. Suvorov, On the Destructive Action of Cavitation, *J. Appl. Phys.,* Vol. 15, pp. 495–506, 1944.

40. M. Rattray Jr., Perturbation Effects in Cavitation Bubble Dynamics, Ph.D. thesis, California Institute of Technology, Pasadena, Calif., 1951.

41. A. Shima, The Behavior of a Spherical Bubble in the Vicinity of a Solid Wall, *J. Basic Eng., Trans. ASME,* Ser. D, Vol. 90, pp. 75–89, 1968.

42. M. S. Plesset and T. P. Mitchell, On the Stability of the Spherical Shape of a Vapor Cavity in a Liquid, *Quart. Appl. Math.,* Vol. 13, pp. 419–430, 1955.

43. C. F. Naudé and A. T. Ellis, On the Mechanism of Cavitation Damage by Nonhemispherical Cavities Collapsing in Contact with a Solid Boundary, *J. Basic Eng., Trans. ASME, Ser. D,* Vol. 83, pp. 648–656, 1961.

44. D. Y. Hsieh, On the Dynamics of Nonspherical Bubbles, *J. Basic Eng., Trans. ASME, Ser. D,* Vol. 94, pp. 655–665, 1972.

45. Y. Sato and A. Shima, The Collapse of an Initially Spherical Bubble Near a Solid Wall, Rept. Inst. High Speed Mech., Tohoku Univ., Vol. 42, pp. 1–24, 1980.

46. A. Shima and K. Nakajima, the Collapse of a Non-Hemispherical Bubble Attached to a Solid Wall, *J. Fluid Mech.,* Vol. 80, pp. 369–391, 1977.

47. A. Shima and Y. Sato, The Collapse of a Bubble Attached to a Solid Wall, *Ing.-Archiv,* Bd. 48. pp. 85–95, 1979.

48. M. S. Plesset and R. B. Chapman, Collapse of an Initially Spherical Vapour Cavity in the Neighbourhood of a Solid Boundary, *J. Fluid Mech.,* Vol. 47, pp. 283–290, 1971.

49. M. S. Plesset and R. B. Chapman, Collapse of an Initially Spherical Vapor Cavity in the Neighborhood of a Solid Boundary, Div. of Eng. and Appl. Science, California Institute of Technology Rept. 85–49, Pasadena, Calif., June 1970.

50. W. Lauterborn and H. Bolle, Experimental Investigation of Cavitation-Bubble Collapse in

Neighbourhood of a Solid Boundary, *J. Fluid Mech.,* Vol. 72, pp. 391–399, 1975.

51. F. P. Bowden and J. H. Brunton, Deformation of Solids by Liquid Impact at Supersonic Speeds, *Proc. Roy. Soc. London, Ser. A,* Vol. 263, pp. 433–450, 1961.

52. P. A. Lush, Impact of a Liquid Mass on a Perfectly Plastic Solid, *J. Fluid Mech,* Vol. 135. pp. 373–387, 1983.

53. J. E. Field, M. B. Lesser and J. P. Dear, Studies of Two-Dimensional Liquid-Wedge Impact and Their Relevance to Liquid-Drop Impact Problems, *Proc. Roy. Soc. London, Ser. A,* Vol. 401, pp. 225–249, 1985.

54. M. Lenoir, Calcul numérique de l'implosion d'une bulle de cavitation au voisinage d'une paroi ou d'une surface libre, *J. Méchan.,* Vol. 15, pp. 725–751, 1976.

55. Y. Tomita and A. Shima, Mechanisms of Impulsive Pressure Generation and Damage Pit Formation by Bubble Collapse, *J. Fluid Mech.,* Vol. 169, pp. 535–564, 1986.

56. J.-B. G. Hwang and F. G. Hammitt, High-Speed Impact between Curved Liquid Surface and Rigid Flat Surface, *J. Fluids Eng., Trans. ASME,* Ser. I, Vol. 99, 396–404, 1977.

57. R. B. Chapman and M. S. Plesset, Nonlinear Effect in the Collapse of a Nearly Spherical Cavity in a Liquid, *J. Basic Eng., Trans. ASME, Ser. D,* Vol. 94, pp. 142–146, 1972.

58. T. M. Mitchell and F. G. Hammitt, Asymmetric Cavitation Bubble Collapse, *J. Basic Eng., Trans. ASME, Ser. I,* Vol. 95, pp. 29–37, 1973.

59. K. Nakajima and A. Shima, Analysis of the Behavior of a Bubble in a Viscous Incompressible Liquid by Finite Element Method, *Ing.-Archiv,* Bd. 46, S. 21–34, 1977.

60. J. T. Oden and L. C. Wellford Jr., Analysis of Flow of Viscous Fluid by the Finite-Element Method, *AIAA J.,* Vol. 10, pp. 1590–1599, 1972.

61. C. Taylor and P. Hood, A Numerical Solution of the Navier-Stokes Equations Using the Finite Element Technique, *Computers & Fluids,* Vol. 1, pp. 73–100, 1973.

62. B. A. Finlayson and L. E. Scriven, The Method of Weighted Residuals and its Relation to Certain Variational Principles for the Analysis of Transport Processes, *Chem. Eng. Sci.,* Vol. 20, pp. 395–404, 1965.

63. D. H. Norrie and G. dé Vries, *The Finite Element Method,* New York and London, 1973.

64. D. C. Gibson, Cavitation Adjacent to Plane Boundaries, *Proc. 3rd Australasian Conf. on Hydraulics and Fluid Mechanics,* Institution of Engineers, Sydney, pp. 210–214, 1968.

65. D. C. Gibson and J. R. Blake, Growth and Collapse of Vapour Bubbles Near Flexible Boundaries, *Proc. 7th Australasian Conf. on Hydraulics and Fluid Mechanics,* Brisbane, pp. 283–286, 1980.

66. D. C. Gibson and J. R. Blake, The Growth and Collapse of Bubbles Near Deformable Surfaces, *Appl. Sci. Res.,* Vol. 38, pp. 215–224, 1982.

67. G. L. Chahine, Interaction between an Osillating Bubble and a Free Surface, *J. Fluids Eng., Trans. ASME, Ser. I,* Vol. 99, pp. 709–716, 1977.

68. J. R. Blake and D. C. Gibson, Growth and Collapse of a Vapour Cavity near a Free Surface, *J. Fluid Mech.,* Vol. 111, pp. 123–140, 1981.

69. L. Guerri, G. Lucca and A. Prosperetti, A Numerical Method for the Dynamics of Nonspherical Cavitation Bubbles, *Proc. 2nd Int. Colloquim on Drops and Bubbles,* edited by D. H. Le Croissette, Jet Propulsion Laboratory, Monterey, pp. 175–181, 1982.

70. S. Fujikawa, T. Hirochi, H. Takahira and T. Akamatsu, Interactions between Two Slightly Nonspherical Bubbles in a Compressible Liquid, *Proc. Int. Symp. on Cavitation,* edited by H. Murai, Sendai, pp. 55–60, 1986.

71. I. Hansson and K. A. Mørch, The Dynamics of Cavity Clusters in Ultrasonic (Vibratory) Cavitation Erosion, *J. Appl. Phys.,* Vol. 51, pp. 4651–4658, 1980.

72. G. L. Chahine, Pressure Generated by a Bubble Cloud Collapse, *Chem. Eng. Commun.,* Vol. 28, pp. 355–367, 1984.

73. J. R. Blake, B. B. Taib and G. Doherty, Transient Cavities near Boundaries, Part 1. Rigid Boundary, *J. Fluid Mech.,* Vol. 170, pp. 479–497, 1986.

74. J. R. Blake, B. B. Taib and G. Doherty, Transient Cavities near Boundaries, Part 2. Free Surface, *J. Fluid Mech.,* Vol. 181, pp. 197–212, 1987.

SIX

BOUNDARY ELEMENTS IN VISCOUS
FLUID MECHANICS AND RHEOLOGY

M. B. Bush, N. Phan-Thien, and R. I. Tanner

ABSTRACT

Some applications of the boundary element method to linear and nonlinear viscous flow problems are reviewed. We conclude that the method is useful in (1) linear problems; creeping and Oseen flows; (2) medium to high Reynolds number flows; and (3) flows with nonlinear material (or rheological) properties. Finally, we note that external flows and flows with free surfaces are often treated very economically using the boundary element method.

NOMENCLATURE

Vectors and tensors are boldfaced.

∇	gradient operator
1	unit tensor
a	sphere or cylinder radius
A	matrix of coefficients
b	pseudo-body-force vector
c	boundary condition vector
C	specific heat at constant pressure
C_D	drag coefficient

C_L	lift coefficient
c_{ij}	coefficient
d	sphere center to plane distance
$\mathbf{d}$	domain integral vector
$\mathbf{D}$	rate of deformation tensor
D/Dt	material or particle-following derivative
$\mathbf{f}$	body-force vector (per unit mass)
h	separation of cylinder centers
i	$= \sqrt{-1}$
k	thermal conductivity
n	power-law exponent
$\mathbf{n}$	unit normal vector
p	pressure
P	general domain or boundary point
Q	general domain or boundary point
r	distance
R	jet radius
Re	Reynolds number
$\mathbf{s}$	strength of "fictitious" traction layer
S	dimensionless surface tension
t	time
$\mathbf{t}$	traction vector
t_{ij}^*	fundamental traction field
T	temperature
$\mathbf{u}$	velocity or displacement vector
u	average velocity
u_{ij}^*	fundamental displacement or velocity field
U	velocity at a large distance
x	coordinate
$\mathbf{x}$	solution vector
X	dimensionless drag in creeping flow
y	coordinate
z	complex variable or coordinate
Z	axial distance
α	angle
$\dot{\gamma}$	rate of shear
Γ	surface of domain
δ_{ij}	unit tensor
ϵ_{ijk}	alternating tensor
$\boldsymbol{\epsilon}$	non-Newtonian stress tensor
ζ	harmonic function
η	viscosity
θ	harmonic function
λ	Lamé elastic constant or time constant

Λ	Weissenberg number (or dimensionless time) $\lambda U/a$
ν	kinematic viscosity or Poisson's ratio
ρ	density
$\boldsymbol{\sigma}$	stress tensor
Σ_{ijk}^*	fundamental stress field
$\boldsymbol{\tau}$	extra stress tensor
ϕ	Airy stress function
ϕ^*	fundamental potential solution
Φ	viscous dissipation function
Φ^*	fundamental solution to diffusion equation
χ	complex potential
ψ	stream function
$\boldsymbol{\omega}$	vorticity vector
Ω	domain of viscous or elastic medium

1 INTRODUCTION

The numerical analysis of problems in viscous fluid flow is, in principle, able to be accomplished by a variety of techniques. For example, one has finite differences, finite elements, boundary elements, and other methods. Each method has some advantage over the others in certain classes of problems, and here we review the use of boundary element methods in some Newtonian and non-Newtonian (or rheological) fluid mechanics problems.

In comparing the various methods, we note that the finite difference (F.D.) and finite element (F.E.M.) methods automatically find the velocity and pressure fields all over the fluid domain. The boundary element method (B.E.M.) does not necessarily do this. For elasticity problems, it is often absolutely necessary to exhibit the fields everywhere in order to locate, for example, the maximum stress points. In fluid mechanics, we are often most interested in the gross features of the flow, for example, the overall pressure drop or drag, and the precise point-wise details of the fields are of less interest. Under these circumstances, the boundary element method has some advantages, which will be discussed below.

The use of "panel" methods of potential flow analysis, related to the B.E.M., is discussed by Hess and Smith [1] (see also Hunt [2]) and follows classical potential theory. It will not be further discussed here (see, for example, Jaswon and Symm [3] and Brebbia et al. [4]). The work of Rizzo [5] on elasticity in 1965, followed by that of Cruse [6], opened the way to the use of the method in viscous creeping flow in 1975 [7].

Sometimes elastic and incompressible creeping fluid flow problems may be conveniently considered together, and this is done in part of what follows for generality, although the emphasis of the present review is on fluid flow. We now discuss the sets of equations to be solved before tracing the development of the subject further.

2 GOVERNING EQUATIONS

The full set of governing equations consists of the linear momentum equations, continuity equation, constitutive equations, and energy equation [8]:

$$\nabla \cdot \boldsymbol{\sigma} - \rho \frac{D\mathbf{u}}{Dt} + \rho \mathbf{f} = \mathbf{0} \qquad \text{(linear momentum)} \tag{1}$$

$$\nabla \cdot \mathbf{u} = 0 \qquad \text{(continuity)} \tag{2}$$

$$\boldsymbol{\sigma} = -p\mathbf{1} + \boldsymbol{\tau} \qquad \text{(constitutive)} \tag{3}$$

$$k\nabla^2 T - \rho C \frac{DT}{Dt} + \Phi = 0 \qquad \text{(energy)} \tag{4}$$

In the above equations, $\boldsymbol{\sigma}$ is the Cauchy stress tensor, $\mathbf{u}$ is the velocity vector, $\mathbf{f}$ is an arbitrary body force vector (per unit mass), $\boldsymbol{\tau}$ is the "extra stress" tensor, $\mathbf{1}$ is the unit tensor, p is the pressure, T is the temperature, Φ is the viscous dissipation function ($= \boldsymbol{\sigma}{:}\nabla\mathbf{u}$) and ρ, k, C are the fluid density, thermal conductivity, and specific heat capacity (at constant pressure), respectively. These latter parameters are assumed to be independent of temperature. In this report we will consider only isothermal flows and ignore the energy equation. The material derivative D/Dt is defined as

$$\frac{D}{Dt} \equiv \frac{\partial}{\partial t} + \mathbf{u} \cdot \nabla \tag{5}$$

Extra equations will be required to describe $\boldsymbol{\tau}$. Newtonian (constant viscosity, inelastic), generalized Newtonian (shear-thinning, inelastic), and viscoelastic forms will be considered. Most applications of the boundary element method deal with Newtonian liquids, for which we can write

$$\boldsymbol{\tau} = \eta[\nabla\mathbf{u} + (\nabla\mathbf{u})^T] \tag{6}$$

where η is the constant viscosity. Substitution of Eq. (6) into Eq. (3) and Eq. (1) yields the well-known Navier-Stokes equations. The set of equations is completed by prescribing suitable boundary conditions.

The above equations may also be written in terms of derived variables. The vorticity is defined as

$$\boldsymbol{\omega} = \nabla \times \mathbf{u} \tag{7}$$

and on eliminating the pressure from the momentum equations we arrive at the vorticity transport equation. For Newtonian liquids in the absence of body forces this becomes:

$$\frac{\partial \boldsymbol{\omega}}{\partial t} = \nabla \times (\mathbf{u} \times \boldsymbol{\omega}) + \nu\nabla^2\boldsymbol{\omega} \tag{8}$$

where ν is the kinematic viscosity η/ρ.

In planar and axisymmetric flows it is also possible to introduce a scalar stream function, ψ, which satisfies the continuity equation identically. In planar problems this can be defined as

$$\frac{\partial \psi}{\partial y} = u_x \qquad \frac{\partial \psi}{\partial x} = -u_y \qquad (9)$$

The relationship between ψ and ω is often expressed in the form of a Poisson's equation:

$$\nabla^2 \psi = -\omega \qquad (10)$$

where ω represents the single nonzero component of the vorticity vector in planar problems. A special case of interest is that of steady creeping Newtonian flow, for which the momentum equations imply:

$$\nabla^4 \psi = 0 \qquad (11)$$

Boundary integral equation methods have been applied to the solution of all of the above equations, in both primitive variable form (velocity and pressure) and derived variable form. It is the purpose of this report to summarize the various techniques adopted.

3 LINEAR PROBLEMS AND APPLICATIONS

3.1 Direct and Indirect Formulations of B.E.M.

The boundary integral method is of course ideally suited to linear problems, which include the elastostatics problem and Stokes (or creeping) flow. The latter arises in the limit of zero Reynolds number, where inertial effects can be ignored. In Stokes flow problems, one deals with the velocity field rather than, as in elastostatics, the displacement field. However, if one sets the Poisson ratio to 0.5 in order to invoke the incompressibility constraint in elasticity, and then compares the equations for the two cases, one finds they are identical, if velocity and displacement are interchanged. Indeed, any boundary element software written for the solution of elastostatic problems can be used without modification for the solution of creeping flow problems and since the elastostatic case is more general, we briefly discuss it here, before specializing to the incompressible case.

For the elastostatic case, the equations of equilibrium are (no body and inertia forces):

$$\nabla \cdot \boldsymbol{\sigma} = \mathbf{0} \qquad (12)$$

where

$$\boldsymbol{\sigma} = \lambda \nabla \cdot \mathbf{u} \, \mathbf{1} + \eta (\nabla \mathbf{u} + \nabla \mathbf{u}^T) \qquad (13)$$

is the total stress tensor. In Eq. (13), $\mathbf{u}$ is the displacement vector, and λ, η are the Lamé moduli. For Stokes flow problems, the first term on the right-hand side of Eq. (13) becomes the indeterminate pressure, $-p\mathbf{1}$, which arises due to the incompressibility constraint, η is the viscosity of the medium and $\mathbf{u}$ is the velocity field. In the standard boundary-value problem, we wish to solve Eqs. (12)–(13) subject to one of the following boundary conditions on the surface Γ of the elastic domain Ω:

1. $\mathbf{u}$ is prescribed on Γ
2. $\mathbf{t} = \boldsymbol{\sigma} \cdot \mathbf{n}$, where $\mathbf{n}$ is the outward unit normal vector on Γ, is prescribed on Γ
3. $\mathbf{u}$ is prescribed on part of Γ and $\mathbf{t}$ on the remaining part of Γ

The reduction of Eqs. (12)–(13) to a set of boundary integral equations is fairly standard and is based on Somigliana's identity [6]. This procedure yields

$$u_i(P) = \int_\Gamma [u_{ij}^*(P, Q)t_j(Q) - t_{ij}^*(P, Q)u_j(Q)]d\Gamma \tag{14}$$

In Eq. (14), P is a point in the domain Ω, Q is a point on the surface Γ and $[u_{ij}^*(P, Q), t_{ij}^*(P, Q)]$ represents a Kelvin-state (Gurtin [9]); that is, $u_{ij}^*(P, Q)$ denotes the jth component of displacement at Q due to a unit force in the ith direction at P, and $t_{ij}^*(P, Q)$ are the associated components of the surface traction at Q. For example, in three dimensions,

$$u_{ij}^*(P, Q) = \frac{1}{16\pi\eta(1 - v)r}\left[(3 - 4v)\delta_{ij} + \frac{r_i r_j}{r^2}\right] \tag{15}$$

where $\mathbf{r}$ is the position vector of Q relative to P (r is its magnitude) and v is Poisson's ratio. When $v = 0.5$, then Eq. (15) represents a Stokeslet [10]. If P is allowed to approach the boundary Γ from inside Ω, then Eq. (14) becomes

$$c_{ij}u_j(P) + \int_\Gamma t_{ij}^*(P, Q)u_j(Q)d\Gamma = \int_\Gamma u_{ij}^*(P, Q)t_j(Q)d\Gamma \tag{16}$$

where the integrals on the left-hand side of Eq. (16) are to be interpreted as Cauchy principal-value integrals, and c_{ij} depends on the surface at P; for a smooth surface $c_{ij} = \frac{1}{2}\delta_{ij}$.

For boundary conditions of type 1, the resulting integral equations for $\mathbf{t}$ are weakly singular (Fredholm integral equations of the first kind). This comes about because $u_{ij}^*(P, Q)$ is of $O(\ln r)$ in two dimensions, and $O(1/r)$ in three dimensions, and these singularities are integrable. For boundary conditions of type 2, the resulting integral equations for $\mathbf{u}$ are singular (Fredholm integral equations of the second kind). Consequently, for mixed boundary conditions of type 3, not all of the integral equations are singular. Singular integral equations are desirable because they lead to a diagonally dominant system of linear algebraic equations which is always numerically well-conditioned. However, they demand special care during the process of numerical integration.

This boundary integral formulation deals directly with the required unknown (traction and/or displacement) and has now become known as the *direct method* of boundary element formulation. One may note that there is only a single form of the direct method, which is embodied in Eqs. (14) and (16). Once the boundary solution has been obtained, a postprocessing of Eq. (14) [or Eq. (16) with the convention that $c_{ij} = \delta_{ij}$ if $P\epsilon\Omega$] will yield the elastic field (or viscous field) in Ω.

An alternative approach to boundary integral formulation, the so-called *indirect method*, was put forward by Massonnet [11], and later by Altiero and Sikarskie [12]. In this method the region Ω of interest is embedded in another region of the same material for which the Green's function is known. A layer of "body forces," or "fictitious tractions," is then applied to the embedded body, and the strength of this layer is adjusted to produce the desired solution in Ω. If we denote the strength of the layer by s, then the elastic field throughout the medium is determined by (principle of superposition)

$$\begin{cases} u_i(P) = \int_\Gamma u_{ij}^*(Q, P)s_j(Q)d\Gamma \\ \\ \sigma_{ij}(P) = \int_\Gamma \Sigma_{ijk}^*(Q, P)s_k(Q)d\Gamma \qquad P\epsilon\Omega \end{cases} \qquad (17)$$

where $\Sigma_{ijk}^*(Q, P)$ are the associated stress components produced by $u_{ik}^*(Q, P)$. Note that $t_{ik}^*(Q, P) = \Sigma_{ijk}^*(Q, P)n_j(P)$ for P on Γ. As before, one lets P approach Γ from inside Ω to obtain

$$\begin{cases} \int_\Gamma u_{ij}^*(Q, P)s_j(Q)d\Gamma = u_i(P) \\ \\ c_{ij}s_j(P) + \int_\Gamma t_{ij}^*(Q, P)s_j(Q)d\Gamma = t_i(P) \end{cases} \qquad (18)$$

where the integrals on the left-hand side of Eq. (18) are Cauchy principal value integrals and, again, c_{ij} depends on the surface at P.

As with the direct formulation, this indirect formulation leads to a set of weakly singular integral equations for boundary conditions of type 1, and, for boundary conditions of type 2, a set of singular integral equations for the unknown strength s. Once s has been found, Eq. (17) can be used to find the elastic (or viscous) field inside Ω.

Another type of indirect formulation, which uses a distribution of dislocations (discontinuity in displacement) on the surface of the embedded region of interest, has been suggested by Louat [13] and Lardner [14]. More importantly, Maiti et al. [15] and Altiero and Gavazza [16] have shown that various indirect formulations (including the one embodied in Eq. (18)) can be directly derived from

Somigliana's identity [6]. This should dispel the notion that indirect methods have no rigorous mathematical foundation. Furthermore, Altiero and Gavazza [16] demonstrated that, with proper choice of either fictitious tractions or fictitious dislocations, one always ends up with a set of singular integral equations. A proper implementation and comparison between various indirect methods is lacking at present. One should note that the indirect method is not a recent development. Eshelby [17] has used the method to determine the elastic field of a crack under an anti-plane loading and the method is extremely popular in slender body theory in Stokes flow [18–20]; a summary is given in Chwang and Wu [10].

3.2 Stokes Flow

An important contribution to the development of the technique in fluid mechanics is the work reported by Youngren and Acrivos [7]. This appears to be the first application of the method to viscous flow problems. The work deals with three-dimensional Stokes flow past bodies of arbitrary shape and highlights an important advantage of the boundary element method over conventional finite element and finite difference methods when applied to such external flow problems. That is, the boundary conditions at infinity can be satisfied exactly, and only the boundary of the body needs to be discretized. No artificial far field boundaries or special "infinite" elements are required, and the size of the computational problem is considerably reduced. This feature had previously been exploited in external potential flow problems by the panel method developed for aerodynamic configurations [1,2].

For the problem of viscous flow past a body (or bodies) with uniform flow at infinity, Youngren and Acrivos [7] found that the general equation Eq. (16) could be simplified to yield

$$U_i = \int_\Gamma u_{ij}^*(P, Q) t_j(Q) d\Gamma \tag{19}$$

where Γ represents the body surface and U_i is the velocity at infinity. Numerical solution of Eq. (19) will yield the surface traction forces and hence the net forces acting on the body. Once the surface traction forces are known, the velocity field and stress field in the fluid domain can also be computed using a similar form to Eq. (19) and introducing additional integral equations for the pressure. Equation (19) is also valid for a body undergoing either a translational or rotational motion, or both (Tran-Cong and Phan-Thien [21]). In these cases, the left-hand side of Eq. (19) should be replaced by the relevant velocity of the body.

Finally, Youngren and Acrivos also noted that numerical treatment of Eq. (19) can be further simplified in axisymmetric problems by analytical integration in the angular direction. The integral equation then reduces to one-dimensional form. This is equivalent to utilizing an axisymmetric form of the Stokeslet; this is the flow field due to a unit load distributed evenly around a ring. The result of this integration is the introduction of elliptic integral functions [22–23]; however, the advantages of reducing the dimension of the integral equation by one overshadows any complication due to the introduction of the transcendental functions.

The use of Eq. (19) to evaluate the net forces acting on a body can be contrasted with the conventional finite element method. This yields the pressure and viscous components of the stress tensor separately; the latter resulting from differentiation of the computed velocity field. Thus, not only is the use of Eq. (19) much more convenient, since no domain discretization is required, but it is also able to yield accurate results with relatively few boundary elements. Youngren and Acrivos demonstrated this feature by computing the forces acting on a number of axisymmetric and three dimensional bodies. The solution variables were assumed to be constant on each boundary element. The element geometry was chosen to fit the geometry of the body being modeled. For the case of axisymmetric flow past spheroids with a wide range of aspect ratios, it was found that the order of 10 equally sized elements (20 simultaneous equations) yielded the drag to within 1% of the true value.

The results are reproduced in Fig. 1, where the effect of element number on accuracy can be seen clearly. The convergence of the computed drag and stress fields with the number of elements used on this and other examples are fully reported and the reader is referred to reference [7] for details. This work was later adapted to model fluid drops suspended in a liquid of a different viscosity to that of the drop [24–25].

More recently the present authors have applied Eq. (16) to the solution of a wider class of Stokes flow problems [26–29]. (See also Milthorpe and Tanner, 1980 [73].) This work again involves the use of boundary elements on which the velocity and traction are taken to be constant; however, in this case the elements

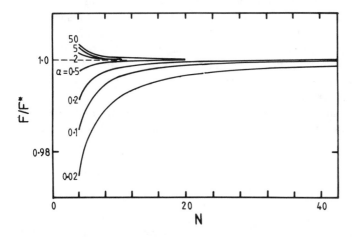

Figure 1 Ratio of numerically calculated (F) and analytic (F*) drags for spheroids $x_1^2 + (x_2^2 + x_3^2)/a^2 = 1$ in axisymmetric flow with N equally spaced points. Redrawn from Youngren and Acrivos [7], with permission of Cambridge University Press.

are geometrically linear. Much of our attention has been focused on the industrially important extrusion problem. This is the process in which a liquid is forced under pressure from a reservoir into the atmosphere through a die. The behavior of the extruded fluid (extrudate) is of interest. The idealized axisymmetric extrusion problem is shown in Fig. 2. The die is assumed to be infinitely long. In the absence of surface tension effects and surface drag, the traction forces on the surface are taken to be zero. However, the free surface is a streamline and an iterative scheme must be used to determine the free surface geometry satisfying both sets of conditions. This is achieved by first applying the traction boundary conditions to an assumed free surface shape (usually $R(z) = R_0$ throughout) and computing the velocity field on the surface. Since the assumed shape is not the required streamline, a net flux of fluid across the surface will occur. The shape of the surface is then updated accordingly and the process repeated until no change in shape is necessary. The boundary element method is a particularly economical technique for this class of problem since the update procedure requires knowledge of the boundary solution only. A domain discretization scheme is therefore wasteful.

The extrusion problem is further complicated by the presence of a stress concentration at the exit lip, where the boundary conditions change suddenly from no-slip conditions in the die to stress-free conditions on the free surface. A finer discretization of elements in the region of the exit lip must be employed to account for this behavior. Another feature of this problem that must be considered is the length of the entrance and exit regions of the computational domain (i.e., Z_{max} and Z_{min}). These must not be so short as to compromise the assumption of a long die and free jet. Experience has shown that $Z_{max} = 3R_0$ and $Z_{min} = -2R_0$ are sufficient [26]. Further lengthening does not result in any significant change to the computed solution.

The computed free surface shape for a circular die is shown in Fig. 3, together with other computational and experimental results. The finite element results were obtained using 2928 degrees of freedom. The boundary element solution required just 39 elements, or 78 degrees of freedom. Convergence was achieved after 4 iterations. We see close agreement between all sets of results. In Fig. 4 we also show the computed tensile stress on the exit plane. The stress concentration near the exit lip is clearly observed.

Another feature of interest is the effect of rounding the exit lip. In the solution presented above the exit lip was assumed to be sharp. However, this is not likely

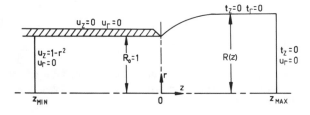

Figure 2 The axisymmetric free jet problem showing computational domain and prescribed boundary conditions.

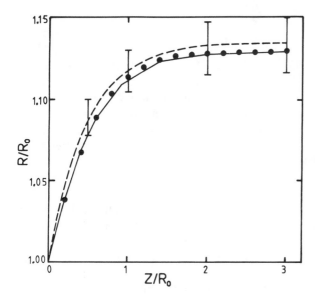

Figure 3 Comparison of jet shape computed using boundary elements (—) with finite elements (●) and the measurements (— — I — —) of Batchelor and Horsfall [30]. The experiments were conducted under conditions of negligible Reynolds number and surface tension.

to be the case in practice. Tanner et al. [28] have used the boundary element method to shed some light on this problem. The significant conclusion drawn from this work is that rounding has little effect on the results, except for the case of large lip radius to tube radius (> 0.2). This is indeed fortunate since detailed modeling of the exit lip is then not required.

Finally we have also extended this work into three dimensions. In this case the elements used were triangular planes. The technique has been employed to study the problem of extrusion from a square orifice [29].

While Eq. (16) is a convenient way to solve Stokes flow problems, it is certainly not the only formulation possible. Several authors have approached the problem by formulating solutions in terms of derived variables. In particular, the biharmonic equation for stream function in planar flows [Eq. (11)] has attracted interest.

Kelmanson [31–32] applied the boundary integral equation method to the solution of Eq. (11) by writing it in the form

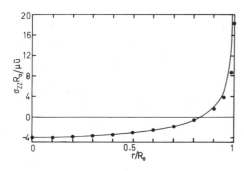

Figure 4 Dimensionless tensile stress $\sigma_{zz}R_0/\eta\bar{u}$ across jet exit plane computed using finite elements (●) and boundary elements (—). See [26].

$$\nabla^2 \psi = -\omega \tag{20}$$

$$\nabla^2 \omega = 0 \tag{21}$$

where ω is the vorticity [see Eq. (7)]. (A scheme of this form was previously outlined, without numerical results, by Fairweather et al. [74].) A biharmonic reciprocal theorem and a harmonic reciprocal theorem can then be used to give rise to a set of two coupled integral equations for ψ, ω and their derivatives along the surface normal vector. Mir-Mohamad-Sadegh and Rajagopal [33] also developed a coupled set of integral equations. However, their equations differ from those of Kelmanson since they used instead the relation

$$\psi = r^2 \zeta + \theta \tag{22}$$

where r is the distance measured from the origin of coordinates and ζ, θ are harmonic functions. Their final equations represent an indirect formulation, since the variables of interest (e.g., ψ) do not appear explicitly in the boundary integrals. Instead, the strength of the source point distribution is given by the equations; the stream function can be computed after this has been obtained. Bezine and Bonneau [34] also considered the solution of the biharmonic equation. A biharmonic reciprocal theorem was used to yield a boundary integral equation relating ψ to its normal derivative, the shear stress and the normal derivative of vorticity on the boundary. A second equation in these quantities was obtained by differentiation.

The formulations described above can be used to compute the stream function in the general flow situation and have been applied to the flow through channels containing obstacles, projections and depressions [33–34]. Kelmanson* has also applied the technique to the planar free surface jet flow problem (the planar form of the axisymmetric jet described earlier) [35]. Instead of updating the free surface in the manner described previously, Kelmanson uses an analytic approximation to the shape of the free surface, based on an expansion for the geometry in the limit of high surface tension. The stream function is specified on the free surface (together with an equation relating vorticity to tangential velocity on the surface), allowing the unknown quantities on an assumed free surface geometry to be computed. The parameters in the analytic free surface geometry equations are then adjusted iteratively until the stress boundary conditions are satisfied and further iterations result in no significant change in free surface geometry. Although this free surface update scheme involves significantly more computation than the technique described earlier, it exhibits sufficient convergence in surface tension dominated problems within only a few iterations. This is to be expected since the analytic expression adopted for the shape is based on an asymptotic form at high surface tension; only a small correction to this initial shape is then required. Reddy and Tanner [36] have incorporated surface tension effects into a finite element solution of axisymmetric jets, using a simpler update scheme. In this the extra

---

*Similar work is said to have been published by Bonneau and Bézine [75]. We have not seen this reference.

surface stresses due to surface tension are estimated for a given surface geometry and these used as boundary conditions. The velocity field is then computed and the free surface shape updated as before. In contrast to the findings of Kelmanson, this technique exhibits slow convergence at high values of surface tension.

In Fig. 5 we have reproduced Kelmanson's results for the shape of the free surface at various values of the dimensionless surface tension parameter S. These were obtained using 280 boundary elements, although comparisons using 70 and 140 elements show little variation. Also shown in Fig. 5 are our own boundary element results, for the case $S = 0$, which were generated using the velocity-traction formulation with 58 elements. (Little change in the solution was observed when the calculation was repeated using 108 elements). No analytic surface shape was assumed. All sets of results were obtained using the 'constant' elements previously described. The computed shape for $S = 0$ and Kelmanson's results at small S differ significantly, although the asymptotic swell at large x_1 agrees to within 0.85%. This difference is most likely to be due to the analytic form of the free surface shape adopted by Kelmanson.

An interesting feature of the solutions to this problem, with $S = 0$, is that most finite element solutions predict a greater degree of swelling than do the boundary element solutions. Kelmanson's results range between 17.72 and 17.87% (for $S = 0.001$ with 70 to 280 elements) while our own boundary element results predict 17.67% swelling. These can be compared with $19.0 \pm 0.2\%$ reported by Tanner [37]. This last result was obtained by taking a variety of published finite element solutions; plotting the predicted expansion ratio against the inverse of the number of degrees of freedom (D.O.F.) and then extrapolating to $1/(\text{D.O.F.}) = 0$. A similar situation is observed in the axisymmetric problem. This phenomenon may be due to the way in which each method approximates the stress singularity at the exit lip, a feature of significant importance in this problem. The matter may be resolved by analytic modeling of the singularity; an innovation already introduced into the boundary element method by Kelmanson [31,32]. However, the effects of this on the free surface geometry are yet to be explored.

Coleman [38] has also considered boundary integral formulations in terms of derived variables. However, in this case the stream function, ψ, and the Airy stress function, ϕ, are utilized. While the stream function satisfies the continuity equation exactly the stress function satisfies the momentum equations exactly; and the boundary integral equations then deal with the constitutive equations. Coleman

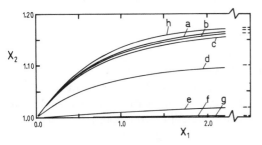

Figure 5 Free surface geometry for various values of the dimensionless surface tension, S [35]: (a) 0.001, (b) 0.01, (c) 0.1, (d) 1.0, (e) 10.0, (f) 100.0, (g) 1000.0. Also shown are our own boundary element results at $S = 0$ (h). The x_1 coordinate is measured from the exit plane and the x_2 coordinate is measured from the centerline. The die wall lies at $x_2 = 1$.

defines a complex potential $\chi = \phi + 2\eta\psi i$ where η is the fluid viscosity, and $i = \sqrt{-1}$. This leads to

$$\frac{\partial^2\chi}{\partial z^2} = 0 \tag{23}$$

where $z = x + yi$. A contour integral representation of this equation in terms of $\partial\chi/\partial z$ and $\partial\chi/\partial\bar{z}$ can then be developed. These variables are expressible in terms of velocity and force components. Thus, one may prescribe these physical quantities as boundary conditions and solve for the unknown values. This formulation was applied to the classic "stick-slip" problem; a problem similar to the free jet problem except that the section of boundary downstream of the exit lip is not "free" but constrained to a fixed geometry. Coleman has also extended the novel contour integral formulation to handle non-Newtonian liquids [39].

The foregoing discussion indicates that there are a wide variety of boundary integral formulations suitable for Stokes flow problems, using either velocity-traction (primitive) or derived variables. It would not be appropriate to select one approach as superior to the others, since each has its own advantages. If the stream function is specifically required then the derived variable approach is advantageous. If the solution is obtained in terms of primitive variables then the stream function must be computed by solution of Eq. (10). The presence of the vorticity in this equation implies the existence of a domain integral in the final integral formulation. Furthermore, this integral involves velocity gradients; however these may be removed by integration by parts to yield a new domain integral involving velocity only.

On the other hand for a problem in which boundary forces are of primary interest, in computation of the net drag and lift forces on a body, for example, then the primitive variable formulation may be more suitable. This approach is also equally applicable to planar and three-dimensional problems, whereas the derived variable approach (involving the stream function) is limited to planar or axisymmetric flows. To our knowledge there has been no attempt to apply derived variable boundary integral formulations to axisymmetric problems, although this should be possible using the Stokes stream function [8].

Finally, there also appear to be no applications of the method to time-dependent Stokes flows, using either formulation. A variety of techniques have been developed for boundary integral modeling of elastodynamic and transient thermal problems [4] and these techniques should be applicable to Stokes flows.

3.3 Oseen Flow

Youngren and Acrivos [7] demonstrated the efficiency with which the problem of creeping motion of a body in an infinite expanse of fluid can be treated using boundary integral equation methods. The effects of inertia can be included by formulating a solution to the complete set of momentum equations [Eq. (1)]. However, this will require the presence of a domain integral and use of an iterative scheme to cope with the nonlinearity so introduced. In some cases, particularly at low Reynolds numbers, it may be suitable to consider an Oseen linearization of Eq. (1). All the advantages of a boundary element solution to a linear problem can then be realized.

Oseen's approximation results if we replace the nonlinear term, $\mathbf{u} \cdot \nabla \mathbf{u}$, with

$$\mathbf{u} \cdot \nabla \mathbf{u} \rightarrow U \frac{\partial \mathbf{u}}{\partial x_1} \tag{24}$$

where U is the velocity of the fluid at infinity, which is assumed to be in the x_1 direction. Far from the body this approximation is valid, but close to the body it is clearly invalid. However, close to the body the viscous effects dominate the inertia effects and the inaccuracy has less consequence. The resulting set of linear differential equations, the Oseen equations, can be converted to a set of boundary integral equations using the fundamental singular solution, the "Oseenlet" [40]. The final integral equation, for the case of steady flow past a rigid particle in the absence of body forces, takes the form [41]:

$$U\delta_{1i} = \int_\Gamma u_{ij}^0(P, Q)t_j(Q)d\Gamma \tag{25}$$

where δ_{1i} is the Kronecker delta, $u_{ij}^0(P, Q)$ represents the Oseenlet and the remaining terms have their previous meanings. Note that Eq. (25) is the direct analogy of Eq. (19) and retains all the advantages of such a representation.

Equation (25) has been applied to the solution of a variety of planar Oseen flow examples [41]; two numerical solutions are reproduced below. No attempt has yet been made to model axisymmetric or three-dimensional Oseen flows. Once again constant boundary elements have been used throughout. The first example is the familiar case of flow past a circular cylinder of infinite length, with the flow direction perpendicular to the cylinder axis. The problem was solved using equal length elements distributed around the circumference of the circular cross section. The drag force was computed using 20, 30, 40 and 50 elements for Reynolds numbers (based on cylinder diameter) in the range 0.4 to 10.0. It was found that the solution varied by less than 0.55% between the simulations based on 20 elements and 50 elements. The computed drag force is shown in Fig. 6, in addition to other experimental, analytical and numerical results. We see that there is close agreement between the present results and other solutions of Oseen's equations (where these are valid), and that the Oseen drag is a reasonably good first approximation to the true drag. Kaplun's asymptotic expansion for the complete momentum equations of course yields a better approximation, but only in the range Re < 1.

The second example is the case of two parallel cylinders shown in Fig. 7. The problem was solved in the range $0.1 < \mathrm{Re} < 1$, using 20 elements on each cylinder. In Figs. 8 and 9, the computed drag and lift forces on the cylinder C_1 are compared with the asymptotic expansions of Fujikawa [46]. Close agreement can be seen, except in the range Re > 0.7, where the expansion formulas (correct

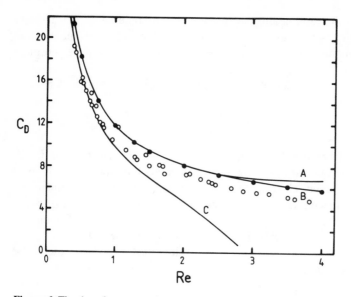

Figure 6 The drag force on a circular cylinder: A, Tomotika and Aoi [42] (asymptotic expansion); B, Tomotika and Aoi [43] (numerical); •, boundary elements; o, Tritton [44] (experimental); C, Kaplun [44] (asymptotic expansion). (After Bush [41].) A and B are for Oseen flow.

to the order of Re^{-1}) become invalid. The boundary element solutions in this range agree with the solutions reported by Yano and Kieda [47]. These authors employed a form of the classical singularity method, in which the singularities are distributed throughout the bodies and the strengths chosen to approximately satisfy the no-slip conditions on the surface.

3.4 Half-Space Problems

Half-space problems, in which the domain of interest is the semi-infinite region, are among those that can be conveniently solved by the B.E.M. technique. Here the relevant singularity solution is either the Mindlin kernel [48] in which the

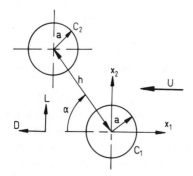

Figure 7 Flow past two equal circular cylinders C_1 and C_2 of radius a, separated by a distance h.

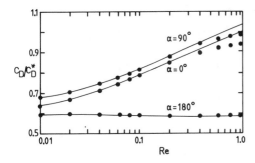

Figure 8 Drag coefficient for cylinder C_1 in Oseen flow with $h/a = 20$. C_D^* is the coefficient for a single circular cylinder: (—), Fujikawa [46]; ●, boundary elements. See [41].

boundary of the half-space is traction free, or the image system of the Kelvin state [49–53] in which the boundary of the half-space is displacement free. Numerical implementations of these kernels have been considered by Telles and Brebbia [54], for the former case, and by Tran-Cong and Phan-Thien [21], for the latter case. Tested cases include a distribution of traction over a finite part of the boundary of the half space, the rigid flat punch indented into the half space, a circular hole near a straight boundary which is under symmetric loading [54], and a sphere undergoing a translational and rotational motion near a rigid plane surface [21]. These latter problems are fully three-dimensional. In Figs. 10 and 11 the computed drag forces, $X = \mathrm{Drag}/(6\pi\eta Ua)$, for a sphere of radius a translating with a velocity U normal and parallel to the plane wall of the semi-infinite domain are compared with the exact solution of Lee and Leal [55]. In all cases the standard boundary element formulation, Eq. (16), was found to be inferior to the simplified theorem (19), especially when the sphere is very near to the plane wall. At $d/a = 1.1$, where d is the distance from the sphere center to the plane wall, and for a sphere translating normal to the plane wall, the standard boundary element formulation underestimates the exact result by a factor of 4.8 whereas the simplified formulation leads to an error of only 3.3%. In both cases, constant elements and the same mesh discretization were employed (a total of 84 elements and 3 sym-

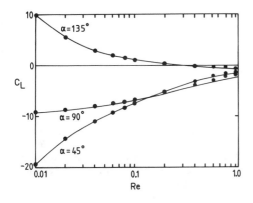

Figure 9 Lift coefficient for cylinder C_1 in Oseen flow with $h/a = 20$. (—), Fujikawa [46]; ●, boundary elements. See [41].

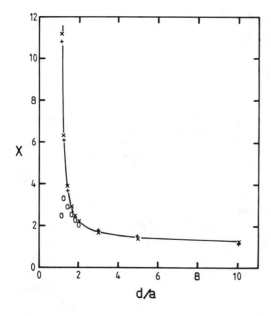

Figure 10 Dimensionless drag on a sphere translating normal to a rigid semi-infinite plane. The computations employed 84 boundary elements: +, two planes of symmetry, Eq. (19); ×, three planes of symmetry, Eq. (19); O, three planes of symmetry, Eq. (16); —, analytical results [55]. See [21].

metry planes). The poor result of the standard boundary element method is mainly due to the inaccuracy in evaluating the integral of $t_{ij}^*(P, Q)$, especially when the sphere is very near to the plane wall; in this region the stress field varies rapidly. The use of higher-order elements may improve the observed performance of the method.

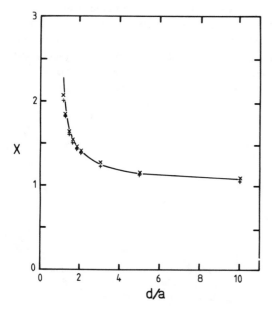

Figure 11 Dimensionless drag on a sphere translating parallel to a rigid semi-infinite plane: +, 108 elements, one symmetry plane; ×, 216 elements, one symmetry plane; —, analytical results [55]. See [21].

4 NONLINEAR PROBLEMS AND APPLICATIONS

4.1 Navier-Stokes Equations

While Oseen's approximation may be used to provide an estimate of the net forces acting on a body as previously described, the approximation is only valid in this flow situation. For more general flow geometry we must consider the complete set of nonlinear momentum equations. The nonlinear nature of these equations necessitates an iterative solution technique. However, of more importance is the appearance of domain integrals to account for the nonlinear terms. The presence of the domain integral implies a domain discretization, however the purpose of this discretization is to numerically approximate the domain integrals and it does not result in an increase in the number of equations to be solved simultaneously at each iteration; this still only depends on the boundary discretization employed.

The velocity-traction formulation has been extended to include inertia terms by Brebbia and Wrobel [56] and Bush and Tanner [26]. The first step is to regard the nonlinear inertia terms in Eq. (1) as "pseudo-body forces" and rewrite the momentum equation as

$$\mathbf{\nabla} \cdot \mathbf{\sigma} + \rho\mathbf{b} = 0 \tag{26}$$

where $\mathbf{b} = \mathbf{f} - \mathbf{u} \cdot \mathbf{\nabla}\mathbf{u}$ for steady state problems (we will restrict attention here to steady problems). The integral equation representation can then be obtained by forming a weighted residual statement including Eq. (26), the continuity equation and the boundary conditions:

$$\int_\Omega \left(\frac{\partial\sigma_{ij}}{\partial x_j} + \rho b_i \right) u_i^* d\Omega + \int_\Omega \frac{\partial u_j}{\partial x_j} p^* d\Omega = \int_{\Gamma_2} (t_j - \bar{t_j}) u_j^* d\Gamma - \int_{\Gamma_1} (u_j - \bar{u_j}) t_j^* d\Gamma \tag{27}$$

where $\bar{t_j}$ are traction boundary conditions specified on the section of boundary represented by Γ_2 and $\bar{u_j}$ are velocity boundary conditions specified on Γ_1. Mixed conditions representing combinations of velocity and traction may exist over another portion of the boundary. However, for simplicity we can include this portion in Γ_1 and Γ_2 simultaneously and write $\Gamma = \Gamma_1 + \Gamma_2$. The terms u_j^*, p^*, and t_j^* are weighting functions. Since Eq. (27) will vanish for any solution satisfying Eq. (26), the continuity equation and the boundary conditions, it must be true for arbitrary (*) fields when the unstarred field is the true solution to the problem in hand.

If the term involving σ_{ij} is integrated by parts twice and the (*) field identified with the Stokeslet (i.e., u_j^*, p^*, and t_j^* represent the velocity, pressure and traction fields due to a unit force applied at a point in an infinite medium) then it is possible to show that [26,56]:

$$c_{ij}u_j(P) = \int_\Gamma t_j(Q)u_{ij}^*(P, Q)d\Gamma - \int_\Gamma u_j(Q)t_{ij}^*(P, Q)d\Gamma + \rho\int_\Omega b_j(Q)u_{ij}^*(P, Q)d\Omega \tag{28}$$

where all symbols have their previous meanings and it is implied that $t_j(Q) = \bar{t_j}$

on Γ_2 and $u_j(Q) = \overline{u}_j$ on Γ_1. Equation (28) is a simple extension of Eq. (16) to include the nonlinear convective terms. An iterative scheme will be required to find its solution.

The numerical solution to Eq. (28) is obtained by dividing the boundary into boundary elements as before and dividing the domain into "domain cells" to allow for numerical approximation of the domain integrals. The resulting set of simultaneous equations will take the form:

$$\mathbf{Ax} = \mathbf{c} + \mathbf{d} \tag{29}$$

where $\mathbf{x}$ is the solution vector (vector of unknown boundary values of velocity and traction), $\mathbf{c}$ is a vector incorporating the prescribed boundary conditions, and $\mathbf{d}$ represents the domain integral. Iterative solution will proceed by first setting $\mathbf{d} = \mathbf{0}$ and evaluating the Stokes flow solution in the usual way. Equation (28) is then reapplied to compute the corresponding velocity distribution in the domain, allowing an estimate of $\mathbf{d}$ to be obtained. Equation (29) is then solved again and the process repeated until satisfactory convergence has been achieved. Note that during this process the coefficient matrix does not change (unless the geometry also changes, as in a free surface problem) and does not need to be assembled and eliminated each time, if a record of the original elimination process is retained.

The iterative process described above effectively applies a correction to the Stokes flow solution. The process is not expected to converge at high Reynolds numbers, since instability will occur once the inertia terms begin to dominate the solution; the method is only suitable for low Reynolds number solutions [0(10)] [26]. This technique has been applied to the problem of flow in a wedge and flow past a sphere [26]. As a further example we include here the problem of an axisymmetric free jet for Reynolds numbers up to 4.2 (based on tube diameter), allowing a comparison to be made with experimental data. The discretization employed is shown in Fig. 12. Once again constant boundary elements are used and a linear variation of velocity is assumed on each triangular domain cell. The free surface was adjusted during each iterative cycle; five iterations were required at Re = 4.2. The computed free jet shapes are shown in Fig. 13, together with other numerical and experimental results. Close agreement can be seen.

Solutions to the Navier-Stokes equations for general flow have also been formulated in terms of derived variables. This formulation has been employed for some time by Wu and co-workers [59] and applied in particular to the problem

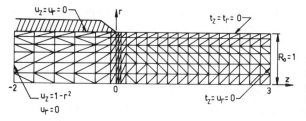

Figure 12 Grid pattern and boundary conditions for the axisymmetric free jet problem; 41 boundary elements, 276 domain cells.

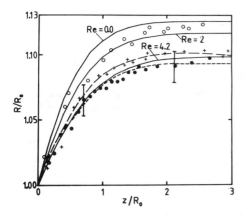

Figure 13 The axisymmetric jet profiles: —, boundary element results; – –, finite element results [57]; ●, experimental results [57] with typical error bars (Re = 4.1); – – –, experimental results [58]; experimental results (+, Re = 4.3; ○, Re = 1.5) of Mr. C. W. Butler, private communication (1986).

of flow past aerofoils and other obstacles, a class of problems to which the method is particularly well suited. In contrast to the previous formulation, solutions at high Reynolds numbers are possible. Derived variable formulations have also been applied to internal flow problems [60–63].

The formulation concentrates on Eqs. (7)–(8) together with the continuity equation, Eq. (2). When boundary conditions are included, this set of equations fully describes the flow problem. Equations (2) and (7) describe the kinematics of the problem, in that they relate velocity to vorticity, and Eq. (8) describes the kinetics of the flow. Note that these equations apply to general three-dimensional problems. In this case the vorticity can be written in tensor notation as

$$\omega_i = \epsilon_{ijk} \frac{\partial u_k}{\partial x_j} \tag{30}$$

where ϵ_{ijk} is the alternating tensor.

The integral equation representation of the kinematic part of the problem can be written as [4,59]:

$$u_i(P) = \int_\Omega \epsilon_{ijk}\omega_j(Q) \frac{\partial \phi^*}{\partial x_k} (P, Q)d\Omega - \int_\Gamma u_j(Q)n_j(Q) \frac{\partial \phi^*}{\partial x_i} (P, Q)d\Gamma$$

$$+ \int_\Gamma \epsilon_{ijk}\epsilon_{jmn}u_m(Q)n_n(Q) \frac{\partial \phi^*}{\partial x_k} (P, Q)d\Gamma \tag{31}$$

where ϕ^* is the fundamental potential solution; in three dimensions this is $\phi^* = 1/(4\pi r)$, where r is the distance between P and Q. Equation (31) applies if P lies in the domain Ω, and allows computation of the velocity at any point P in Ω if the boundary values of velocity and the distribution of vorticity in Ω are known. The most important aspect of Eq. (31) is that the domain integral is zero in regions of flow that are vorticity free. This is of greatest consequence in external flow problems, in which case it is possible to show that Eq. (31) reduces to

$$u_i(P) = \int_\Omega \epsilon_{ijk}\omega_j(Q) \frac{\partial \phi^*}{\partial x_k}(P, Q)d\Omega + U_i \tag{32}$$

where U_i is the velocity at infinity. In such an external flow problem the vorticity is zero far from the body, and the domain integral in Eq. (32) therefore implies integration over a limited region surrounding the body. Note however that the far field boundary conditions are automatically incorporated and no fictitious far field boundary is implied.

The integral representation of the vorticity transport equation takes the form:

$$\omega_i(P, t_f) = \int_\Omega \omega_i(Q, t_0)\Phi^*(P, Q, t_f, t_0)d\Omega$$

$$+ \int_{t_0}^{t_f} dt \int_\Omega \left(\omega_j(Q, t) \frac{\partial u_i}{\partial x_j}(Q, t) \right.$$

$$\left. - u_j(Q, t) \frac{\partial \omega_i}{\partial x_j}(Q, t) \right) \Phi^*(P, Q, t_f, t_0)d\Omega \, dt$$

$$- v \int_{t_0}^{t_f} dt \int_\Gamma \left(\omega_i(Q, t) \frac{\partial \Phi^*}{\partial x_j}(P, Q, t_f, t) \right.$$

$$\left. - \Phi^*(P, Q, t_f, t) \frac{\partial \omega_i}{\partial x_j}(Q, t) \right) n_j(Q)d\Gamma \, dt \tag{33}$$

where $t = t_0$ is the initial time, $t = t_f$ is the final time and $\Phi^*(P, Q, t_f, t)$ is the time dependent fundamental solution of the diffusion equation (see for example [4]). Note that the first integral in Eq. (33) represents the initial conditions in the problem. The second integral may be further integrated by parts to remove the velocity and vorticity gradients [59], and this can be further specialized for the case of external flow past a stationary body.

Equations (31) and (33) are the complete integral representation of the problem. If steady state solutions are required then the problem can be stepped through time until transients have died out. Alternatively, Eq. (33) can be replaced by an equivalent time independent integral equation [4]. Instead of the integral formulation of the kinetics, the vorticity transport equation has also been treated using conventional finite element on finite difference techniques.

The numerical solution is obtained using Eqs. (31) and (33) [or the finite difference or finite element equivalents of Eq. (33)] by evaluating the vorticity distribution at the end of a given time step and evaluating the corresponding velocity distribution, before proceeding to the next time step. As previously indicated only the velocity and vorticity distribution within the viscous region of flow need be considered. This implies that the domain discretization must be adapted during the solution process to account for the changing vorticity distribution. A number of innovations have been introduced by Wu and co-workers, including the treatment of compressible and turbulent flow. The reader is referred to reference [59] for details.

The power of this approach is illustrated in Figs. 14 and 15. The first figure shows the steady state pressure distribution on a circular cylinder at a Reynolds number of 40, based on diameter (from Wu and Rizk [64]). The second figure shows the drag coefficient as a function of Reynolds number (from Wu [59]). The close agreement between computed and measured results is encouraging.

Although the technique is highly suited to external flow problems, it is certainly not limited to this class of problem and several authors have used the above formulation, and alternative derived variable formulations, to successfully solve a variety of internal flow problems [60–64].

4.2 Non-Newtonian Problems: Generalized Newtonian and Viscoelastic Liquids

Boundary element techniques have recently been applied to non-Newtonian fluid mechanics problems; inelastic and elastic non-Newtonian behavior have been examined [67–69]. In this section we outline the general integral formulation for such fluids and demonstrate its application to the generalized Newtonian fluid.

The set of equations governing the motion of non-Newtonian fluids is given by Eqs. (1)–(4), in addition to constitutive equations relating the extra stress τ to the strain field. These constitutive equations may simply relate the fluid viscosity to the instantaneous rate-of-strain (the generalized Newtonian fluid) or they may consist of a set of differential or integral equations relating the stress field to the strain history of the fluid (the elastic, or memory fluid). In either case we can develop an integral equation representation of the momentum and continuity equations by first breaking the extra stress tensor τ into Newtonian and non-Newtonian components in the form:

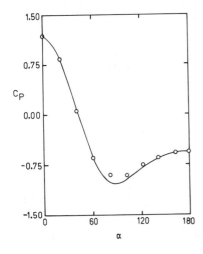

Figure 14 Pressure distribution C_p on a circular cylinder as a function of the angle α measured from the stagnation point: —, boundary elements; O, experimental data [65]. Redrawn from Wu and Rizk [64], with permission from Springer-Verlag, Berlin.

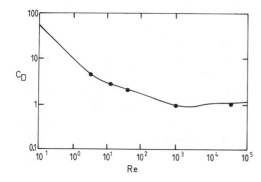

Figure 15 Drag coefficient for a circular cylinder: ●, computed results; —, experimental data. Redrawn from Wu [59], with permission from Applied Science Publishers, London.

$$\boldsymbol{\sigma} = -p\mathbf{1} + 2\eta_N \mathbf{D} + \boldsymbol{\epsilon} \qquad (34)$$

where η_N is an arbitrary constant Newtonian viscosity, $\mathbf{D}$ is the rate-of-strain tensor ($\mathbf{D} = \frac{1}{2}[\nabla\mathbf{u} + (\nabla\mathbf{u})^T]$) and $\boldsymbol{\epsilon}$ is the non-Newtonian component of the extra stress. When this is substituted into Eq. (1) we obtain

$$\nabla \cdot \boldsymbol{\sigma}_N + \nabla \cdot \boldsymbol{\epsilon} - \rho \frac{D\mathbf{u}}{Dt} + \rho\mathbf{f} = \mathbf{0} \qquad (35)$$

where $\boldsymbol{\sigma}_N$ represents the Newtonian component of the total stress tensor: $\boldsymbol{\sigma}_N = -p\mathbf{1} + 2\eta_N\mathbf{D}$. We will restrict discussion here to steady problems. In this case we may then combine the non-Newtonian components, the inertia terms and $\mathbf{f}$ to form a pseudo-body force, $(\rho\mathbf{b} + \nabla \cdot \boldsymbol{\epsilon})$ where, as before, $\mathbf{b} = \mathbf{f} - \mathbf{u} \cdot \nabla\mathbf{u}$:

$$\nabla \cdot \boldsymbol{\sigma}_N + (\rho\mathbf{b} + \nabla \cdot \boldsymbol{\epsilon}) = \mathbf{0} \qquad (36)$$

The development of an integral equation then follows the process outlined for the treatment of Eq. (26). However, in this case we are manipulating only the Newtonian component of the total stress tensor and the boundary traction terms that are generated in the process represent the Newtonian component of the total traction force. That is, we generate $\mathbf{t}_N = \boldsymbol{\sigma}_N \cdot \mathbf{n}$ where

$$\mathbf{t} = \boldsymbol{\sigma} \cdot \mathbf{n} = \mathbf{t}_N + \boldsymbol{\epsilon} \cdot \mathbf{n} \qquad (37)$$

We can recover the total traction by integrating the domain integral containing $\nabla \cdot \boldsymbol{\epsilon}$ by parts. The resulting equation can be written as

$$c_{ij}u_j(P) = \int_\Gamma t_j(Q)u_{ij}^*(P, Q)d\Gamma$$
$$- \int_\Gamma u_j(Q)t_{ij}^*(P, Q)d\Gamma$$
$$+ \int_\Omega \left(\rho b_j(Q)u_{ij}^*(P, Q) - \epsilon_{jk}(Q)\frac{\partial u_{ij}^*}{\partial x_k} \right)d\Omega \qquad (38)$$

In Eq. (38) it is implied that the Stokeslet acts in an infinite medium of viscosity

η_N. When combined with the additional equations describing ϵ this can be used to solve general steady non-Newtonian fluid mechanics problems. The starting point for iteration is generally the Newtonian stress and velocity field. This equation has been used to successfully solve a number of elastic and inelastic fluid flow problems [67–69].

The generalized Newtonian constitutive equation can be written in the form

$$\sigma = -p\mathbf{1} + 2\eta(\dot{\gamma})\mathbf{D} \tag{39}$$

where $\dot{\gamma} = \sqrt{(2D_{ij}D_{ij})}$ is the "shear rate" and $\eta(\dot{\gamma})$ is the shear-rate dependent viscosity. This model is not capable of representing memory effects, but merely expresses the viscosity as a function of the instantaneous shear rate. Equation (39) may be rewritten in the form:

$$\sigma = -p\mathbf{1} + 2\eta_N\mathbf{D} + 2[\eta(\dot{\gamma}) - \eta_N]\mathbf{D} \tag{40}$$

which is the form required for the integral equation solution. A variety of different models for $\eta(\dot{\gamma})$ have been proposed [70]; the one used in this work is the Carreau model:

$$\eta(\dot{\gamma}) = \eta_\infty + (\eta_0 - \eta_\infty)[1 + (\lambda\dot{\gamma})^2]^{(n-1)/2} \tag{41}$$

where η_0 and η_∞ are the zero-shear rate and infinite-shear rate asymptotes, λ is a time constant and n is the power-law index. This model accurately represents shear-thinning behavior, in which η is a decreasing function of $\dot{\gamma}$.

Although the choice of η_N in Eq. (40) is arbitrary, experience with simple test problems indicates that the best choice for η is the minimum viscosity in the problem. This ensures that the magnitude of the difference $(\eta(\dot{\gamma}) - \eta_N)$ is small in regions where $\dot{\gamma}$ is large. The size of the product $(\eta(\dot{\gamma}) - \eta_N)D_{jk}$ is then limited and errors due to numerical approximation in the domain are minimized.

We illustrate the technique by reproducing results obtained for the drag force acting on a sphere in creeping motion through an infinite expanse of fluid. An equation similar to Eq. (38) also applies to such axisymmetric problems, however an additional term due to the axisymmetric hoop stress is also included. See Bush and Phan-Thien [67] for details. The results are illustrated in Fig. 16, where the dimensionless drag force $X = \text{DRAG}/(6\pi\eta_0 Ua)$ (U = sphere velocity, a = radius) is plotted against dimensionless velocity (or Weissenberg number) $\Lambda = \lambda U/a$ for two test fluids. The numerical technique has accurately predicted the drag reduction observed in shear-thinning fluids. The upper and lower bounds based on simple power-law theory [72] are valid at high values of Λ and provide extra validation of the computed results.

The method has also been applied to highly elastic liquids [68–69]. In this case we must consider an additional set of differential or integral constitutive equations which describe the extra-stress tensor τ in a viscoelastic fluid [39,69]. The method has been applied with considerable success to the extrusion problem discussed above. In this problem the difficulty in securing convergence with other

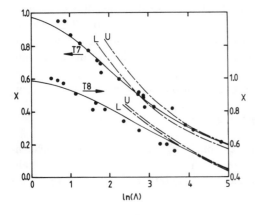

Figure 16 Dimensionless drag, x, on a sphere in two different Carreau model fluids (T7,T8): —, boundary element results; •, experiments [71]; U, L, upper and lower bounds based on simple power-law theory [72]. See reference [67].

methods (F.E.M., F.D.) is high. Using the boundary element method coupled with a differential constitutive model the swelling of extrudates up to 1.8 times the die diameter have been demonstrated. For details of the method of calculation see [68,69].

5 CONCLUSION

We have reviewed the applicability of the boundary element method to various viscous flow problems and one sees that it is well matched to certain classes of problems. For the solution of linear problems, particularly external flow problems and free surface problems, the method is outstanding. For the Navier-Stokes equations the method of Wu appears to offer the possibility of convergence at high Reynolds numbers; the simpler Stokes iteration schemes are confined to medium Reynolds numbers. The method also performs quite well for the generalized Newtonian fluid and outstandingly well in the nonlinear viscoelastic extrusion problem. (The reason for the latter success is not yet completely understood). We therefore conclude that for the class of problems discussed above, and probably for others not yet explored, the boundary element method is an excellent alternative to other computational methods.

ACKNOWLEDGMENTS

This work was supported by the Australian Research Grants Scheme at The University of Sydney and at The University of Western Australia. This support is gratefully acknowledged. Some of this research was carried out using low-cost computer time supplied by the Western Australian Regional Computing Centre.

REFERENCES

1. J. L. Hess and A. M. O. Smith, Calculation of Potential Flow about Arbitrary Bodies, in *Progress in Aero. Sciences,* Vol. 8, ed. D. Kuchemann, Pergamon, Oxford, 1967.
2. B. Hunt, The mathematical basis and numerical principles of the boundary integral method for incompressible potential flow over 3-D aerodynamic configurations, in *Numerical Methods in Applied Fluid Dynamics,* B. Hunt (ed.), Academic Press, U.S.A., 1980.
3. M. A. Jaswon and G. T. Symm, *Integral Equation Methods in Potential Theory and Elastostatics,* Academic Press, London, 1977.
4. C. A. Brebbia, J. C. F. Telles and L. C. Wrobel, *Boundary Element Techniques—Theory and Applications in Engineering,* Springer-Verlag, Berlin, 1984.
5. F. J. Rizzo, An integral equation approach to boundary value problems in classical elastostatics, *Q. J. Appl. Math.,* 25 (1967) 83–95.
6. T. A. Cruse, Numerical solutions in three dimensional elastostatics, *Int. J. Solids Struct.,* 5 (1969) 1295–1274.
7. G. K. Youngren and A. Acrivos, Stokes flow past a particle of arbitrary shape: a numerical method of solution, *J. Fluid Mech.,* 69 (1975) 377–403.
8. G. K. Batchelor, *An Introduction to Fluid Dynamics,* Cambridge University Press, Cambridge, 1967.
9. M. E. Gurtin, *Handbuch der Physik,* Band VIa/2, ed. C. Truesdell, Springer-Verlag, Berlin, 1972.
10. A. T. Chwang and T. Y. Wu, Hydromechanics of low-Reynolds-number flow. Part 2. Singularity method for Stokes flows, *J. Fluid Mech.,* 67 (1975) 787–815.
11. C. E. Massonnet, Numerical use of integral procedures, in *Stress Analysis: Recent Developments in Numerical and Experimental Methods,* ed. O. C. Zeinkiewicz and G. S. Holister, Wiley, London, 1965.
12. N. J. Altiero and D. L. Sikarskie, An integral equation method applied to penetration problems in rock mechanics, in *Boundary-Integral Equation Method: Computational Applications in Applied Mechanics,* ed. T. A. Cruse and F. J. Rizzo, *ASME AMD,* Vol. 11, New York, 1975.
13. N. Louat, Solution of boundary value problems in plane strain, *Nature,* 196 (1962) 1081–1082.
14. R. W. Lardner, Dislocation layers and boundary value problems of plane elasticity, *Quart. J. Mech. Appl. Maths,* 25 (1972) 45–61.
15. M. Maiti, B. Das and S. S. Palit, Somigliana's method applied to plane problems of elastic half-spaces, *J. Elasticity,* 6 (1976) 429–439.
16. N. J. Altiero and S. D. Gavazza, On a unified boundary-integral equation method, *J. Elasticity,* 10 (1980), 1–9.
17. J. D. Eshelby, The elastic field of a crack extending non-uniformly under general anti-plane loading, *J. Mech. Phys. Solids,* 17 (1969) 177–199.
18. J. R. Blake, Singularities of viscous flow. Part II: Applications to slender body theory, *J. Eng. Mathematics,* 8 (1974) 113–124.
19. N. J. De Mestre and D. F. Katz, Stokes flow about a sphere attached to a slender body, *J. Fluid Mech.,* 64 (1974) 817–826.
20. D. F. Katz, J. R. Blake and S. L. Paveri-Fontana, On the movement of slender bodies near plane boundaries at low Reynolds number, *J. Fluid Mech.,* 72 (1975) 529–540.
21. T. Tran-Cong and N. Phan-Thien, Boundary element solution for half-space elasticity of Stokes problems with a no-slip boundary, *J. Comp. Mech.,* 1(1986) 259–268.
22. T. A. Cruse, D. W. Snow and R. B. Wilson, Numerical solutions in axisymmetric elasticity, *Comp. Struct.,* 7 (1977) 445–451.
23. T. Kermanidis, A numerical solution for axially symmetrical elasticity problems, *Int. J. Solids Struct.,* 11 (1975) 493–500.
24. G. K. Youngren and A. Acrivos, On the shape of a gas bubble in a viscous extensional flow, *J. Fluid Mech.,* 76 (1976) 433–442.

25. J. M. Rallison and A. Acrivos, A numerical study of the deformation and burst of a viscous drop in an extensional flow, *J. Fluid Mech.*, 89 (1978) 191–200.

26. M. B. Bush and R. I. Tanner, Numerical solution of viscous flows using integral equation methods, *Int. J. Numerical Methods in Fluids*, 3 (1983) 71–92.

27. N. Phan-Thien, C. J. Goh and M. B. Bush, Viscous flow through corrugated tube by boundary element method, *J. Applied Math. Physics (ZAMP)*, 36 (1985) 475–480.

28. R. I. Tanner, H. Lam and M. B. Bush, The separation of viscous jets, *Phys. Fluids*, 28 (1985) 23–25.

29. M. B. Bush and N. Phan-Thien, Three dimensional viscous flows with a free surface: flow out of a long square die, *J. Non-Newtonian Fluid Mech.*, 18 (1985) 211–218.

30. J. Batchelor and F. Horsfall, *Die Swell in Elastic and Viscous Fluids*, Rubber and Plastics Research Association of Great Britain Report 189, 1971.

31. M. A. Kelmanson, An integral equation method for the solution of singular slow flow problems, *J. Comput. Physics*, 51 (1983) 139–158.

32. M. A. Kelmanson, Modified integral equation solution of viscous flows near sharp corners, *Computers Fluids*, 11 (1983) 307–324.

33. A. Mir-Mohamad-Sadegh and K. R. Rajagopal, The flow of a non-Newtonian fluid past projections and depressions, *J. Appl. Mech.*, 47 (1980) 485–488.

34. G. Bezine and D. Bonneau, Integral equation method for the study of two dimensional Stokes flows, *Acta Mechanica*, 41 (1981) 197–209.

35. M. A. Kelmanson, Boundary integral equation solution of viscous flows with a free surface, *J. Eng. Mathematics*, 17 (1983) 329–343.

36. K. R. Reddy and R. I. Tanner, Finite element solution of viscous jet flows with surface tension, *Computers Fluids*, 6 (1978) 83–91.

37. R. I. Tanner, *Engineering Rheology*, Clarendon Press, Oxford, 1985.

38. C. J. Coleman, A contour integral formulation of plane creeping Newtonian flow, *Q. J. Mech. Applied Math.*, 34 (1981) 453–464.

39. C. J. Coleman, On the use of boundary integral methods in the analysis of non-Newtonian fluid flow, *J. Non-Newtonian Fluid Mech.*, 16 (1984) 347–355.

40. L. Rosenhead, *Laminar Boundary Layers*, Oxford University Press, London, 1963.

41. M. B. Bush, Modelling two dimensional flow past arbitrary cylindrical bodies using boundary element formulations, *Applied Math. Modelling*, 7 (1983) 386–394.

42. S. Tomotika and T. Aoi, An expansion formula for the drag on a circular cylinder moving through a viscous fluid at small Reynolds numbers, *Q. J. Mech. Applied Math.*, 4 (1951) 401–406.

43. S. Tomotika and T. Aoi, The steady flow of a viscous fluid past a sphere and a circular cylinder at small Reynolds numbers, *Q. J. Mech. Applied Math.*, 3 (1950) 140–161.

44. D. J. Tritton, Experiments on the flow past a circular cylinder at low Reynolds numbers, *J. Fluid Mech.*, 6 (1959) 547–567.

45. M. Van Dyke, *Perturbation Methods in Fluid Mechanics*, Academic Press, London, 1964.

46. H. Fujikawa, The forces acting on two circular cylinders of arbitrary radii in a uniform stream at low values of Reynolds numbers, *J. Phys. Soc. Japan*, 11 (1956) 690–701.

47. H. Yano and A. Kieda, An approximate method for solving two-dimensional low-Reynolds-number flow past arbitrary cylindrical bodies, *J. Fluid Mech.*, 97 (1980) 157–179.

48. R. D. Mindlin, Force at a point in the interior of a semi-infinite solid, *Physics*, 7 (1936) 195–202.

49. H. A. Lorentz, Ein Allgemeiner Satz, die Bewegung Einer Reibenden Flüssigkeit Betreffend, Nebst Einigen Anwendungen Desselben, *Abh. Theor. Phys. Leipzig*, 1 (1907) 23.

50. J. Blake, A note on the image system for a Stokeslet in a no-slip boundary, *Proc. Camb. Phil. Soc.*, 70 (1971) 303–310.

51. U. B. C. O. Ejike, Boundary effects due to body forces and body couples in the interior of a semi-infinite elastic solid, *Int. J. Eng. Sci.*, 8 (1970) 909–924.

52. H. Hasimoto and O. Sano, Stokeslets and eddies in creeping flow, *Ann. Rev. Fluid Mech.*, 12 (1980) 335–365.

53. N. Phan-Thien, On the image system for the Kelvin state, *J. Elasticity*, 13 (1983) 231–235.
54. J. C. F. Telles and C. A. Brebbia, Boundary element solution for half-space problems, *Int. J. Solid Structures*, 17 (1978) 1149–1158.
55. S. H. Lee and L. G. Leal, Motion of a sphere in the presence of a plane interface, Part 2: an exact solution in bipolar coordinates, *J. Fluid Mech.*, 98 (1980) 193–224.
56. C. A. Brebbia and L. C. Wrobel, The boundary element method, in *Computer Methods in Fluids;* K. Morgan, C. Taylor and C. A. Brebbia (eds.), Pentech Press, London, 1980.
57. R. L. Gear, M. Keentok, J. F. Milthorpe and R. I. Tanner, On the shape of low Reynolds number jets, *Phys. Fluids,* 26 (1983) 7–9.
58. S. L. Goren and S. Wronski, The shape of low-speed capillary jets of Newtonian liquids, *J. Fluid Mech.*, 25 (1966) 185–198.
59. J. C. Wu, Problems of general viscous flow, in *Developments in Boundary Element Methods 2*, P. K. Banerjee and R. P. Shaw (eds.), Applied Science Publishers, London, 1982.
60. J. C. Wu and M. M. Wahbah, Numerical solution of viscous flow equations using integral representations, Lecture Notes in Physics, 59, Springer-Verlag, Berlin, 1976.
61. P. Skerget, A. Alujevic and C. A. Brebbia, The solution of the Navier-Stokes equations in terms of vorticity-velocity variables by boundary elements, in *Boundary Elements IV,* Proc. 6th Int. Conf., Queen Elizabeth 2 (Southampton to New York), 1984, C. A. Brebbia (ed), Springer-Verlag, Berlin, 1984.
62. P. Skerget and A. Alujevic, The solution of the Navier-Stokes equations in terms of vorticity-velocity variables by the boundary element method, *Z. Angew. Math. Mech. (ZAMM),* 65 (1985) T245–T248.
63. K. Onishi, T. Kuroki and M. Tanaka, An application of boundary element method to incompressible laminar viscous flows, *Eng. Analysis,* 1 (1984) 122–127.
64. J. C. Wu and Y. M. Rizk, Integral-representation approach for time-dependent viscous flows, Lecture Notes in Physics, 90, Springer-Verlag, Berlin, 1979.
65. A. S. Grove, F. H. Shair, E. E. Petersen and A. Acrivos, An experimental investigation of the steady separated flow past a circular cylinder, *J. Fluid Mech.*, 33 (1964) 60–80.
66. H. Schlichting, *Boundary Layer Theory,* McGraw-Hill, New York, 1968.
67. M. B. Bush and N. Phan-Thien, Drag force on a sphere in creeping motion through a Carreau model fluid, *J. Non-Newtonian Fluid Mech.*, 16 (1984) 303–313.
68. M. B. Bush, J. F. Milthorpe and R. I. Tanner, Finite element and boundary element methods for extrusion computations, *J. Non-Newtonian Fluid Mech.*, 16 (1984) 37–51.
69. M. B. Bush, R. I. Tanner and N. Phan-Thien, A boundary element investigation of extrudate swell, *J. Non-Newtonian Fluid Mech.*, 18 (1985) 143–162.
70. R. B. Bird, R. C. Armstrong and O. Hassager, *Dynamics of Polymeric Fluids,* Vol. 1, J. Wiley and Sons, U.S.A., 1977.
71. R. B. Chhabra and P. H. T. Uhlerr, Creeping motion of sphere through shear-thinning elastic fluids described by the Carreau viscosity equation, *Rheol. Acta,* 19 (1980) 187–195.
72. Y. I. Cho and J. P. Hartnett, Drag coefficients of a slowly moving sphere in non-Newtonian fluids, *J. Non-Newtonian Fluid Mech.*, 12 (1983) 243–247.
73. J. F. Milthorpe and R. I. Tanner, Boundary element methods for free surface viscous flows, in *Proc. 7th Australian Hydraulics and Fluid Mechanics Conf.*, Brisbane (1980) 103–106.
74. C. Fairweather, F. J. Rizzo, D. J. Shippy and Y. S. Wu, On the numerical solution of two-dimensional potential problems by an improved boundary integral equation method, *J. Comp. Phys.*, 31 (1979) 96–112.
75. D. Bonneau and G. Bezine, Etude des écoulements bidimensionels à surface libre en régime de Stokes par la méthode des équations intégrales, in 2ème Congrès International sur les méthodes numériques dans les Sciences de l'Ingénieur, Chatenay Malabry, Dunod (1980).

SEVEN

COMPUTATION OF CONFINED
VORTICAL FLOWS

Egon Krause

ABSTRACT

Internal vortical flows, in particular formation and destruction of vortices of flows in vessels, are discussed. Because of the great variety of such flows, the examples considered here are restricted to cylindrical and spherical boundary shapes. Computational concepts for momentum and vorticity transport are discussed in the frame of finite-difference approaches. Several examples of flow predictions demonstrate some characteristic properties of such flows and exhibit some of the difficulties arising in their computation.

NOMENCLATURE

D^2	differential operator, defined in Eq. (5)
f_1, f_2	damping coefficients
r	radial coordinate
Re	Reynolds number
s	Lagrangian particle path
Sr	Strouhal number
t	time
u, v, w	velocity components
$\mathbf{v}$	velocity vector
Γ	circulation

ζ	vorticity component
θ	colateral angle
ρ	density
σ'	Stokes stress tensor
ϕ	function, defined in Eq. (1)
ψ	stream function
ω	vorticity vector

1 INTRODUCTION

Because of the presence of walls, confined flows always carry vorticity. What is, however, striking, is that they can generate vortices, which have the same size as the vessel the flow is contained in; further, that they may not be able to attain a steady state, even when there is no change in the boundary conditions in the course of time. It is clear that such flows pose a challenging problem for computational work, and, aside from the interest in practical applications, it is of great interest to know whether numerical solutions of the conservation laws of fluid mechanics correctly predict flow variations due to time-dependent in- and outflow; to relative motion of the walls or parts of them with respect to each other; due to sharp corners on the bounding surfaces, or due to pressure and temperature variations.

If formation and destruction of vortices in the flow must be envisaged, the corresponding length and time scales must be known. Spatial and temporal resolution must guarantee that the radial variation of the flow properties in the vortices are computed with sufficient accuracy and that the development of the flow is correctly predicted in time. If, for example, artificial damping of the numerical solution is too large, vortex formation may not take place at all, or the decay of already existing vortices may be too fast.

The time scale is also of great importance for transitional states between two flow modes. For example, if the flow is in the transition from a mode, in which a certain number of vortices dominate, to another, in which the number of vortices is decreased or increased, the characteristic transition time must be known. If the temporal resolution is not appropriate the transition may not be predicted by the numerical solution or may be false in the sense that the direction of the transition may be inversed.

Discussion of transition from one flow mode to another as just described does not include transition from laminar to turbulent flow. The latter as well as time-dependent turbulent internal flow motion will not be reviewed here. Too many uncertainties and the lack of experimental data for validation of already existing theories do not encourage one to attempt a review. It can, however, be expected that the next few years will furnish a lot more experimental information through time-dependent wall-pressure measurements and laser velocimetry. Similar expectations can be extended to the computational side. Numerical flow simulation through direct or large-eddy simulation, see, for example [1], or by the random vortex method [2], seem to be potential although competing tools for analysis of internal flows. One of the greatest handicaps of the methods just mentioned is the enormous numerical effort still necessary even for model problems.

Time dependent internal flow problems are in many cases three-dimensional in nature. The introduction of symmetry conditions, as for example axial or equatorial symmetry, may cause such severe a deviation from the actual flow behavior that the prediction is far from reality.

Computation of external flows by finite-difference techniques is often plagued by high-Reynolds-numbers restrictions, and artificial damping must be employed. Since there are no safe guidelines for determining the damping coefficients, it is not known how much of the actual local flow behavior is suppressed in the numerical solution.

Despite their enormous technical importance, internal flows seem to have received much less attention in the past than did external flows. One obvious reason is that investigation methods were not available for detailed flow studies. Numerical tools to describe complex boundary shapes and to devise suitable grids became the subject of systematic investigations only after enough computational speed and storage capacity were available [3,4]. Almost all known methods of solution have been used for the computation of internal flows: finite difference, finite element, spectral methods, and others. Description and application of these techniques are therefore not discussed here. The reader interested in the details of the solution techniques is referred to textbooks on this subject, as, for example, Peyret's and Taylor's *Computational Methods for Fluid Flow* [5].

This chapter sets out to describe some of the characteristic aspects of vorticity transport. Because of the large variety of flow modes observed, only two geometrical shapes will be considered in the two sections that follow: cylindrical and spherical vessels. Flows in spherical vessels are discussed first. Some remarks with regard to future work conclude the chapter.

2 MATHEMATICAL FORMULATION

Flows in containers of cylindrical or spherical shapes are naturally computed from solutions of the conservation equations for mass, momentum, and energy, written in cylindrical or spherical coordinates, respectively. The simplest case that can be considered is axially symmetric incompressible flow. Then the pressure can be eliminated from the equations of motion by introducing a stream function and combining two of the momentum equations to the transport equation of the vorticity component in the meridional plane. If the three velocity components u, v, w designate the radial, circumferential, and meridional velocity component, respectively, they can be defined in terms of the stream function ψ, a function ϕ, the radial coordinate r, and the colateral angle θ:

$$u = -\frac{1}{r^2 \sin \theta} \frac{\partial \psi}{\partial \theta} \qquad w = \frac{1}{r \sin \theta} \frac{\partial \psi}{\partial r} \qquad v = \frac{\phi}{r \sin \theta} \qquad (1)$$

With this notation the momentum equation for the circumferential direction takes on the following form:

$$\text{Sr} \frac{\partial \phi}{\partial t} + u \frac{\partial \phi}{\partial r} + \frac{w}{r} \frac{\partial \phi}{\partial \theta} = \frac{1}{\text{Re}} D^2 \phi \tag{2}$$

and the transport equation for the vorticity component ζ in the meridional plane is

$$\text{Sr} \frac{\partial \zeta}{\partial t} + u \frac{\partial \zeta}{\partial r} + \frac{w}{r} \frac{\partial \zeta}{\partial \theta} - \frac{\partial \zeta}{\partial r} (u + w \cot \theta)$$

$$+ \frac{2\phi}{r^3 \sin \theta} \left(\frac{\partial \phi}{\partial \theta} - \frac{\partial \phi}{\partial r} r \cot \theta \right) = \frac{1}{\text{Re}} D^2 \zeta \tag{3}$$

Equations (1)–(3) are written in dimensionless form. The quantities Sr and Re are the Strouhal number and the Reynolds number, respectively, based on characteristic length and time scales of the problem considered. The stream function ψ is related to the vorticity component ζ through the Poisson equation

$$D^2 \psi = \zeta \tag{4}$$

where the differential operator D^2 is defined as

$$D^2 \equiv \frac{\partial^2}{\partial r^2} + \frac{1}{r^2} \frac{\partial^2}{\partial \theta^2} - \frac{\cot\theta}{r^2} \frac{\partial}{\partial \theta} \tag{5}$$

and the vorticity component is

$$\zeta = \left[\frac{\partial (wr)}{\partial r} - \frac{\partial u}{\partial \theta} \right] \sin \theta \tag{6}$$

The system of Eqs. (1)–(6) can readily be reduced to cylindrical coordinates.

The time development of the velocity field must be obtained by integrating the momentum equation for the circumferential direction and the vorticity transport equation. For each time level, the Poisson equation for the stream function, Eq. (4) must be solved. Hence either Dirichlet or Neumann boundary conditions must be described on the inner surface of, for example, a hollow sphere, which is assumed to be filled with fluid. Either ψ itself is specified on the boundary, or its derivative in the tangential direction; thereby either radial in- or out-flow is prescribed, or, if the equatorial plane is considered, a symmetry condition.

The momentum equation for the circumferential direction and the vorticity transport equation are parabolic in nature but reduce to elliptic equations for steady flow conditions. Both equations require initial and boundary conditions. While the boundary condition for the circumferential velocity component is simply given by the no-slip condition, there are no natural boundary conditions for the vorticity component ζ. The values of ζ on the boundary must be obtained from the Poisson equation for the stream function, Eq. (4). When finite-difference solutions are used, these result in a one-sided approximation for the second derivative in the radial direction. Depending on what representation is being used, the quantitative as well as the qualitative behavior of the solution may be changed.

The influence of the finite-difference representations of the boundary conditions on the solution was discussed in detail in [6]. The knowledge of this influence of the discretization on the solution is of particular interest, as the solution itself depends on the magnitude on the Reynolds number and on the Strouhal number. A change in Reynolds number can also cause a change in the number of vorticities appearing in the flow.

As mentioned earlier, finite-difference solutions for the conservation equations may require artificial damping in order to retain diagonal dominance of the difference equations. So far, damping terms of second and fourth order have been used. While they are necessary to ensure numerical stability, they may falsify the kinematics of the flow, in particular that of vortices, as can easily be demonstrated [7]. For that purpose the momentum equations are written in vector form with the damping terms included:

$$\frac{d\mathbf{V}}{dt} = -\frac{1}{\rho}\nabla p - \frac{1}{\rho}\nabla\sigma' + f_1\nabla^2\mathbf{V} + f_2\nabla^4\mathbf{V} \tag{7}$$

and f_1 and f_2 designate damping coefficients, which must be "suitably" chosen. Since the rate of change of the local circulation Γ can be expressed by the line integral over the material derivative of $\mathbf{v}$, it follows from Eq. (6) that

$$\frac{d\Gamma}{dt} = -\oint_c \frac{dp}{\rho} - \oint_c \frac{1}{\rho}(\nabla\sigma')\,d\mathbf{s} + \oint_c f_1(\nabla^2\mathbf{V})d\mathbf{s} + \oint_c f_2(\nabla^4\mathbf{V})d\mathbf{s} \tag{8}$$

It is clear from this equation that the last two integrals, containing the damping terms, must be small in comparison to the others, if the rate of change of the local circulation is not to be falsified by the damping terms. Note that Γ is closely related to ϕ, introduced earlier.

Although the pressure does not appear in the vorticity-stream-function formulation, it certainly does influence the distribution of vorticity and the formation of vortices in the flow. This is born out of the following consideration. Let $\boldsymbol{\omega}$ denote the vorticity vector; then for three-dimensional incompressible flow the vorticity transport equation reads

$$\frac{d\boldsymbol{\omega}}{dt} = (\boldsymbol{\omega}\nabla)\mathbf{V} + \nu\nabla^2\boldsymbol{\omega} \tag{9}$$

According to this equation, the time rate of change is solely determined by the velocity and the vorticity distribution in the flow field. However, the pressure is implicitly contained in the first term on the right-hand side of Eq. (8). Integration of Eq. (7), the momentum equation, with respect to time yields, if inserted into Eq. (9):

$$\frac{d\boldsymbol{\omega}}{dt} = -(\boldsymbol{\omega}\nabla)\int_s \frac{1}{\rho V^2}(\nabla p + \nabla\sigma')(\mathbf{V}d\mathbf{s}) + \nu\nabla^2\boldsymbol{\omega} \tag{10}$$

where $\mathbf{s}$ is the Lagrangian particle path. The integral over $\mathbf{s}$ reflects a "history effect" of the pressure and the viscous forces on the local time rate of change of the vorticity vector.

Equation (10) was written for incompressible flow. It follows from the Bjerkness theorem [8] that for compressible flows there is a direct influence of the density (or pressure and temperature field) on the rate of change of the vorticity. The vorticity transport equation then reads for inviscid flow

$$\frac{d\boldsymbol{\omega}}{dt} = \frac{\boldsymbol{\omega}}{\rho}\frac{d\rho}{dt} + (\boldsymbol{\omega}\nabla)\mathbf{V} - \nabla x \left(\frac{1}{\rho}\nabla p\right) \tag{11}$$

It is seen that the time rate of change of the density can influence the vorticity distribution as can local density gradients. There may be substantial differences due to density variations of the local circulation, and one cannot generally conclude that the formation and destruction of vortices is the same as in incompressible flow. With this in mind, a few typical examples of internal spherical flows are considered next.

3 SPHERICAL FLOWS

Consider a hollow sphere filled with an incompressible fluid. If the sphere is rotated about a fixed axis, momentum is transferred to the fluid through viscous forces. If the angular velocity is held constant, the fluid will eventually rotate like a rigid body about the axis of rotation. During the spin-up, however, secondary motion sets in, and a vortex ring is being formed on each side of the equator. The direction of the secondary flow motion is from the pole towards the equator, and in the opposite direction in parts of the fluid close to the axis of rotation. The secondary flow first increases, and then after a certain time decays fast with time. If such a flow is to be computed, the time steps have to be very small in the vicinity of the center of the sphere, and computations can be very tedious. In [9], a computation of the secondary flow is reported by Bartels. He determined the flow pattern with a finite-difference solution of the equations of motion, given in the previous section for a reference Reynolds number Re = 100. Figure 1 shows the stream lines of the secondary flow motion, the increase and rapid decrease of the value of the stream function at the center of the vortex for a rotation with constant angular velocity.

Also of great interest is spherical flow motion for time dependent boundary conditions. Then the hollow sphere rotates with an angular velocity changing periodically in time. The resulting flow has been observed in geophysical problems, in particular the magnetic field of the earth seems to be strongly influenced by hydrodynamic motion in the inner of the earth, initiated through its precession [10]. Such a motion can be simulated through rigid body rotation and a perturbation of the angular velocity of the form $R \sin(\omega t)$, if the precession angle is small. For certain fixed ratios of the circular frequency of the perturbation ω, and that of the rigid body rotation, say Ω, eigenmodes exist in the inner sphere, and they are referred to as inertial oscillations [11]. The fluid in the hollow sphere can then oscillate, in particular with certain eigenfrequences, corresponding to the modes just mentioned. The layer of the fluid adjacent to the wall, the Ekman layer, oscillates with the sphere and produces a small volume flux in the radial

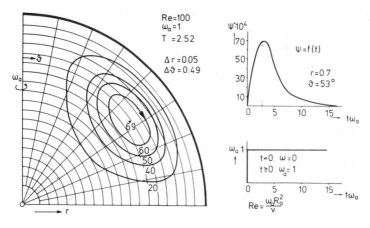

Figure 1 Spin-up in a spherical flow. Development of secondary flow motion in a vortex ring and decay to rigid-body rotation (from [9]).

direction, which is proportional to the square root of the Ekman number $E = \nu / \Omega R^2$. The quantity R is the radius of the sphere. If the frequency of the volume flux coincides with the eigenfrequency, the flow is going to resonate, and large time-dependent variations of the flow properties are being observed [10]. Experimental investigations [12] confirm the existence of such excited states.

In [13], Israeli investigated two eigenmodes for small Ekman numbers ($E \le 2.5 \times 10^{-4}$) with an implicit finite-difference solution. In [9], this investigation was extended to larger Ekman numbers, i.e., $E = O(10^{-2})$ by solving the complete system of equations for a third eigenmode with a finite-difference solution for the vorticity-stream-function formulation. For the above-mentioned order of magnitude of the Ekman number, the Ekman layer looses its boundary-layer characteristics and three flow regions, separated from each other, begin to exist. Figure 2, taken from [9], shows the flow modes for Ekman numbers $E = 10^{-2}$ and $E = 2 \times 10^{-3}$ during the second period after excitation. It can be seen, that for $E = 2 \times 10^{-3}$ at least temporarily the three flow modes appear, while for $E = 10^{-2}$ the flow pattern looks completely different. Aside from the interesting fluid dynamical aspects of the results, Fig. 2 makes clear that computation of such flows must be carried out with extreme care, as the results may strongly depend on the spatial and temporal variation.

Similar complicated flow patterns are found in spherical gaps, which will be considered next. Assume that a second sphere with smaller radius is placed inside a hollow sphere, and further, that they both are coaxially aligned in such a way that the gap between the two spheres has a constant width. Let the gap be filled with fluid and assume that flow can enter the gap at one pole and leave it at the other of the outer sphere. Then, if the inner sphere is allowed to rotate about the common axis of rotation, a great number of steady and unsteady flow modes can be observed. In fact, the variety of flows is so large, that a series of special meetings was initiated in order to study this subject [14] in detail.

Recently, Bühler summarized the state of the art in [15]. He describes primarily those results worked out by Zierep, Sawatzki, Wimmer, and himself at the University of Karlsruhe. Several other research groups are also engaged in research on this subject, and it is impossible to give even a reasonably complete list of references here. The reader is therefore referred to the literature survey given in [15].

A few characteristic features of the flow in narrow gaps are discussed in the following; emphasis is again placed on the fluid dynamic aspects rather than on methodical questions. The main reason for doing so is to illustrate how much care must be exercised, if fluid mechanical problems are to be solved accurately by means of numerical techniques [16].

In [15], Bühler gives an overview over the possible laminar basic flow modes and their transitions. They are indicated in Fig. 3, taken from [15]. Flow instabilities in the gap can be observed if the inner sphere is rotated about the common axis, such that a spherical Couette flow results. Steady and unsteady Taylor vortices can be formed, when a certain critical Reynolds number is exceeded. The flow destabilization is caused by the decrease of the circumferential velocity component in the radial direction, and transition from one mode to another can be initiated through time-dependent boundary conditions.

If mass is added and subtracted through holes in the vicinity of the poles of the outer sphere, the equatorial symmetry is, naturally, destroyed. This spherical source-sink-flow, depicted in the lower part of Fig. 3, experiences the concavity of the wall of the outer sphere and develops Görtler vortices. A spiraling basic flow results when the inner sphere is rotated and mass is added to and subtracted from the flow in the gap through a source at one pole and a sink at the other. The resulting vortex rings are influenced by the angular velocity of the inner sphere and the volume flux of the source-sink flow. The flow can be steady or unsteady.

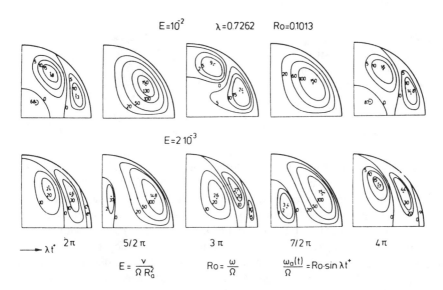

Figure 2 Stream line pattern of a flow in a hollow rotating sphere with precession (from [9]).

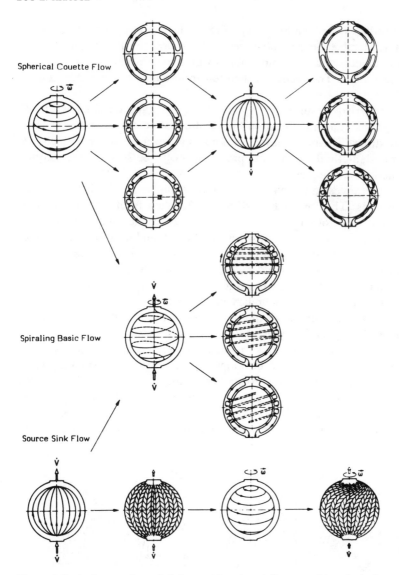

Figure 3 Basic flow modes and their transitions in axially symmetric spherical gaps (from [15]).

Transition from one flow mode to another, as for example from mode I (indicated in the upper left part of Fig. 3) to mode III or IV (in Bühler's nomenclature) has been studied experimentally and numerically by several authors. Figure 4 (again taken from [15]) shows two experimentally determined transition diagrams for the gap width 0.1538 and 0.1779, nondimensionalized with respect to the radius of the inner sphere. The solid lines indicate steady flow conditions of the three modes I, III, and IV. The wavy lines indicate time-dependent flows, with a certain period in the circumferential direction. The wavy modes have been observed experimentally. Bühler reports in [15] that the Taylor vortices of the modes III and IV, as discussed above, clearly show a certain waviness in the circumferential direction; they are no longer axially symmetric.

Equatorial symmetry may be destroyed during transition between two flow modes. For example, transitions from mode III to mode IV and from mode III to mode I are equatorially symmetric, while the transitions from mode I to mode III and from mode IV to mode III are not. As a consequence, flow computations, for which equatorial symmetry is assumed, will not be able to determine the latter two transitions. This has been confirmed in numerical experiments; see for example, [17] and [18].

Another question, which is of equal importance, is whether the flow is steady for large times or not. In order to clarify this point extensive computations were carried out with the solution reported in [9], [19], and [20]. But until now, the experimental and numerical data available do not yield a criterion, according to which steadiness or unsteadiness can be predicted. More recent computations of unsteady flow behavior are reported in [15]. Figures 5 and 6 show the periodic variation of the torque coefficient and the pattern of the secondary stream lines for Re $= 1000$ and a dimensionless volume flux of $\dot{V} = 0.5 \times 10^{-2}$.

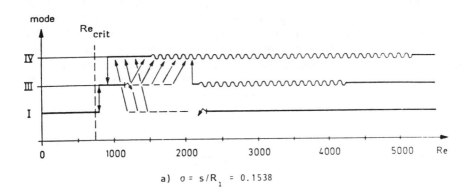

a) $\sigma = s/R_1 = 0.1538$

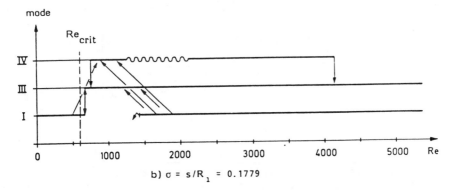

b) $\sigma = s/R_1 = 0.1779$

Figure 4 Experimentally determined transitions (from [15]).

It was first observed in [9] that with increasing number of vortices, the generation and destruction of vortices may not be periodic but may become irregular. Here again, the question of proper resolution must be raised; solutions which do not satisfy certain step size or accuracy conditions as developed in [9] cannot hope to give an accurate prediction of the actual flow behaviour.

Steady flow conditions in spherical gaps were extensively studied by Schrauf in a series on papers, [21–23]. Stimulated by the work of Meyer-Spasche and Keller [24] he used the continuation method of Keller [25] for the solution of the equations of motion to determine the various flow modes for infinitely long transition times. One of the major difficulties encountered in such calculations is the determination of the bifurcation points. Since the finite-difference approximation perturbs the bifurcation, the computed solution branches do not intersect each other as analytically determined branches would. The studies, reported in [21–23] give, for example, insight into the dependence of the solution of the finite-difference equations on the step size. Some of the results reported in [23] show the influence of the gap width on the formation of a single vortex on each side of the equator for constant Reynolds number of Re = 800. It is seen that the one vortex-mode disappears with decreasing gap width, and that eventually the basic flow prevails (Fig. 7).

It is noteworthy that there has been a strong and fruitful interaction between the experimental work in Karlsruhe, directed by Professor J. Zierep, and the numerical work carried in Aachen and Bonn. At Bonn University investigations of this topic were strongly supported by Professors S. Hildebrandt and J. Frehse. The experimental data were indispensable in the verification of the numerical results; but on the other hand, the latter also stimulated new experiments. In the following, flows in cylindrical containers will be discussed.

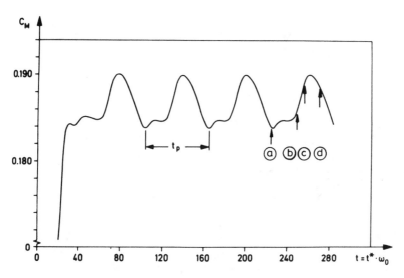

Figure 5 Periodic Torque coefficient computed for Re = 1000, and spherical source-sink flow with dimensionless volume flux $\dot{V} = 0.5 \times 10^{-2}$ (from [15]).

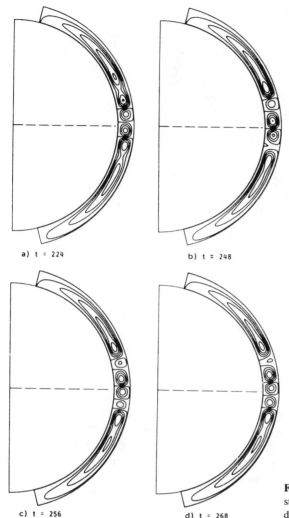

a) t = 224

b) t = 248

c) t = 256

d) t = 268

Figure 6 Stream line pattern of un-steady flow in a spherical gap. Conditions same as in Fig. 5 (from [15]).

4 FLOWS IN CYLINDRICAL CONTAINERS

A special type of cylindrical flows is directly related to those just discussed: formation of Taylor-vortices in cylindrical gaps. This topic has been studied in numerous investigations, and it would be out of question to make an attempt of surveying only the most important results here. The reader interested in this topic is referred to Fasel's and Booz's article in [26]. Instead, only one example of an interesting computational aspect of this problem will be discussed here briefly. This is the following: Assume that a cylindrical gap of finite length and gap width is given, and further, that only a certain part of the inner cylinder is rotating. This part of the cylinder is indicated in Fig. 8 by the thick solid line. The questions to be answered then are: How does the flow and also the numerical solution react

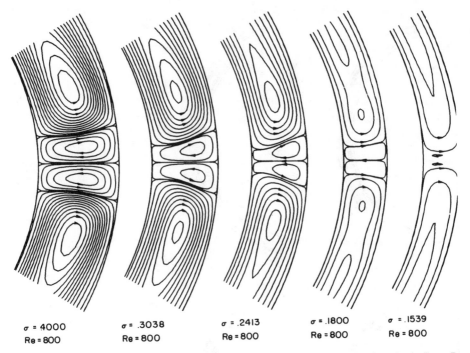

$\sigma = 4000$
Re = 800

$\sigma = .3038$
Re = 800

$\sigma = .2413$
Re = 800

$\sigma = .1800$
Re = 800

$\sigma = .1539$
Re = 800

Figure 7 Influence of the gap width σ on the transition of the one-vortex mode to basic flow. Re = 800 (from [23]).

to the abrupt transition from the nonrotating to the rotating part [27]? What happens when the length of the rotating part of the inner cylinder is decreased? Such a case is depicted in Fig. 8 taken from [28]. In the upper part of the figure the length of the rotating part is seven units, and in the lower part six and a half. It is seen that such a small difference is sufficient to decrease the number of Taylor cells from eight to six. There seems to be a critical length for transition, since the flow computations showed that the number of vortices did not change when the length of the rotating part was reduced from eight to seven. An increase from eight to ten length units created a new pair of vortices. A similar behavior of the

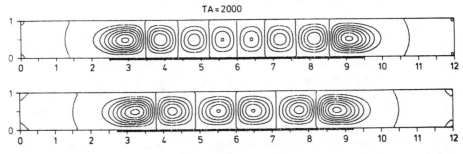

Figure 8 Taylor vortices in a cylindrical gap. Change of length of rotating part (thick solid line) by half a unit reduces number of vortices from eight to six (from [28]).

flow was also observed in experiments [29], and the bifurcation problem is also encountered here. The most recent analysis of bifurcation flows in cylindrical gaps is given in [30].

Another interesting problem is the acceleration or deceleration of a flow in a rotating cylinder, the spin-up or spin-down. Both may be enforced by rotating either the top or the bottom of the container, or, if the container rotates, holding top or bottom plate still for deceleration. As was discussed for spherical flows, secondary flow motion sets in during the transient phase, forming vortex rings as depicted in Fig. 1. Similar behavior can also be expected during the deceleration or acceleration phase of flows in rotating cylinders. One of the most recent experiments on the acceleration of a flow in a cylindrical container through rotation of its bottom plate is reported in [31]. Therein the conditions for vortex breakdown to occur are discussed, and it is reported therein that several authors had computed such flows, but with the exception of Lugt and Haussling [32], had avoided flow conditions for which vortex breakdown was to be expected.

In [33], the slowing down of a vortex in a container was investigated by solving the time-dependent Navier-Stokes equations numerically for an initially prescribed distribution of the circumferential velocity component, and holding top and bottom and the sidewall suddenly still. The initial distribution of the circumferential velocity component was close to rigid body rotation near the center and decayed with r^{-1} with increasing distance from the axis. The first computation was carried out for Re = 100. The results of the integration showed that two ring vortices were generated, dividing the flow into two symmetric parts in the container almost immediately after top and bottom were stopped to rotate. The center of the vortices is first located near the end walls, but moves inward with increasing time, until finally all flow motion is stopped through viscous forces. The flow pattern is completely different, when the Reynolds number is increased to 300. Immediately after stopping, the two vortices form as in the case of Re = 100; but soon thereafter, three small vortex rings are being formed near the axis. This can be seen in Fig. 9, where the stream line patterns of the secondary flow are depicted for the times $t = 1.6$ s and $t = 32$ s after stopping. Shown is only the lower half of the flow.

Figures 10 and 11 show the complete development of the secondary flow in the cylinder, until finally after 200 s the formation and destruction of the smaller vortices near the axis no longer occur. The flow motion eventually comes to a complete stop, when all kinetic energy has been dissipated. Here again, the grid spacing must be carefully chosen if the small vortex rings are to be predicted.

It is interesting to follow a particle on its path; see Fig. 12. The initial coordinates are chosen in such a way that the particle does not get close to the small vortex rings near the axis. The particle moves up in a spiral, until it comes to rest near the outer wall of the cylinder at the end of the motion.

An accelerating flow in a cylindrical container was investigated experimentally in [31]. It is seen in Fig. 13 that the pattern of the secondary flow is similar to the one just discussed. Figure 11 is taken from [31]. Direct comparison of both flows is not possible, since the flow conditions and the dimensions of the containers were not the same as those used in the computations.

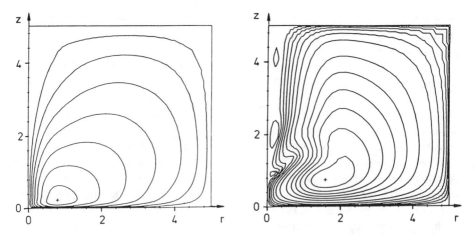

Figure 9 Spin-down in a cylindrical container. Stream line pattern of secondary flow, computed for $t = 7s$ (left), and $t = 32s$ (right), after stopping the bottom plate (from [33]).

Flows in rotating cylinders have found many technical applications as, for example, centrifuges are used for isotope separation. Since the molecular masses of the isotopes differ only by a small margin, the speed of rotation of the centrifuge must be very large in order to make an acceptable efficiency of the separation process possible. As a consequence, the radial variation of the gas density is very large, extending over several orders of magnitude over the radius of the centrifuge. Experimental investigations of such flows are extremely difficult, as a probe introduced into the flow would change the flow pattern substantially and make an analysis of the data difficult. For that reason theoretical investigations of this problem were undertaken by several reseach groups, and the most recent results were reported at the Sixth Workshop on Gases in Strong Rotation in 1985 in Tokyo.

Although the part of the flow near the axis is rarefied, most analyses are based on numerical solutions of the Navier-Stokes equations for compressible rotating flow. One such solution is described in [34]. The integration is carried out for the conservation equations of mass, momentum, and energy without making use of physical simplifications. The solution is based on a linearization in time and a spatial approximate factorization of the difference equations. Further details of the solution can be found in [35] and [36].

One of the main problems of the flow in gas centrifuges is the generation, development, and control of the secondary flow motion, which sets in when the boundary conditions are changed as a function of time. As demonstrated earlier, the secondary flow in the rotating cylinder can be influenced by changing the angular velocity of top or bottom or both with respect to the rotating cylindrical wall. Three time scales can be identified during the spin-up phase [34]. The first is the time the flow requires for developing viscous layers near the walls, the Ekman layers; the second is the spin-up time of the flow; and the third is the time that is necessary to damp out local flow nonuniformities through viscous forces.

Secondary flow motion has been studied extensively for incompressible flow.

Very little work has, however, been done for compressible flows. In the latter, temperature differences may also be employed in order to generate secondary flow motion. For example, a temperature variation in the cylindrical wall may generate countercurrent axial flow in the vicinity of the wall, where most of the mass of the gas is concentrated. The following figures show the stream line pattern of the secondary flow, generated mechanically and thermally. Figure 14 gives an example for a mechanically driven flow as reported in [37] by Merten and Hänel. The bottom of the centrifuge rotates with an angular velocity, which is 5% lower

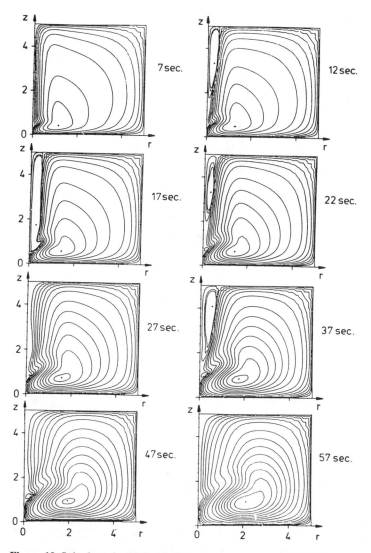

Figure 10 Spin-down in a cylindrical container. Stream line pattern of secondary flow pattern computed for $t = 7$ s to $t = 57$ s. Formation and destruction of small vortex rings near the axis (from [33]).

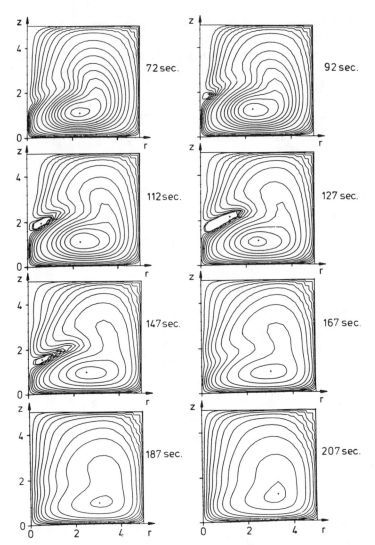

Figure 11 Same as Fig. 10. Secondary flow pattern computed for $t = 72$ sec to $t = 207$ sec (from [33]).

than that of the cylindrical wall. The reference Reynolds number is Re $= 10^4$. The axis of rotation is identical with the z axis on the left side of the figure. Figure 15 shows the pattern of the secondary flow for a linearly varying temperature along the cylindrical wall for the same reference Reynolds number [34]. The temperature of the top is 2% lower, and the temperature of the bottom is 2% higher than that at $Z/L = 0.5$ on the outer wall. It is seen that thermal control of the secondary flow can be very effective, and only small temperature changes are necessary for initiating a countercurrent flow motion. Results have also been

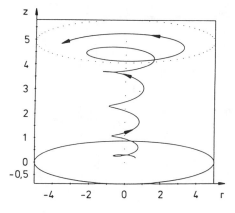

Figure 12 Lagrangian particle path of the flow depicted in Figs. 10 and 11 (from [33]).

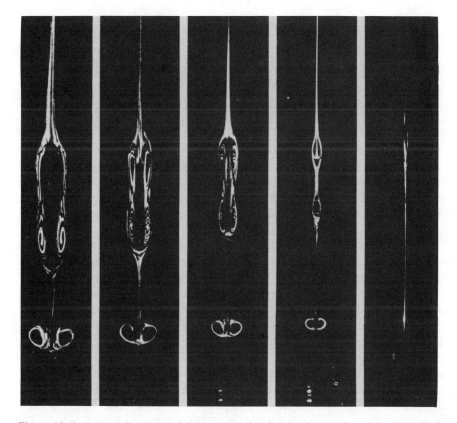

Figure 13 Experimentally observed flow pattern of secondary flow motion during acceleration in a cylindrical container. Forms of vortex breakdown. Note the similarity to the flow pattern of Figs. 10 and 11 (from [31]).

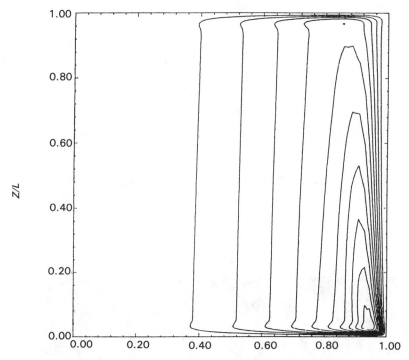

Figure 14 Mechanically driven flow in a gas centrifuge. Numerical solution of the Navier-Stokes equations for compressible flow; Re = 10^4. The angular velocity of the bottom is five percent lower than that of the cylinder (from [37]).

obtained for long centrifuges, with a length/diameter ratio of five, and Reynolds numbers of 10^6 and more, although computation times can be long. In order to keep the computation effort to a minimum, careful adjustment of the time step to the time scales is necessary, as reported in [34].

The last problem to be discussed is the description of the gas motion in the cylinder of piston engines. Large efforts have recently been initiated in order to obtain a better understanding of the flow process, the chemical reactions, and the energy release. Although there is no doubt that turbulent momentum and energy transport play an important role, experimental and numerical studies clearly revealed that large vortical structures can be found in the flow. It should therefore be possible to determine the main flow by solving the Euler equations numerically. This was done by, among others, Henke and Hänel [38] and Henke [39]. In these analyses, the method of approximate factorization was used to compute the flow variables for various inflow conditions. An example of the results obtained so far is shown in the following sequence of figures. The first, Fig. 16, shows the plot of the velocity vectors of an axially symmetric flow in a cylinder of a piston engine. The flow enters the cylinder through a slot, the axis of which is inclined to the axis of the cylinder by an angle $\beta = 30°$. The number of rev-

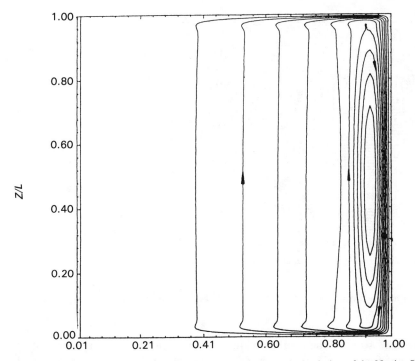

Figure 15 Thermally driven flow in a gas centrifuge. Numerical solution of the Navier-Stokes equations for compressible flow; Re $= 10^4$. The temperature varies linearly along the cylindrical wall, being two percent lower at the top and two percent higher at the bottom (from [34]).

olutions per minute is assumed to be 3000, and the compression ratio as $\epsilon = 11$. In Fig. 16 the vector plot shows the formation of a vortex ring in the cylinder. The crank angle is 180°. Figure 17 shows the vector plot of an axially symmetric flow for a mold piston. During the suction stroke two large ring vortices are being formed. In other computations a circumferential velocity component was superimposed on the flow in the entrance slot in order to simulate swirling motion. The circumferential velocity component was chosen in such a way that the ratio between the circumferential and the axial velocity component remained constant.

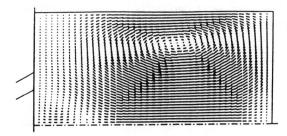

Figure 16 Inviscid flow computed with numerical solution of Euler equations in cylinder of a piston engine. Injection angle $\beta = 30°$; number of revolutions per minute $n = 3000$; compression ration $\epsilon = 11$. Formation of vortex ring (from [30]).

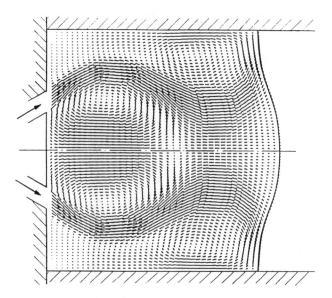

Figure 17 Formation of two vortex rings during the suction stroke in a cylinder with a mould piston.

Figure 18 shows some of the results obtained in the computations. Fig. 18*a* shows the vector plot for a crank angle α = 82° and a piston velocity of 15.6 m/s and Fig. 18*b* the vector plot for a crank angle of α = 180°, for which the piston velocity is zero. In comparison to the flow without swirl (Figs. 19*a* and *b*), the centers of the vortices depicted in Figs. 18*a* and *b* are moved toward the outer wall of the cylinder, and the differences between the two flow fields are clearly evident. As to whether or not the vortex systems survive the compression stroke remains to be answered in future investigations.

Experimental investigations recently carried out at the Aerodynamisches Institut with the light sheet technique in incompressible flow and with Schlieren technique [39] also showed the large vortex structures just discussed. This can be seen in Figs. 20 and 21. In Figs. 20*a* and *b* the vortices are shown for an incompressible flow for a curved and a straight inlet; Figs. 21*a* and *b* show vortex pattern for compressible flow, made visible in a plane cylinder with a step piston.

So far very little is known about the fine structure of these flows. One of the main differences between compressible and incompressible flow consists in the influence of the density variation as discussed in connection with the Bjerkness theorem. The answers to such questions must be left to future investigations.

5 CONCLUSIONS

This chapter summarized some of the results obtained through numerical integration of the equations for viscous and inviscid flows in spherical and cylindrical

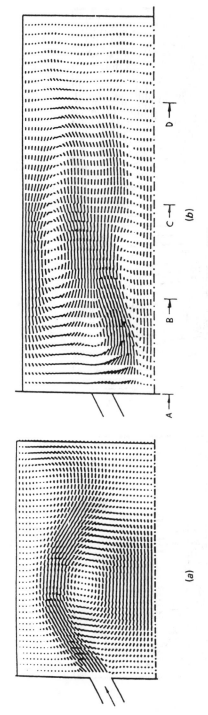

Figure 18 Influence of swirl on the location of the center of the vortices in piston engines (from [38]).

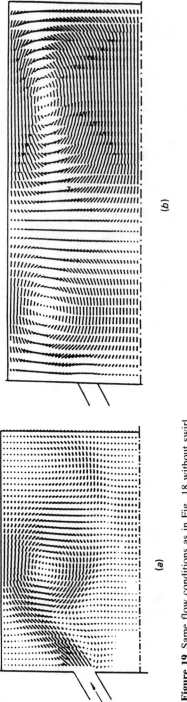

(b)

(a)

Figure 19 Same flow conditions as in Fig. 18 without swirl.

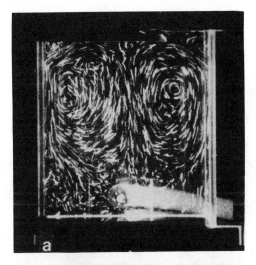

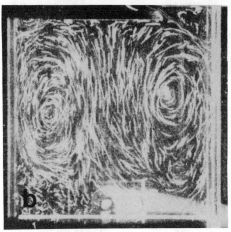

Figure 20 Experimentally observed ring vortex in incompressible flow in a piston engine. Nonaxially symmetric (a) and axially symmetric flow (b) (from &39é).

containers. Primary aim of the investigation was to point to some of the large-scale vortex phenomena which can be observed in such flows and to call attention to some of the difficulties arising in numerical predictions of such flows. It was shown that a steady state for large times may not be attained and that vortex formation during transient phases may develop its own dynamics, in part determined by vortex breakdown and formation of new vortices. It was also demonstrated that numerical analyses of spherical flows and of flows in cylindrical containers can be of great importance in the explanation of geophysical and technical flows, although computations at the present time are still very much handicapped by the available storage capacity and computational speed. Future work should open the door to the analysis of time-dependent three-dimensional flows, which, so far, barely have been touched on.

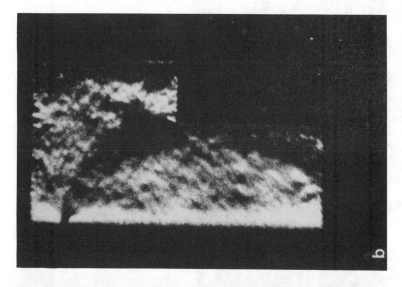

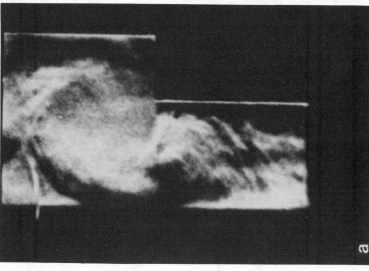

Figure 21 Experimentally observed vortices in compressible flow in a plane cylinder with a stepped piston (from [39]).

REFERENCES

1. P. Moin, I. Kim, Numerical Investigation of Turbulent Channel Flow, *J. Fluid Mech.* 118, pp. 341–377, 1982.

2. A. F. Ghoniem, A. J. Chorin, and A. K. Oppenheim, Numerical Modelling of Turbulent Flow in a Combustion Tunnel, *Philosophical Transactions Royal Society* A 304, pp. 303–325, 1982.

3. I. F. Thompson, T. C. Thames, C. W. Martin, Automatic Numerical Generation of Body-Fitted Curvilinear Coordinate System for Field Containing any Number of Arbitrary Two-Dimensional Bodies, *J. Comput. Phys.* 15, pp. 299–319, 1974.

4. P. R. Eisenman, Grid Generation for Fluid Mechanics Computations, *Ann. Rev. Fluid Mech.*, vol. 17, pp. 487–521, 1985.

5. R. Peyret, T. D. Taylor, *Computational Methods for Fluid Flow*, Springer-Verlag, New York, Heidelberg, Berlin, Springer Series in Computational Physics, 1983.

6. G. Schrauf, Lösungen der Navier-Stokes-Gleichungen für stationäre Strömungen im Kugelspalt, Abhandl. *Aerodyn. Institut RWTH Aachen*, Heft 27, pp. 24–35, 1985.

7. E. Krause, Review of Some Vortex Relations, *Computers & Fluids*, vol. 13, No. 4, pp. 513–515, 1985.

8. V. Bjerkness, Lectures on Hydrodynamics Forces Acting at a Distance (German), Leipzig, 1900–1902.

9. F. Bartels, Rotationssymmetrische Strömungen im Spalt konzentrischer Kugeln, Diss. RWTH Aachen, 1978.

10. K. Stewartson, P. H. Roberts, On the Motion of a Precessing Rigid Body, *J. Fluid Mech.*, vol. 17, pp. 1–20, 1963.

11. H. P. Greenspan, *The Theory of Rotating Fluids*, Cambridge University Press, 1968.

12. K. D. Aldridge, A. Toomre, Axisymmetric Inertial Oscillations of a Fluid in a Rotating Spherical Container, *J. Fluid Mech.*, vol. 37, pp. 307–323, 1969.

13. M. Israeli, Nonliner Motions of a Confined Rotating Fluid, Proceedings of the IUTAM Symposium on Unsteady Boundary Layers, Quebec, 1971.

14. K. Bühler, M. Wimmer, I. Zierep, Taylor Vortex Flow Working Party, Synopsis of Contributions, Fourth Meeting, Karlsruhe, May 20–22, 1985.

15. K. Bühler, Strömungsmechanische Instabilitäten zäher Medien im Kugelspalt, Habilitationsschrift, Universität Friedericiana, Karlsruhe, 1985.

16. E. Krause, Computational Fluid Dynamics: Its Present Status and Future Direction, *Computers & Fluids*, vol. 13, No. 3, pp. 239–269, 1985.

17. E. Krause and F. Bartels, Finite-Difference Solutions of the Navier-Stokes Equations for Axially Symmetric Flows in Spherical Gaps, in: *Approximation Methods for Navier-Stokes Problems*, Proc. IUTAM Symp., Paderborn, pp. 213–222, 1980.

18. G. Schrauf and E. Krause, Symmetric and Asymmetric Taylor Vortices in a Spherical Gap, in V. V. Kozlov (ed.), *The Second IUTAM Symposium on Laminar-Turbulent Transition*, Springer-Verlag, 1984.

19. F. Bartel and E. Krause, Taylor Vortices in Spherical Gaps, in: *IUTAM Symposium on Laminar-Turbulent Transition*, Springer-Verlag, pp. 415–425, 1980.

20. F. Bartels, Taylor Vortices Between Two Concentric Rotating Spheres, *J. Fluid Mech.*, vol. 119, pp. 1–25, 1982.

21. G. Schrauf, Branching of Navier-Stokes Equation in a Spherical Gap, in E. Krause (ed.), *Eighth International Conference on Numerical Methods in Fluid Dynamics*, Lecture Notes in Physics 170, Springer-Verlag, pp. 474–480, 1982.

22. G. Schrauf, Lösungen der Navier-Stokes Gleichungen für stationäre Strömungen im Kugelspalt, Diss. Universität Bonn, 1983, und Abhandl. *Aerodyn. Institut RWTH Aachen*, pp. 24–34, 1985.

23. G. Schrauf, The First Instability in Spherical Taylor-Couette Flow, *J. Fluid Mech.*, vol. 166, pp. 287–303, 1986.

24. R. Meyer-Spasche, H. B. Keller, Computations of the Axisymmetric Flow Between Rotating Cylinders, *J. Comp. Phys.* 35, pp. 100–109, 1980.

25. H. B. Keller, Numerical Solution of Bifurcation and Non-Linear Eigen-Value Problems, in P. Rabinowitz (ed.), *Applications of Bifurcation Theory*, Academic Press, New York, pp. 359–384, 1977.

26. H. Fasel, O. Booz, Numerical Investigation of Supercritical Taylor-Vortex Flow for a Wide Gap. *J. Fluid Mech.*, vol. 138, pp. 21–58, 1984.

27. B. Binninger, Einlaufströmung in einem Ringspalt mit rotierendem Innenzylinder, Diplomarbeit, RWTH Aachen, 1983.

28. H. Eckstein, B. Binninger, Private Communication, 1985.

29. T. A. Cole, Taylor-Vortex Instability and Annulus-Length Effects, *J. Fluid Mech.*, vol. 75, pp. 1–15, 1976.

30. M. Paffrath, Das Taylorproblem und die numerische Behandlung von Verzweigungen, Diss. Univ. Bonn, 1986.

31. M. P. Escudier, Observations of the Flow Produced in a Cylindrical Container by a Rotating Endwall, *Experiments in Fluids*, vol. 2, No. 4, pp. 189–196, 1984.

32. H. I. Lugt, H. I. Haussling, Development of Flow Circulation in a Rotating Tank, *Acta Mechanica* 18, pp. 255–272, 1973.

33. A. Vornberger, Numerische Simulation des Abklingvorganges eines Wirbels zwischen festen Wänden, Diplomarbeit, RWTH Aachen, 1984.

34. A. Merten, D. Hänel, Implicit Solution of the Unsteady Navier-Stokes-Equations for the Flow in a Gas Centrifuge, Proceedings Sixth Workshop on Gases in Strong Rotation, Tokyo, Y. Takashima (ed.), pp. 43–59, 1985.

35. A. Merten, D. Hänel, Navier-Stokes for Compressible Flow in a Rotating Cylinder, H. Viviand (ed.), *Proceedings Fourth GAMM-Conference Num. Meth. in Fluid Mechanics*, Vieweg, vol. 5, Stuttgart, pp. 197–206, 1982.

36. A. Merten, D. Hänel, Implicit Solution of the Navier-Stokes Equations for the Flow in a Centrifuge, E. Rätz (ed.), *Proceedings Fourth Workshop on Gases in Strong Rotation*, Oxford, pp. 479–488, 1981.

37. H. Henke, D. Hänel, Numerical Simulation of Gas Motion in Piston Engines, *Proceedings Ninth International Conference on Numerical Methods in Fluid Dynamics*, Lecture Notes in Physics, Springer Verlag, vol. 218, pp. 267–271, 1985.

38. H. Henke, Lösung der Euler-Gleichungen mit der Methode der angenäherten Faktorisierung, Diss. RWTH Aachen, 1986.

39. H. Weiß, B. Binninger, E. Krause, W. Limberg, Visualisierung des Einströmvorganges im Zylinder von Verbrennungsmototen, Sonderforschungsbereich 224 "Motorische Verbrennung," RWTH Aachen, Kolloquium, Proceedings, pp. 129–141, 1985.

EIGHT

THE ALTERNATING GROUP EXPLICIT (AGE) ITERATIVE METHOD TO SOLVE PARABOLIC AND HYPERBOLIC PARTIAL DIFFERENTIAL EQUATIONS

D. J. Evans and M. S. Sahimi

ABSTRACT

In this chapter the alternating group explicit (AGE) iterative method is applied to a variety of problems involving parabolic and hyperbolic partial differential equations. The method is shown to be extremely powerful and flexible and affords its users many advantages.

1 INTRODUCTION

The ADI method was developed to deal with two-dimensional parabolic (and elliptic) problems and the solutions were obtained *implicitly* in the horizontal and vertical directions. The method could then be extended for applications to higher-dimensional problems. Thus we find that the method has no analogue for the one-dimensional case.

We will, however, show that it is possible to derive another method, the analysis of which is analogous to the ADI scheme. Initially we present the method for one-dimensional problems and then extend its implementation to higher dimensional ones. This *iterative* method employs the fractional splitting strategy which is applied alternately at each half (intermediate) time step on tridiagonal systems of difference schemes and which has proved to be stable. Its rate of convergence is governed by the acceleration parameter r. The accuracy of this method is, in general, comparable if not better than that of the GE class of problems as well as other existing schemes (Evans and Abdullah, 1983).

2 ALTERNATING GROUP EXPLICIT METHOD TO SOLVE SECOND ORDER PARABOLIC EQUATIONS WITH DIRICHLET BOUNDARY CONDITIONS

Consider the following second order parabolic equation,

$$\frac{\partial U}{\partial t} = \frac{\partial^2 U}{\partial x^2} \qquad 0 \leq x \leq 1 \qquad 0 < t \leq T \tag{2.1a}$$

subject to the initial-boundary conditions

$$U(x, 0) = f(x) \qquad 0 < x < 1$$

$$U(0, t) = g(t) \qquad 0 < t \leq T \tag{2.1b}$$

and

$$U(1, t) = h(t)$$

a uniformly spaced network whose mesh points are $x_i = i\Delta x$, $t_j = j\Delta t$ for $i = 0$, 1, ..., m, $m + 1$ and $j = 0, 1, \ldots, n$, $n + 1$ is used with $\Delta x = 1/(m + 1)$, $\Delta t = T/(n + 1)$ and $\lambda = \Delta t/(\Delta x)^2$, the mesh ratio. The real line $0 \leq x \leq 1$ is thus divided as illustrated in Fig. 1.

A weighted approximation to the differential Eq. (2.1a) at the point $(x_i, t_{j+1/2})$ is given by

$$-\lambda\theta u_{i-1,j+1} + (1 + 2\lambda\theta)u_{i,j+1} - \lambda\theta u_{i+1,j+1}$$

$$= \lambda(1 - \theta)u_{i-1,j} + [1 - 2\lambda(1 - \theta)]u_{ij}$$

$$+ \lambda(1 - \theta)u_{i+1,j} \qquad i = 1, 2, \ldots, m \tag{2.2}$$

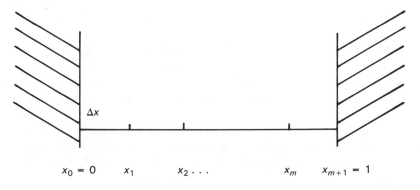

$$x_0 = 0 \qquad x_1 \qquad x_2 \ldots \qquad x_m \qquad x_{m+1} = 1$$

Figure 1 Division of the real line, $0 \leq x \leq 1$.

This approximation can be displayed in a more compact matrix form as

$$
\begin{bmatrix}
a & b & & & & \\
c & a & b & & & \\
& c & a & b & \bigcirc & \\
& & \ddots & \ddots & \ddots & \\
& \bigcirc & & c & a & b \\
& & & & c & a
\end{bmatrix}_{(m \times m)}
\begin{bmatrix}
u_1 \\
u_2 \\
\vdots \\
\vdots \\
u_{m-1} \\
u_m
\end{bmatrix}_{j+1}
=
\begin{bmatrix}
f_1 \\
f_2 \\
\vdots \\
\vdots \\
f_{m-1} \\
f_m
\end{bmatrix}
\qquad (2.3)
$$

i.e.,

$$
\mathbf{Au} = \mathbf{f} \qquad (2.4a)
$$

where

$$
\left.
\begin{aligned}
& c = -\lambda\theta \qquad a = 1 + 2\lambda\theta \qquad b = -\lambda\theta \\
& f_1 = \lambda(1 - \theta)(u_{0j} + u_{2j}) + \lambda\theta u_{0,j+1} + [1 - 2\lambda(1 - \theta)]u_{1,j} \\
& f_i = \lambda(1 - \theta)(u_{i-1,j} + u_{i+1,j}) + [1 - 2\lambda(1 - \theta)]u_{ij} \\
& \qquad i = 2, 3, \ldots, m - 2, m - 1 \\
& f_m = \lambda(1 - \theta)(u_{m-1,j} + u_{m+1,j}) + [1 - 2\lambda(1 - \theta)]u_{mj} \\
& \qquad + \lambda(1 - \theta)u_{m+1,j} + \lambda\theta u_{m+1,j+1} \\
& \mathbf{u} = (u_{1,j+1}, u_{2,j+1}, \ldots, u_{m,j+1})^T \quad \text{and} \quad \mathbf{f} = (f_1, f_2, \ldots, f_m)^T
\end{aligned}
\right\} \qquad (2.4b)
$$

We note that $\mathbf{f}$ is a column vector of order m consisting of the boundary values as well as known $\mathbf{u}$ values at time level j while $\mathbf{u}$ are the values at time level $(j + 1)$ which we seek. We also recall that Eq. (2.4a) corresponds to the fully implicit, the Crank-Nicolson, the Douglas and the classical explicit methods when θ takes the values $1, \frac{1}{2}, \frac{1}{2} - 1/12\lambda$ and 0 with accuracies of the order $O[(\Delta x)^2 + \Delta t]$, $O[(\Delta x)^2 + (\Delta t)^2]$, $O[(\Delta x)^4 + (\Delta t)^2]$ and $O[(\Delta x)^2 + \Delta t]$, respectively.

Let us first assume that we have an *even number of intervals* (corresponding to an odd number of internal points, i.e., m odd) on the real line $0 \le x \le 1$. We can then perform the following splitting of the coefficient matrix A:

$$
A = G_1 + G_2 \qquad (2.5)
$$

where

$$
G_1 =
\begin{bmatrix}
a/2 & & & & & & & \\
& a/2 & b & & & & & \\
& c & a/2 & & & & & \\
& & & a/2 & b & & & \\
& & & c & a/2 & \bigcirc & & \\
& & & & \bigcirc & \ddots & & \\
& & & & & & a/2 & b \\
& & & & & & c & a/2
\end{bmatrix}_{(m \times m)}
\qquad (2.6)
$$

and

$$G_2 = \begin{bmatrix} a/2 & b & & & & & \\ c & a/2 & & & & & \\ & & a/2 & b & & & \\ & & c & a/2 & & & \\ & & & & \bigcirc & & \\ & & & \bigcirc & & a/2 & b \\ & & & & & c & a/2 \\ & & & & & & a/2 \end{bmatrix}_{(m \times m)} \tag{2.7}$$

It is assumed that the following conditions are satisfied:
  (i) $G_1 + rI$ and $G_2 + rI$ are nonsingular for any $r > 0$,
  (ii) for any vectors $\mathbf{f}_1$ and $\mathbf{f}_2$ and for any $r > 0$, the systems

$$(G_1 + rI)\mathbf{u}_1 = \mathbf{f}_1$$

and
$$(G_2 + rI)\mathbf{u}_2 = \mathbf{f}_2 \tag{2.8}$$

are more easily solved in explicit form since they consist of only (2×2) sub-systems.

Thus, with these conditions, system (2.4a) becomes

$$(G_1 + G_2)\mathbf{u} = \mathbf{f} \tag{2.9}$$

The alternating group explicit (AGE) iteration consists of writing Eq. (2.9) as a pair of equations:

$$(G_1 + rI)\mathbf{u} = (rI - G_2)\mathbf{u} + \mathbf{f}$$

and
$$(G_2 + rI)\mathbf{u} = (rI - G_1)\mathbf{u} + \mathbf{f} \tag{2.10}$$

The AGE method using *the Peaceman and Rachford variant* for the *stationary case* ($r = $ constant) is given by

$$(G_1 + rI)\mathbf{u}^{(p+1/2)} = (rI - G_2)\mathbf{u}^{(p)} + \mathbf{f}$$

and
$$(G_2 + rI)\mathbf{u}^{(p+1)} = (rI - G_1)\mathbf{u}^{(p+1/2)} + \mathbf{f} \qquad p \geq 0 \tag{2.11}$$

where $\mathbf{u}^{(0)}$ is a starting approximation and r are positive constants called acceleration parameters whose values are chosen to maximize the rate of convergence. We now seek to analyze the convergence properties of the AGE method. From Eq (2.11) we can write

$$\mathbf{u}^{(p+1)} = M(r)\mathbf{u}^{(p)} + \mathbf{q}(r) \qquad p \geq 0 \tag{2.12}$$

where the *AGE iteration matrix* is given by

$$M(r) = (G_2 + rI)^{-1}(rI - G_1)(G_1 + rI)^{-1}(rI - G_2) \tag{2.13}$$

If $\mathbf{e}$ denotes the error vector and $\mathbf{U}$ the exact solution of Eq. (2.1a), then $\mathbf{e}^{(p)} =$

$\mathbf{u}^{(p)} - \mathbf{U}$ and $\mathbf{e}^{(p+1)} = M(r)\mathbf{e}^{(p)}$. Hence, we have

$$\mathbf{e}^{(p)} = M^p(r)\,\mathbf{e}^{(0)} \qquad p \geq 1 \tag{2.14}$$

We now prove the following theorem.

Theorem 2.1 If G_1 and G_2 are real positive definite matrices, and if $r > 0$, then

$$\rho[M(r)] < 1 \tag{2.15}$$

PROOF If we let $\tilde{M}(r) = (G_2 + rI)M(r)(G_2 + rI)^{-1}$, then by a similarity transformation, $M(r)$ and $\tilde{M}(r)$ have the same eigenvalues. Hence, from Eq. (2.13) we find that

$$\rho[M(r)] = \rho[\tilde{M}(r)]$$

$$\leq \|\tilde{M}(r)\|$$

$$\leq \|(rI - G_1)(G_1 + rI)^{-1}\| \, \|(rI - G_2)(G_2 + rI)^{-1}\| \tag{2.16}$$

where $\rho[M(r)]$ is the spectral radius of $M(r)$. But since G_1 and G_2 are symmetric and $(rI - G_1)$ commutes with $(G_1 + rI)^{-1}$, then in the L_2 norm we have

$$\|(rI - G_1)(G_1 + rI)^{-1}\|_2 = \rho[(rI - G_1)(G_1 + rI)^{-1}]$$

$$= \max_{1 \leq i \leq m} \left| \frac{r - \mu_i}{r + \mu_i} \right| \tag{2.17}$$

where μ_i are the eigenvalues of G_1. But since G_1 is positive definite, its eigenvalues are positive. Therefore

$$\|(rI - G_1)(G_1 + rI)^{-1}\|_2 < 1 \tag{2.18}$$

Similarly,

$$\|(rI - G_2)(G_2 + rI)^{-1}\|_2 < 1$$

and we have

$$\rho[M(r)] = \rho[\tilde{M}(r)] \leq \|M(r)\|_2 < 1 \tag{2.19}$$

and convergence is assured. We note that to establish the condition (2.15) for unsymmetric matrices G_1 and G_2 may require us to evaluate directly the eigenvalues of $M(r)$, which can be difficult.

It is possible to determine the optimum parameter $\hat{r}$ such that the bound for $\rho[M(r)]$ is minimized. To investigate this we assume that G_1 and G_2 are real positive definite matrices and that bounds for their eigenvalues μ_i and η_i are available, i.e.,

$$0 < \alpha \leq \mu_i, \eta_i \leq \beta \tag{2.20}$$

In the L_2 norm, Eq. (2.16) implies

$$\rho[M(r)] \leq \rho[(rI - G_1)(G_1 + rI)^{-1}]\rho[(rI - G_2)(G_2 + rI)^{-1}]$$

$$= \left(\max_{1 \leq i \leq m} \left| \frac{r - \mu_i}{r + \mu_i} \right| \right) \left(\max_{1 \leq i \leq m} \left| \frac{r - \eta_i}{r + \eta_i} \right| \right)$$

$$\leq \left(\max_{\alpha \leq z \leq \beta} \left| \frac{r - z}{r + z} \right| \right)^2 = \phi(\alpha, \beta; r) \qquad (2.21)$$

But $(r - z)/(r + z)$ is an increasing function of z. Therefore, we find that

$$\max_{\alpha \leq z < \beta} \left| \frac{r - z}{r + z} \right| = \max \left(\left| \frac{r - \alpha}{r + \alpha} \right|, \left| \frac{r - \beta}{r + \beta} \right| \right) \qquad (2.22)$$

When $r = \sqrt{\alpha\beta}$, we have

$$\left| \frac{r - \alpha}{r + \alpha} \right| = \left| \frac{r - \beta}{r + \beta} \right| = \frac{\sqrt{\beta} - \sqrt{\alpha}}{\sqrt{\alpha} + \sqrt{\beta}} \qquad (2.23)$$

For $0 < r < \sqrt{\alpha\beta}$, we obtain

$$\left| \frac{r - \beta}{r + \beta} \right| - \frac{\sqrt{\beta} - \sqrt{\alpha}}{\sqrt{\alpha} + \sqrt{\beta}} = \frac{2\sqrt{\beta}(\sqrt{\alpha\beta} - r)}{(r + \beta)(\sqrt{\alpha} + \sqrt{\beta})} > 0$$

i.e.,

$$\left| \frac{r - \beta}{r + \beta} \right| > \frac{\sqrt{\beta} - \sqrt{\alpha}}{\sqrt{\alpha} + \sqrt{\beta}} \qquad (2.24)$$

Similarly, for $\sqrt{\alpha\beta} < r$, we get

$$\left| \frac{r - \alpha}{r + \alpha} \right| - \frac{\sqrt{\beta} - \sqrt{\alpha}}{\sqrt{\alpha} + \sqrt{\beta}} = \frac{2\sqrt{\alpha}(r - \sqrt{\alpha\beta})}{(r + \alpha)(\sqrt{\alpha} + \sqrt{\beta})} > 0$$

i.e.,

$$\left| \frac{r - \alpha}{r + \alpha} \right| > \frac{\sqrt{\beta} - \sqrt{\alpha}}{\sqrt{\alpha} + \sqrt{\beta}} \qquad (2.25)$$

Hence, using Eqs. (2.21)–(2.25), we deduce that the bound $\phi(\alpha, \beta; r)$ for $\rho[M(r)]$ is minimized when $r = \hat{r} = \sqrt{\alpha\beta}$ and $\rho[M(\hat{r})] \leq \phi(\alpha, \beta; \hat{r}) = [(\sqrt{\beta} - \sqrt{\alpha})/(\sqrt{\alpha} + \sqrt{\beta})]^2$.

For an efficient implementation of the AGE algorithm, it is essential to vary the acceleration parameters r_p from iteration to iteration—the nonstationary case. This will result in an improvement in the rate of convergence of the AGE method when the Peaceman-Rachford variant is employed. The Peaceman-Rachford formula (2.11) will now become

$$(G_1 + r_{p+1}I)\mathbf{u}^{(p+1/2)} = (r_{p+1}I - G_2)\mathbf{u}^{(p)} + \mathbf{f}$$

and $\quad(G_2 + r_{p+1}I)\mathbf{u}^{(p+1)} = (r_{p+1}I - G_1)\mathbf{u}^{(p+1/2)} + \mathbf{f} \qquad p \geq 0 \qquad (2.26)$

The best values of r_p can be ascertained provided G_1 and G_2 are *commutative*—a property which is not possessed by our model problem. However, these matrices commute if the boundary conditions are *periodic* and of order 4, that is, when the conditions (2.1*b*) are replaced by

$$u(0, t) = u(1, t) \qquad \frac{\partial U}{\partial x}(0, t) = \frac{\partial U}{\partial x}(1, t) \qquad (2.27)$$

For the general application of Eq. (2.26), $(r_p > 0)$ we assume first of all that the positive definite matrices G_1 and G_2 commute. Thus, G_1 and G_2 have a common set of (orthonormal) eigenvectors. Let $(\mu_j, \mathbf{v}_j)_{j=1}^m$ and $(\eta_j, \mathbf{v}_j)_{j=1}^m$ be the eigensystem of G_1 and G_2, respectively.

For p iterations of Eq. (2.26), relation (2.13) yields, for $1 \leq j \leq m$,

$$\left[\prod_{i=1}^{p} M(r_i)\right]\mathbf{v}_j = \left(\prod_{i=1}^{p} \frac{(r_i - \mu_j)}{(r_i + \mu_j)}\frac{(r_i - \eta_j)}{(r_i + \eta_j)}\right)\mathbf{v}_j \qquad (2.28)$$

It follows that $\prod_{i=1}^{p} M(r_i)$ is symmetric and, therefore,

$$\left\|\prod_{i=1}^{p} M(r_i)\right\|_2 = \rho\left[\prod_{i=1}^{p} M(r_i)\right]$$

$$= \max_{1 \leq j \leq m} \prod_{i=1}^{p} \left|\frac{r_i - \mu_j}{r_i + \mu_j}\right|\left|\frac{r_i - \eta_j}{r_i + \eta_j}\right| < 1 \qquad (2.29)$$

and convergence of the iterative process is achieved. Now it is clear that

$$\max_{1 \leq j \leq m} \prod_{i=1}^{p} \left|\frac{r_i - \mu_j}{r_i + \mu_j}\right|\left|\frac{r_i - \eta_j}{r_i + \eta_j}\right| \leq \max_{1 \leq j \leq m} \prod_{i=1}^{p} \left|\frac{r_i - \mu_j}{r_i + \mu_j}\right| \max_{1 \leq j \leq m} \prod_{i=1}^{p} \left|\frac{r_i - \eta_j}{r_i + \eta_j}\right|$$

$$\leq \left(\max_{\alpha \leq z \leq \beta} \prod_{i=1}^{p} \left|\frac{r_i - z}{r_i + z}\right|\right)^2 \qquad (2.30)$$

where we have used the bounds for the eigenvalues given by Eq. (2.20). Hence,

$$\rho\left[\prod_{i=1}^{p} M(r_i)\right] \leq \max_{\alpha \leq z \leq \beta} |R_p(z)|^2 = \phi(\alpha, \beta; r_1, \ldots, r_p) \qquad (2.31)$$

where

$$R_p(z) = \prod_{i=1}^{p} \frac{r_i - z}{r_i + z}$$

The difficulty of determining the optimum parameters by minimizing $\phi(\alpha,$

$\beta; r_1, \ldots, r_p)$ has led a number of authors to devise alternative sequences. For example, the parameters used by Peaceman and Rachford (1955) are

$$\hat{r}_j = \beta\left(\frac{\alpha}{\beta}\right)^{(2j-1)/(2p)} \qquad j = 1, 2, \ldots, p \qquad (2.32)$$

from which we obtain the result

$$\rho\left[\prod_{i=1}^{p} M(r_i)\right] \leq \left[\frac{1 - (\alpha/\beta)^{1/(2p)}}{1 + (\alpha/\beta)^{1/(2p)}}\right]^2 \qquad (2.33)$$

Wachpress (1966), on the other hand, solved the minimization problem analytically in terms of elliptic functions and arrived at the result

$$\hat{r}_j = \beta\left(\frac{\alpha}{\beta}\right)^{(j-1)/(p-1)} \qquad p \geq 2 \qquad j = 1, 2, \ldots, p \qquad (2.34)$$

It must be noted that the requirement that G_1 and G_2 be commutative can be very restrictive indeed. However, in far more general situations we can expect convergence of the AGE iteration for a *fixed acceleration parameter* without the condition $G_1 G_2 = G_2 G_1$.

We conclude this section by considering the case when we have an *odd number of intervals* (corresponding to an even number of internal points, i.e., m even) on $0 \leq x \leq 1$. The coefficient matrix A will be split as in Eq. (2.5) but G_1 and G_2 now take the form

$$(2.35)$$

and

$$(2.36)$$

All the preceding conditions and arguments regarding convergence and the choice of the optimal acceleration parameter for both the stationary and nonstationary cases remain valid.

3 VARIANTS OF THE AGE SCHEME AND ITS COMPUTATION

Many variants of the basic Peaceman-Rachford scheme can be proposed. For example, we have, on modifying the second stage of Eq. (2.26) (the nonstationary case),

$$(G_1 + r_{p+1}I)\mathbf{u}^{(p+1/2)} = (r_{p+1}I - G_2)\mathbf{u}^{(p)} + \mathbf{f}$$

and $\quad (G_2 + r_{p+1}I)\mathbf{u}^{(p+1)} = [G_2 - (1 - \omega)r_{p+1}I]\mathbf{u}^{(p)} + (2 - \omega)r_{p+1}\mathbf{u}^{(p+1/2)}$

$$(3.1)$$

where ω is a parameter. For $\omega = 0$, we have the Peaceman-Rachford scheme (2.26), and for $\omega = 1$, we obtain the scheme due to Douglas and Rachford (1956). For G_1 and G_2 symmetric and positive definite and with a *fixed acceleration parameter $r > 0$*, the resulting *generalized AGE scheme* is convergent for any $0 \leq \omega \leq 2$. As we shall see in a later paper, a natural extension of the AGE algorithm is to implement it on higher dimensional boundary value problems using the Douglas-Rachford variant.

For the purpose of computation, we shall now attempt to derive equations that are satisfied at each intermediate (half-time) level. For *the Peaceman-Rachford variant*, in particular, and with fixed parameter r, the *AGE method* can be applied to determine $\mathbf{u}^{(p+1/2)}$ and $\mathbf{u}^{(p+1)}$ implicitly by

$$(G_1 + rI)\mathbf{u}^{(p+1/2)} = (rI - G_2)\mathbf{u}^{(p)} + \mathbf{f}$$

and $\quad (G_2 + rI)\mathbf{u}^{(p+1)} = (rI - G_1)\mathbf{u}^{(p+1/2)} + \mathbf{f}$

$$(3.2)$$

or explicitly by

$$\left.\begin{array}{l}\mathbf{u}^{(p+1/2)} = (G_1 + rI)^{-1}[(rI - G_2)\mathbf{u}^{(p)} + \mathbf{f}] \\[2mm] \mathbf{u}^{(p+1)} = (G_2 + rI)^{-1}[(rI - G_1)\mathbf{u}^{(p+1/2)} + \mathbf{f}]\end{array}\right\}$$

and

$$(3.3)$$

If we assume m to be odd *(even number of intervals)* and if we write

$$\hat{G} = \begin{pmatrix} r_2 & b \\ c & r_2 \end{pmatrix}$$

$$(3.4)$$

where

$$r_2 = r + \frac{a}{2}$$

$$(3.5)$$

then from Eqs. (2.6) and (2.7) we have

$$(G_1 + rI) = \begin{bmatrix} r_2 & & & & & \\ & \hat{\hat{G}} & & & & \\ & & \hat{G} & & & \\ & & & \circ & & \\ & & & & \circ & \\ & & & & & \hat{\hat{G}} \end{bmatrix}_{(m \times m)} \tag{3.6}$$

and

$$(G_2 + rI) = \begin{bmatrix} \hat{G} & & & & & \\ & \hat{G} & & & & \\ & & \circ & & & \\ & & & \circ & & \\ & & & & \hat{G} & \\ & & & & & r_2 \end{bmatrix}_{(m \times m)} \tag{3.7}$$

It is clear that $(G_1 + rI)$ and $(G_2 + rI)$ are block diagonal matrices. All the diagonal elements except the first {or the last for $(G_2 + rI)$ are (2×2)} submatrices. Therefore, $(G_1 + rI)$ and $(G_2 + rI)$ can be easily inverted by merely inverting their block diagonal entries. Hence,

$$(G_1 + rI)^{-1} = \begin{bmatrix} 1/r_2 & & & & \\ & \hat{\hat{G}}^{-1} & & & \\ & & \hat{G}^{-1} & \circ & \\ & & \circ & & \\ & & & & \hat{\hat{G}}^{-1} \end{bmatrix}_{(m \times m)} \tag{3.8}$$

$$= \frac{1}{\Delta} \begin{bmatrix} \Delta/r_2 & & & & \\ & r_2 & -b & & \\ & -c & r_2 & & \\ & & & \circ & \\ & & \circ & & r_2 & -b \\ & & & & -c & r_2 \end{bmatrix}_{(m \times m)} \tag{3.9}$$

where
$$\Delta = r_2^2 - bc \tag{3.10}$$

Similarly, we obtain

$$(G_2 + rI)^{-1} = \frac{1}{\Delta} \begin{bmatrix} r_2 & -b & & & & \\ -c & r_2 & & & & \\ & & r_2 & -b & & \\ & & -c & r_2 & \circ & \\ & & & \circ & & \\ & & & & r_2 & -b \\ & & & & -c & r_2 \\ & & & & & & \Delta/r_2 \end{bmatrix}_{(m \times m)} \tag{3.11}$$

From Eq. (3.3), $\mathbf{u}^{(p+1/2)}$ and $\mathbf{u}^{(p+1)}$ are given by

$$
\begin{bmatrix} u_1^{(p+1/2)} \\ u_2^{(p+1/2)} \\ u_3^{(p+1/2)} \\ \vdots \\ \\ u_{m-1}^{(p+1/2)} \\ u_m^{(p+1/2)} \end{bmatrix} = \frac{1}{\Delta}
\begin{bmatrix}
\Delta/r_2 & & & & & \\
& r_2 & -b & & & \\
& -c & r_2 & & \bigcirc & \\
& & & & & \\
& & \bigcirc & & r_2 & -b \\
& & & & -c & r_2
\end{bmatrix}
$$

$$
\times
\begin{bmatrix}
r_1 u_1^{(p)} - b u_2^{(p)} + f_1 \\
-c u_1^{(p)} + r_1 u_2^{(p)} + f_2 \\
r_1 u_3^{(p)} - b u_4^{(p)} + f_3 \\
-c u_3^{(p)} + r_1 u_4^{(p)} + f_4 \\
\vdots \\
r_1 u_{m-2}^{(p)} - b u_{m-1}^{(p)} + f_{m-2} \\
-c u_{m-2}^{(p)} + r_1 u_{m-1}^{(p)} + f_{m-1} \\
r_1 u_m^{(p)} + f_m
\end{bmatrix}
\qquad (3.12)
$$

and

$$
\begin{bmatrix} u_1^{(p+1)} \\ u_2^{(p+1)} \\ u_3^{(p+1)} \\ \vdots \\ u_{m-2}^{(p+1)} \\ u_{m-1}^{(p+1)} \\ u_m^{(p+1)} \end{bmatrix} = \frac{1}{\Delta}
\begin{bmatrix}
r_2 & -b & & & & \\
-c & r_2 & & & & \\
& & r_2 & -b & & \\
& & -c & r_2 & \bigcirc & \\
& & & & & \\
& & \bigcirc & & r_2 & -b \\
& & & & -c & r_2 \\
& & & & & \Delta/r_2
\end{bmatrix}
$$

$$
\times
\begin{bmatrix}
r_1 u_1^{(p+1/2)} + f_1 \\
r_1 u_2^{(p+1/2)} - b u_3^{(p+1/2)} + f_2 \\
-c u_2^{(p+1/2)} + r_1 u_3^{(p+1/2)} + f_3 \\
r_1 u_4^{(p+1/2)} - b u_5^{(p+1/2)} + f_4 \\
-c u_4^{(p+1/2)} + r_1 u_5^{(p+1/2)} + f_5 \\
\vdots \\
r_1 u_{m-1}^{(p+1/2)} - b u_m^{(p+1/2)} + f_{m-1} \\
-c u_{m-1}^{(p+1/2)} + r_1 u_m^{(p+1/2)} + f_m
\end{bmatrix}
\qquad (3.13)
$$

where $\qquad r_1 = r - \dfrac{a}{2} \qquad r_2 = r + \dfrac{a}{2} \qquad$ and $\qquad \Delta = r_2^2 - bc \qquad$ (3.14)

The *alternating implicit* nature of the (2×2) groups in Eq. (3.2) is shown in Fig. 2 where the implicit/explicit values are given on the forward/backward levels for sweeps on the ($p + 1/2$)th and ($p + 1$)th levels.

The corresponding explicit expressions for the AGE equations are obtained by carrying out the multiplications in Eqs. (3.12) and (3.13). Thus, we have

(i) at level ($p + 1/2$)

$$\left.\begin{array}{l} u_1^{(p+1/2)} = \dfrac{r_1 u_1^{(p)} - b u_2^{(p)} + f_1}{r_2} \\[2em] u_i^{(p+1/2)} = \dfrac{A u_{i-1}^{(p)} + B u_i^{(p)} + C u_{i+1}^{(p)} + D u_{i+2}^{(p)} + E_i}{\Delta} \\[2em] \text{and} \\[1em] u_{i+1}^{(p+1/2)} = \dfrac{\tilde{A} u_{i-1}^{(p)} + \tilde{B} u_i^{(p)} + \tilde{C} u_{i+1}^{(p)} + \tilde{D} u_{i+2}^{(p)} + \tilde{E}_i}{\Delta} \end{array}\right\} \begin{array}{l} i = 2, 4, \ldots, m - 1 \end{array} \right\} \text{(3.15a)}$$

where

$$\left.\begin{array}{l} A = -cr_2 \quad B = r_1 r_2 \quad C = -br_1 \quad E_i = r_2 f_i - b f_{i+1} \\[1em] D = \begin{cases} 0 & \text{for } i = m - 1 \\ b^2 & \text{otherwise} \end{cases} \\[1.5em] \tilde{A} = c^2 \quad \tilde{B} = -cr_1 \quad \tilde{C} = r_1 r_2 \quad \tilde{E}_i = r_2 f_{i+1} - c f_i \\[1em] \tilde{D} = \begin{cases} 0 & \text{for } i = m - 1 \\ -br_2 & \text{otherwise} \end{cases} \end{array}\right\} \text{(3.15b)}$$

and

with the computational molecules shown in Fig. 3.

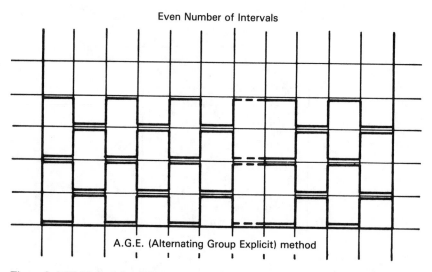

Even Number of Intervals

A.G.E. (Alternating Group Explicit) method

Figure 2 AGE Method (implicit).

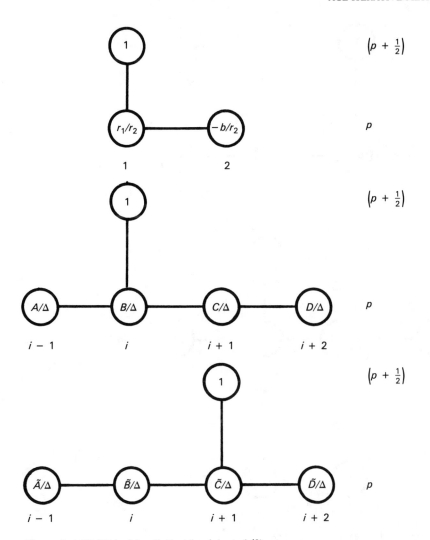

Figure 3 AGE Method (explicit) at level $(p + 1/2)$.

(ii) at level $(p + 1)$

$$
\left.
\begin{aligned}
u_i^{(p+1)} &= \frac{Pu_{i-1}^{(p+1/2)} + Qu_i^{(p+1/2)} + Ru_{i+1}^{(p+1/2)} + Su_{i+2}^{(p+1/2)} + T_i}{\Delta} \\[2mm]
u_{i+1}^{(p+1)} &= \frac{\tilde{P}u_{i-1}^{(p+1/2)} + \tilde{Q}u_i^{(p+1/2)} + \tilde{R}u_{i+1}^{(p+1/2)} + \tilde{S}u_{i+2}^{(p+1/2)} + \tilde{T}_i}{\Delta} \\[2mm]
& i = 1, 3, \ldots, m - 2 \\[2mm]
u_m^{(p+1)} &= \frac{-cu_{m-1}^{(p+1/2)} + r_1 u_m^{(p+1/2)} + f_m}{r_2}
\end{aligned}
\right\} \quad (3.16a)
$$

and

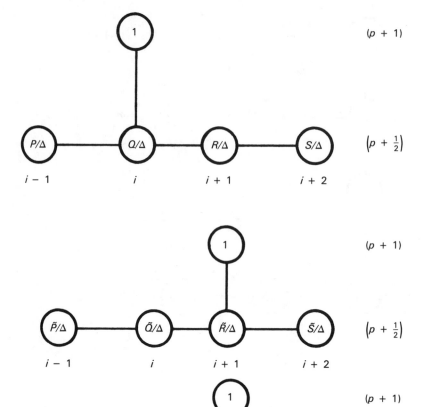

Figure 4 AGE Method (explicit) at level $(p + 1)$.

where

$$P = \begin{cases} 0 & \text{for } i = 1 \\ -cr_2 & \text{for } i \neq 1 \end{cases} \quad Q = r_1 r_2 \quad R = -br_1 \quad S = b^2$$

$$T_i = r_2 f_i - b f_{i+1}$$

and

$$\tilde{P} = \begin{cases} 0 & \text{for } i = 1 \\ c^2 & \text{for } i \neq 1 \end{cases} \quad \tilde{Q} = -cr_1 \quad \tilde{R} = Q = r_1 r_2$$

$$\tilde{S} = -br_2 \quad \tilde{T}_i = -cf_i + r_2 f_{i+1}$$

(3.16b)

with its computational molecules given by Fig. 4.

For the generalized AGE scheme (3.1) with fixed acceleration parameter r, the relevant equations at level $(p + 1/2)$ remain the same as in Eqs. (3.15a) and (3.15b). The equations at level $(p + 1)$ are, however, now replaced by

$$
\begin{bmatrix} u_1^{(p+1)} \\ u_2^{(p+1)} \\ u_3^{(p+1)} \\ \vdots \\ u_{m-2}^{(p+1)} \\ u_{m-1}^{(p+1)} \\ u_m^{(p+1)} \end{bmatrix}
= \frac{1}{\Delta}
\begin{bmatrix}
r_2 & -b & & & & & \\
-c & r_2 & & & & & \\
& & r_2 & -b & & & \\
& & -c & r_2 & & & \\
& & & & \bigcirc & & \\
& & & \bigcirc & & r_2 & -b \\
& & & & & -c & r_2 \\
& & & & & & \Delta/r_2
\end{bmatrix}
$$

$$
\times \begin{bmatrix}
r_3 u_1^{(p)} + b u_2^{(p)} + r_4 u_1^{(p+1/2)} \\
c u_1^{(p)} + r_3 u_2^{(p)} + r_4 u_2^{(p+1/2)} \\
r_3 u_3^{(p)} + b u_4^{(p)} + r_4 u_3^{(p+1/2)} \\
c u_3^{(p)} + r_3 u_4^{(p)} + r_4 u_4^{(p+1/2)} \\
\vdots \\
r_3 u_{m-2}^{(p)} + b u_{m-1}^{(p)} + r_4 u_{m-2}^{(p+1/2)} \\
c u_{m-2}^{(p)} + r_3 u_{m-1}^{(p)} + r_4 u_{m-1}^{(p+1/2)} \\
r_3 u_m^{(p)} + r_4 u_m^{(p+1/2)}
\end{bmatrix}
\qquad (3.17)
$$

where r_1, r_2 and Δ are given by Eq. (3.14) and $r_3 = (a/2)(1 - \omega)r$ and $r_4 = (2 - \omega)r$. This leads to

$$
\left. \begin{aligned}
u_i^{(p+1)} &= \frac{P u_i^{(p+1/2)} + Q u_{i+1}^{(p+1/2)} + R u_i^{(p)} + S u_{i+1}^{(p)}}{\Delta} \\[2mm]
u_{i+1}^{(p+1)} &= \frac{\tilde{P} u_i^{(p+1/2)} + \tilde{Q} u_{i+1}^{(p+1/2)} + \tilde{R} u_i^{(p)} + \tilde{S} u_{i+1}^{(p)}}{\Delta}
\end{aligned} \right\} \; i = 1, 3, \ldots, m-2
$$

and

$$
u_m^{(p+1)} = \frac{r_3 u_m^{(p)} + r_4 u_m^{(p+1/2)}}{r_2}
$$

$$(3.18)$$

where

$$
\left. \begin{aligned}
P &= r_2 r_4 & Q &= -b r_4 & R &= r_2 r_3 - bc & S &= b(r_2 - r_3) \\
\tilde{P} &= -c r_4 & \tilde{Q} &= P = r_2 r_4 & \tilde{R} &= c(r_2 - r_3) & \\
&& \text{and} \quad \tilde{S} &= r_2 r_3 - bc &&
\end{aligned} \right\} \quad (3.19)
$$

and the computational molecules are given by Fig. 5.

Finally, let us now consider the case when *m is even (corresponding to an odd number of intervals)*. We shall then have, from Eqs. (2.35) and (2.36),

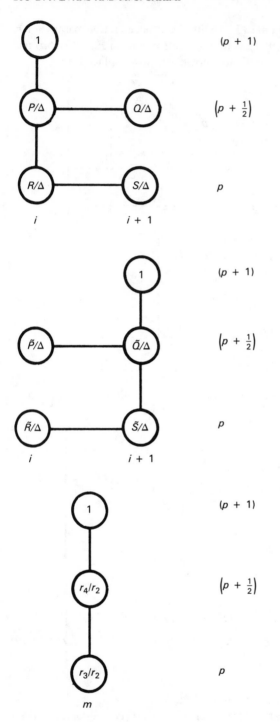

Figure 5 Generalized AGE method [explicit at level $(p + 1)$].

$$(G_1 + rI)^{-1} = \begin{bmatrix} \hat{G}^{-1} & & & & \\ & \hat{G}^{-1} & & & \\ & & & \bigcirc & \\ & & \bigcirc & & \\ & & & & \hat{G}^{-1} \end{bmatrix}_{(m \times m)} \tag{3.20}$$

$$= \frac{1}{\Delta} \begin{bmatrix} r_2 & -b & & & & & & \\ -c & r_2 & & & & & & \\ & & r_2 & -b & & & & \\ & & -c & r_2 & & & & \\ & & & & & & \bigcirc & \\ & & & & \bigcirc & & & \\ & & & & & & r_2 & -b \\ & & & & & & -c & r_2 \end{bmatrix}_{(m \times m)} \tag{3.21}$$

and

$$(G_2 + rI)^{-1} = \begin{bmatrix} 1/r_2 & & & & \\ & \hat{G}^{-1} & & & \\ & & & \bigcirc & \\ & & \bigcirc & & \\ & & & \hat{G}^{-1} & \\ & & & & 1/r_2 \end{bmatrix}_{(m \times m)} \tag{3.22}$$

The Peaceman-Rachford variant in its implicit form (3.2) can be pictorially represented by Fig. 6.

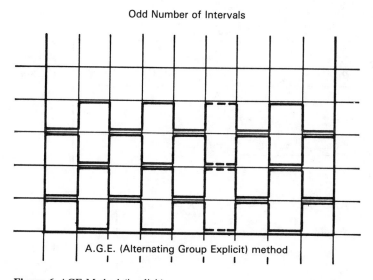

Odd Number of Intervals

A.G.E. (Alternating Group Explicit) method

Figure 6 AGE Method (implicit).

By means of Eqs. (2.35)–(2.36), (3.3), and (3.20)–(3.22), we obtain the following explicit expressions for the AGE scheme:

(i) *at level* $(p + 1/2)$

$$u_i^{(p+1/2)} = \frac{Au_{i-1}^{(p)} + Bu_i^{(p)} + Cu_{i+1}^{(p)} + Du_{i+2}^{(p)} + E_i}{\Delta}$$

and

$$u_{i+1}^{(p+1/2)} = \frac{\tilde{A}u_{i-1}^{(p)} + \tilde{B}u_i^{(p)} + \check{C}u_{i+1}^{(p)} + \tilde{D}u_{i+2}^{(p)} + \tilde{E}_i}{\Delta}$$

$$i = 1, 3, \ldots, m - 1 \qquad (3.23a)$$

where

$$A = \begin{cases} 0 & \text{for } i = 1 \\ -cr_2 & \text{otherwise} \end{cases} \qquad B = r_1 r_2 \qquad C = -br_1$$

$$D = \begin{cases} 0 & \text{for } i = m - 1 \\ b^2 & \text{otherwise} \end{cases} \qquad E_i = r_2 f_i - bf_{i+1}$$

and

$$\tilde{A} = \begin{cases} 0 & \text{for } i = 1 \\ c^2 & \text{otherwise} \end{cases} \qquad \tilde{B} = -cr_1 \qquad \tilde{C} = r_1 r_2$$

$$\tilde{D} = \begin{cases} 0 & \text{for } i = m - 1 \\ -br_2 & \text{otherwise} \end{cases} \qquad \tilde{E}_i = -cf_i + r_2 f_{i+1}$$

$$(3.23b)$$

(ii) *at level* $(p + 1)$

$$u_1^{(p+1)} = \frac{r_1 u_1^{(p+1/2)} - bu_2^{(p+1/2)} + f_1}{r_2}$$

$$u_i^{(p+1)} = \frac{Pu_{i-1}^{(p+1/2)} + Qu_i^{(p+1/2)} + Ru_{i+1}^{(p+1/2)} + Su_{i+2}^{(p+1/2)} + T_i}{\Delta}$$

$$u_{i+1}^{(p+1)} = \frac{\tilde{P}u_{i-1}^{(p+1/2)} + \tilde{Q}u_i^{(p+1/2)} + \tilde{R}u_{i+1}^{(p+1/2)} + \tilde{S}u_{i+2}^{(p+1/2)} + \tilde{T}_i}{\Delta}$$

$$(3.24a)$$

$$i = 2, 4, \ldots, m - 2$$

and

$$u_m^{(p+1)} = \frac{-cu_{m-1}^{(p+1/2)} + r_1 u_m^{(p+1/2)} + f_m}{r_2}$$

where

$$P = -cr_2 \qquad Q = r_1 r_2 \qquad R = -br_1$$

$$S = b^2 \qquad T_i = r_2 f_i - bf_{i+1}$$

and

$$\tilde{P} = c^2 \qquad \tilde{Q} = -cr_1 \qquad \tilde{R} = Q = r_1 r_2$$

$$\tilde{S} = -br_2 \qquad \tilde{T}_i = -cf_i + r_2 f_{i+1}$$

$$(3.24b)$$

For *the generalized AGE scheme* (3.1) with fixed acceleration parameter r, the same equations in (3.23a) still apply for level $(p + 1/2)$. At level $(p + 1)$, we have the following equations:

$$u_1^{(p+1)} = \frac{r_3 u_1^{(p)} + r_4 u_1^{(p+1/2)}}{r_2}$$

$$\left. \begin{array}{l} u_i^{(p+1)} = \dfrac{P u_i^{(p)} + Q u_{i+1}^{(p)} + R u_i^{(p+1/2)} + S u_{i+1}^{(p+1/2)}}{\Delta} \\[3mm] u_{i+1}^{(p+1)} = \dfrac{(\tilde{P} u_i^{(p)} + \tilde{Q} u_{i+1}^{(p)} + \tilde{R} u_i^{(p+1/2)} + \tilde{S} u_i^{(p+1/2)})}{\Delta} \end{array} \right\} \; i = 2, 4, \ldots, m - 2$$

$$u_m^{(p+1)} = \frac{r_3 u_m^{(p)} + r_4 u_m^{(p+1/2)}}{r_2}$$

$$\tag{3.25a}$$

where

$$\left. \begin{array}{ll} P = r_2 r_3 - bc & Q = b(r_2 - r_3) \\[2mm] R = r_2 r_4 & S = -b r_4 \\[2mm] \tilde{P} = c(r_2 - r_3) & \tilde{Q} = r_2 r_3 - bc \\[2mm] \tilde{R} = -c r_4 & \tilde{S} = r_2 r_4 \end{array} \right\} \qquad (3.25b)$$

and

The AGE algorithm is completed *explicitly* by using the required equations at levels $(p + 1/2)$ and $(p + 1)$ in *alternate sweeps* along all the points in the interval $(0,1)$ until a specified convergence criterion is satisfied.

4 AGE METHOD TO SOLVE SECOND ORDER PARABOLIC EQUATIONS WITH PERIODIC BOUNDARY CONDITIONS

Consider the following second order parabolic equation,

$$\frac{\partial U}{\partial t} = \frac{\partial^2 U}{\partial x^2} + k(x, t) \qquad 0 \le x \le 1 \qquad 0 < t \le T \tag{4.1a}$$

subject to the initial condition

$$U(x, 0) = f(x) \qquad 0 < x < 1$$

and the boundary condition

$$U(0, t) = U(1, t) \qquad \frac{\partial U}{\partial x}(0, t) = \frac{\partial U}{\partial x}(1, t) \tag{4.1b}$$

As in Section 2, a weighted approximation to the differential equation (4.1a) at the point $(x_i, t_{j+1/2})$ is

$$-\lambda\theta u_{i-1,j+1} + (1 + 2\lambda\theta)\, u_{i,j+1} - \lambda\theta u_{i+1,j+1} = \lambda(1 - \theta)u_{i-1,j}$$

$$+ [1 - 2\lambda(1 - \theta)]u_{ij} + \lambda(1 - \theta)u_{i+1,j} + \Delta t k_{i,j+1/2}$$

$$\text{for } i = 1, 2, \ldots, m \tag{4.2}$$

When written in the matrix form (2.3) and taking account of (4.1b) we obtain the system (2.4) where

$$c = -\lambda\theta \qquad a = (1 + 2\lambda\theta) \qquad b = -\lambda\theta$$

$$f_1 = [1 - 2\lambda(1 - \theta)]u_{1j} + \lambda(1 - \theta)u_{2j} + \lambda(1 - \theta)u_{mj} + \Delta t k_{1,j+1/2}$$

$$f_i = \lambda(1 - \theta)u_{i-1,j} + [1 - 2\lambda(1 - \theta)]u_{ij} + \lambda(1 - \theta)u_{i+1,j} + \Delta t k_{i,j+1/2}$$

$$i = 2, \ldots, m - 1$$

$$f_m = \lambda(1 - \theta)u_{1j} + \lambda(1 - \theta)u_{m-1,j} + [1 - 2\lambda(1 - \theta)]u_{mj} + \Delta t k_{m,j+1/2}$$

To implement the AGE algorithm, we split the coefficient matrix A as in Eq. (2.5) with A given by

$$A = \begin{bmatrix} a & b & & & & c \\ c & a & b & & & \\ & c & a & b & & \\ & & & \ddots & & \\ & & & c & a & b \\ b & & & & c & a \end{bmatrix} \tag{4.3}$$

and to ascertain the forms of G_1 and G_2, we treat two different cases of m as before.

(a) m even (even number of points since $u_0 = u_m$ at every level with $x_0 = 0$ and $x_m = 1$). We have

$$G_1 = \begin{bmatrix} a/2 & b & & & & \\ c & a/2 & & & & \\ & & a/2 & b & & \\ & & c & a/2 & & \\ & & & & a/2 & b \\ & & & & c & a/2 \end{bmatrix}_{(m \times m)} \tag{4.4}$$

$$G_2 = \begin{bmatrix} a/2 & & & & & c \\ & a/2 & b & & & \\ & c & a/2 & & & \\ & & & a/2 & b & \\ & & & c & a/2 & \\ b & & & & & a/2 \end{bmatrix}_{(m \times m)} \tag{4.5}$$

and

$$(G_1 + rI)^{-1} = \frac{1}{\Delta} \begin{bmatrix} r_2 & -b & & & & & \\ -c & r_2 & & & & & \\ & & r_2 & -b & & & \\ & & -c & r_2 & \bigcirc & & \\ & & & & & & \\ & & \bigcirc & & & & \\ & & & & & r_2 & -b \\ & & & & & -c & r_2 \end{bmatrix}_{(m \times m)}$$ (4.6)

and

$$(G_2 + rI)^{-1} = \frac{1}{\Delta} \begin{bmatrix} r_2 & & & & & & -c \\ & r_2 & -b & & & & \\ & -c & r_2 & & & & \\ & & & & \bigcirc & & \\ & & & \bigcirc & r_2 & -b & \\ & & & & -c & r_2 & \\ -b & & & & & & r_2 \end{bmatrix}_{(m \times m)}$$ (4.7)

Hence, using Eqs. (3.1) and (4.4)–(4.7) we obtain the following equations for the computation of *the generalized AGE scheme*.

(i) at level $(p + 1/2)$

$$\begin{bmatrix} u_1^{(p+1/2)} \\ u_2^{(p+1/2)} \\ u_3^{(p+1/2)} \\ \vdots \\ \vdots \\ u_{m-1}^{(p+1/2)} \\ u_m^{(p+1/2)} \end{bmatrix} = \frac{1}{\Delta} \begin{bmatrix} r_2 & -b & & & & & \\ -c & r_2 & & & & & \\ & & r_2 & -b & & & \\ & & -c & r_2 & \bigcirc & & \\ & & & & & & \\ & & \bigcirc & & & & \\ & & & & & r_2 & -b \\ & & & & & -c & r_2 \end{bmatrix}$$

$$\times \begin{bmatrix} r_1 u_1^{(p)} - c u_m^{(p)} + f_1 \\ r_1 u_2^{(p)} - b u_3^{(p)} + f_2 \\ -c u_2^{(p)} + r_1 u_3^{(p)} + f_3 \\ r_1 u_4^{(p)} - b u_5^{(p)} + f_4 \\ -c u_4^{(p)} + r_1 u_5^{(p)} + f_5 \\ \vdots \\ r_1 u_{m-2}^{(p)} - b u_{m-1}^{(p)} + f_{m-2} \\ -c u_{m-2}^{(p)} + r_1 u_{m-1}^{(p)} + f_{m-1} \\ -b u_1^{(p)} + r_1 u_m^{(p)} + f_m \end{bmatrix}$$ (4.8)

which leads to

$$u_i^{(p+1/2)} = \frac{Au_{i-1}^{(p)} + Bu_i^{(p)} + Cu_{i+1}^{(p)} + Du_{i+2}^{(p)} + E_i}{\Delta}$$

$$\left.\begin{array}{r} 1, 3, \ldots, m-1 \end{array}\right.$$

and

$$u_{i+1}^{(p+1/2)} = \frac{\tilde{A}u_{i-1}^{(p)} + \tilde{B}u_i^{(p)} + \tilde{C}u_{i+1}^{(p)} + \tilde{D}u_{i+2}^{(p)} + \tilde{E}_i}{\Delta}$$

$$\left.\begin{array}{r}\\\\\\\\\end{array}\right\}(4.9a)$$

with $\quad u_0^{(p)} = u_m^{(p)} \quad$ and $\quad u_{m+1}^{(p)} = u_1^{(p)}$

where

$$A = -cr_2 \quad B = r_1r_2 \quad c = -br_1 \quad D = b^2 \quad E_i = r_2f_i - bf_{i+1}$$

and

$$\left.\begin{array}{r}\\\\\end{array}\right\}(4.9b)$$

$$\tilde{A} = c^2 \quad \tilde{B} = -cr_1 \quad \tilde{C} = r_1r_2 \quad \tilde{D} = -br_2 \quad \tilde{E}_i = -cf_i + r_2f_{i+1}$$

(ii) at level (p + 1)

$$\begin{bmatrix} u_1^{(p+1)} \\ u_2^{(p+1)} \\ \vdots \\ \vdots \\ \vdots \\ u_{m-1}^{(p+1)} \\ u_m^{(p+1)} \end{bmatrix} = \frac{1}{\Delta} \begin{bmatrix} r_2 & & & & & -c \\ & r_2 & -b & & & \\ & -c & r_2 & & & \\ & & & \bigcirc & & \\ & & & \bigcirc & r_2 & -b \\ & & & & -c & r_2 \\ -b & & & & & r_2 \end{bmatrix}$$

$$\times \begin{bmatrix} r_3u_1^{(p)} + cu_m^{(p)} + r_4u_1^{(p+1/2)} \\ r_3u_2^{(p)} + bu_3^{(p)} + r_4u_2^{(p+1/2)} \\ cu_2^{(p)} + r_3u_3^{(p)} + r_4u_3^{(p+1/2)} \\ r_3u_4^{(p)} + bu_5^{(p)} + r_4u_4^{(p+1/2)} \\ cu_4^{(p)} + r_3u_5^{(p)} + r_4u_5^{(p+1/2)} \\ \vdots \\ r_3u_{m-2}^{(p)} + bu_{m-1}^{(p)} + r_4u_{m-2}^{(p+1/2)} \\ cu_{m-2}^{(p)} + r_3u_{m-1}^{(p)} + r_4u_{m-1}^{(p+1/2)} \\ bu_1^{(p)} + r_3u_m^{(p)} + r_4u_m^{(p+1/2)} \end{bmatrix} \quad (4.10)$$

This gives

$$\underline{u_i^{(p+1)} = (Pu_{i-1}^{(p)} + Qu_i^{(p)} + Ru_{i-1}^{(p+1/2)} + Su_i^{(p+1/2)})}$$
$$\Delta$$

and

$$\left. \begin{array}{l} \\ \\ \underline{u_{i+1}^{(p+1)} = (\tilde{P}u_{i+1}^{(p)} + \tilde{Q}u_{i+2}^{(p)} + \tilde{R}u_{i+1}^{(p+1/2)} + \tilde{S}u_{i+2}^{(p+1/2)})} \\ \Delta \end{array} \right\} \begin{array}{l} i = 1, 3, \ldots, m-1 \end{array} \quad (4.11a)$$

with $u_0 = u_m$ and $u_1 = u_{m+1}$ at both alternating levels
where

$$\begin{array}{llll} & P = c(r_2 - r_3) & Q = r_2 r_3 - bc & R = -cr_4 & S = r_2 r_4 \\ \text{and} & \tilde{P} = r_2 r_3 - bc & \tilde{Q} = b(r_2 - r_3) & \tilde{R} = r_2 r_4 & \tilde{S} = -br_4 \end{array} \right\} \quad (4.11b)$$

  (b) m odd (odd number of points). We have

$$G_1 = \begin{bmatrix} a/2 & & & & & & \\ & a/2 & b & & & & \\ & c & a/2 & & & & \\ & & & a/2 & b & & \\ & & & c & a/2 & \bigcirc & \\ & & & & & \ddots & \\ & & & \bigcirc & & a/2 & b \\ & & & & & c & a/2 \end{bmatrix}_{(m \times m)} \quad (4.12)$$

$$G_2 = \begin{bmatrix} a/2 & b & & & & & c \\ c & a/2 & & & & & \\ & & a/2 & b & & & \\ & & c & a/2 & & & \\ & & & & & \bigcirc & \\ & & & & \bigcirc & \ddots & \\ b & & & & & a/2 \end{bmatrix}_{(m \times m)} \quad (4.13)$$

and

$$(G_1 + rI)^{-1} = \frac{1}{\Delta} \begin{bmatrix} \Delta/r_2 & & & & & & \\ & r_2 & -b & & & & \\ & -c & r_2 & & & & \\ & & & r_2 & -b & & \\ & & & -c & r_2 & \bigcirc & \\ & & & & & \ddots & \\ & & & \bigcirc & & r_2 & -b \\ & & & & & -c & r_2 \end{bmatrix} \quad (4.14)$$

and

$$(G_2 + rI)^{-1} = \frac{1}{\Delta(\Delta_1)}$$

$$\times \begin{bmatrix} r_2\Delta & -b\Delta & & & & & & -c\Delta \\ -c\Delta & \Delta^2/r_2 & & & & & & c^2\Delta/r_2 \\ & & r_2\Delta_1 & -b\Delta_1 & & & & \\ & & -c\Delta_1 & r_2\Delta_1 & & & & \\ & & & & \bigcirc & & & \\ & & & & & r_2\Delta_1 & -b\Delta_1 & \\ & & & & \bigcirc & -c\Delta_1 & r_2\Delta_1 & \\ -b\Delta & b^2\Delta/r_2 & & & & & & \Delta^2/r_2 \end{bmatrix} \quad (4.15)$$

where
$$\Delta_1 = r_2^2 - 2bc \quad (4.16)$$

We shall now derive the AGE equations using Eqs. (3.1) and (4.12)–(4.15).
  (i) *level* $(p + 1/2)$

$$\begin{bmatrix} u_1^{(p+1/2)} \\ u_2^{(p+1/2)} \\ \vdots \\ \vdots \\ \vdots \\ u_{m-1}^{(p+1/2)} \\ u_m^{(p+1/2)} \end{bmatrix} = \frac{1}{\Delta} \begin{bmatrix} \Delta/r_2 & & & & \\ & r_2 & -b & & \\ & -c & r_2 & & \\ & & & \bigcirc & \\ & & & \bigcirc & \\ & & & & r_2 & -b \\ & & & & -c & r_2 \end{bmatrix}$$

$$\times \begin{bmatrix} r_1 u_1^{(p)} - b u_2^{(p)} - c u_m^{(p)} + f_1 \\ -c u_1^{(p)} + r_1 u_2^{(p)} + f_2 \\ r_1 u_3^{(p)} - b u_4^{(p)} + f_3 \\ -c u_3^{(p)} + r_1 u_4^{(p)} + f_4 \\ \vdots \\ r_1 u_{m-2}^{(p)} - b u_{m-1}^{(p)} + f_{m-2} \\ -c u_{m-2}^{(p)} + r_1 u_{m-1}^{(p)} + f_{m-1} \\ r_1 u_m^{(p)} + f_m - b u_1^{(p)} \end{bmatrix} \quad (4.17a)$$

Hence, we obtain

$$
\left.\begin{array}{l}
u_1^{(p+1/2)} = \dfrac{r_1 u_1^{(p)} - b u_2^{(p)} - c u_m^{(p)} + f_1}{r_2} \\[2ex]
u_i^{(p+1/2)} = \dfrac{A u_{i-1}^{(p)} + B u_i^{(p)} + C u_{i+1}^{(p)} + D u_{i+2}^{(p)} + E_i}{\Delta} \\[2ex]
\qquad\qquad i = 2, 4, \ldots, m-3, m-1
\end{array}\right\} \quad (4.17b)
$$

and

$$
\left.u_{i+1}^{(p+1/2)} = \dfrac{\tilde{A} u_{i-1}^{(p)} + \tilde{B} u_i^{(p)} + \tilde{C} u_{i+1}^{(p)} + \tilde{D} u_{i+2}^{(p)} + \tilde{E}_i}{\Delta}\right\}
$$

with $u_1^{(p)} = u_{m+1}^{(p)}$

where

$$
\left.\begin{array}{l}
A = -cr_2 \quad B = r_1 r_2 \quad C = -br_1 \quad D = b^2 \quad E_i = r_2 f_i - b f_{i+1} \\[2ex]
\tilde{A} = c^2 \quad \tilde{B} = -cr_1 \quad \tilde{C} = r_1 r_2 \quad \tilde{D} = -br_2 \quad \tilde{E}_i = -c f_i + r_2 f_{i+1}
\end{array}\right\} \quad (4.17c)
$$

and

(ii) level (p + 1)

$$
\begin{bmatrix}
u_1^{(p+1)} \\
u_2^{(p+1)} \\
\vdots \\
\vdots \\
u_{m-1}^{(p+1)} \\
u_m^{(p+1)}
\end{bmatrix}
= \frac{1}{\Delta(\Delta_1)}
\left[\begin{array}{cc:cc:c:cc}
r_2\Delta & -b\Delta & & & & & -c\Delta \\
-c\Delta & \Delta^2/r_2 & & & & & c^2\Delta/r_2 \\ \hdashline
& & r_2\Delta_1 & -b\Delta_1 & & & \\
& & -c\Delta_1 & r_2\Delta_1 & & & \\ \hdashline
& & & & \bigcirc & & \\ \hdashline
& & & \bigcirc & & r_2\Delta_1 & -b\Delta_1 \\
& & & & & -c\Delta_1 & r_2\Delta_1 \\ \hdashline
-b\Delta & b^2\Delta/r_2 & & & & & \Delta^2/r_2
\end{array}\right]
$$

$$
\times
\begin{bmatrix}
r_3 u_1^{(p)} + b u_2^{(p)} + c u_m^{(p)} + r_4 u_1^{(p+1/2)} \\
c u_1^{(p)} + r_3 u_2^{(p)} + r_4 u_2^{(p+1/2)} \\
r_3 u_3^{(p)} + b u_4^{(p)} + r_4 u_3^{(p+1/2)} \\
c u_3^{(p)} + r_3 u_4^{(p)} + r_4 u_4^{(p+1/2)} \\
\vdots \\
r_3 u_{m-2}^{(p)} + b u_{m-1}^{(p)} + r_4 u_{m-2}^{(p+1/2)} \\
c u_{m-2}^{(p)} + r_3 u_{m-1}^{(p)} + r_4 u_{m-1}^{(p+1/2)} \\
b u_1^{(p)} + r_3 u_m^{(p)} + r_4 u_m^{(p+1/2)}
\end{bmatrix}
\qquad (4.18)
$$

which leads to

$$
u_1^{(p+1)} = \frac{P_1 u_1^{(p)} + P_2 u_2^{(p)} + P_3 u_m^{(p)} + P_4 u_1^{(p+1/2)} + P_5 u_2^{(p+1/2)} + P_6 u_m^{(p+1/2)}}{\Delta_1}
$$

$$
u_2^{(p+1)} = \frac{Q_1 u_1^{(p)} + Q_2 u_2^{(p)} + Q_3 u_m^{(p)} + Q_4 u_1^{(p+1/2)} + Q_5 u_2^{(p+1/2)} + Q_6 u_m^{(p+1/2)}}{\Delta_1}
$$

$$
\left. \begin{array}{l}
u_i^{(p+1)} = \dfrac{P u_i^{(p)} + Q u_{i+1}^{(p)} + R u_i^{(p+1/2)} + S u_{i+1}^{(p+1/2)}}{\Delta} \\[4mm]
u_{i+1}^{(p+1)} = \dfrac{\tilde{P} u_i^{(p)} + \tilde{Q} u_{i+1}^{(p)} + \tilde{R} u_i^{(p+1/2)} + \tilde{S} u_{i+1}^{(p+1/2)}}{\Delta}
\end{array} \right\} \quad i = 3, 5, \ldots, m-2
$$

$$
u_m^{(p+1)} = \frac{R_1 u_1^{(p)} + R_2 u_2^{(p)} + R_3 u_m^{(p)} + R_4 u_1^{(p+1/2)} + R_5 u_2^{(p+1/2)} + R_6 u_m^{(p+1/2)}}{\Delta_1}
$$

$$(4.19a)$$

where

$$
\begin{aligned}
&P_1 = r_2 r_3 - 2bc \qquad P_2 = b(r_2 - r_3) \qquad P_3 = c(r_2 - r_3) \qquad P_4 = r_2 r_4 \\
& P_5 = -br_4 \quad P_6 = -cr_4 \\[2mm]
&Q_1 = c(r_2 - r_3) \qquad Q_2 = -bc + \frac{r_3 \Delta}{r_2} \qquad Q_3 = -c^2 \frac{r_2 - r_3}{r_2} \qquad Q_4 = -cr_4 \\[2mm]
& Q_5 = \frac{r_4 \Delta}{r_2} \quad Q_6 = \frac{r_4 c^2}{r_2} \\[2mm]
&P = r_2 r_3 - bc \qquad Q = b(r_2 - r_3) \qquad R = r_2 r_4 \qquad S = -br_4 \\
&\tilde{P} = c(r_2 - r_3) \qquad \tilde{Q} = r_2 r_3 - bc \qquad \tilde{R} = -cr_4 \qquad \tilde{S} = r_2 r_4 \\[2mm]
&R_1 = b(r_2 - r_3) \qquad R_2 = b^2 \frac{r_3 - r_2}{r_2} \qquad R_3 = -bc + \frac{r_3 \Delta}{r_2} \qquad R_4 = -br_4 \\[2mm]
& R_5 = \frac{b^2 r_4}{r_2} \quad R_6 = \frac{\Delta r_4}{r_2}
\end{aligned}
$$

$$(4.19b)$$

The iterative process is continued for each alternate sweep until convergence is reached.

5 AGE METHOD TO SOLVE THE DIFFUSION–CONVECTION EQUATION

Consider the following diffusion-convection equation

$$\frac{\partial U}{\partial t} = \epsilon \frac{\partial^2 U}{\partial x^2} - k \frac{\partial U}{\partial x} \tag{5.1}$$

with its Dirichlet boundary conditions at $x = 0, 1$ specified by Eq. (2.1b). At the point $(x_i, t_{j+1/2})$, the derivatives in Eq. (5.1) are approximated by generalized finite differences (Evans and Abdullah, 1983). Thus we are led to the following generalized formula,

$$-(E\theta_2 + K\alpha_3)u_{i-1,j+1} + [1 + E(\theta_1 + \theta_2) - K(\alpha_1 - \alpha_3)]u_{i,j+1}$$

$$- (E\theta_1 - K\alpha_1)u_{i+1,j+1} = (E\theta_4 + K\alpha_2)u_{i-1,j}$$

$$+ [1 - E(\theta_3 + \theta_4) + K(\alpha_4 - \alpha_2)]u_{ij} + (E\theta_3 - K\alpha_4)u_{i+1,j}$$

$$i = 1, 2, \ldots, m \tag{5.2a}$$

where

$$\lambda = \frac{\Delta t}{(\Delta x)^2} \qquad E = \epsilon\lambda \qquad K = \frac{1}{2}k\lambda\Delta x \tag{5.2b}$$

Different choices of the weighting parameters α, θ lead to a variety of finite-difference schemes. As examples, we have

(a) $\theta_1 = \theta_2 = 0$, $\alpha_1 = \alpha_3 = 0$, $\theta_3 = \theta_4 = 1$, $\alpha_2 = \alpha_4 = 1$. Equation (5.2) reduces to

$$u_{i,j+1} = (E + K)u_{i-1,j} + (1 - 2E)u_{ij} + (E - K)u_{i+1,j} \tag{5.3}$$

which is *the classical explicit* scheme with $O[(\Delta x)^2 + \Delta t]$ accuracy.

(b) $\theta_1 = \theta_2 = 1$, $\alpha_1 = \alpha_3 = 1$, $\theta_3 = \theta_4 = 0$, $\alpha_2 = \alpha_4 = 0$ gives the following *fully implicit* scheme:

$$-(E + K)u_{i-1,j+1} + (1 + 2E)u_{i,j+1} - (E - K)u_{i+1,j+1} = u_{ij} \tag{5.4}$$

with $O[(\Delta x)^2 + \Delta t]$ accuracy.

(c) $\theta_1 = \theta_2 = \theta_3 = \theta_4 = 1/2$, $\alpha_2 = \alpha_3 = 1$, $\alpha_1 = \alpha_4 = 0$ gives

$$-\frac{1}{2}(E + 2K)u_{i-1,j+1} + (1 + E + K)u_{i,j+1} - \frac{1}{2}Eu_{i+1,j+1}$$

$$= \frac{1}{2}(E + 2K)u_{i-1,j} + (1 - E - K)u_{ij} + \frac{1}{2}Eu_{i+1,j} \tag{5.5}$$

which is *the Crank-Nicolson scheme with upwinding* with $O[\Delta x + (\Delta t)^2]$ accuracy.

(d) $\theta_1 = \theta_2 = \theta_3 = \theta_4 = 1/2$, $\alpha_1 = \alpha_2 = \alpha_3 = \alpha_4 = 1/2$ gives

$$-\frac{1}{2}(E + K)u_{i-1,j+1} + (1 + E)u_{i,j+1} - \frac{1}{2}(E - K)u_{i+1,j+1}$$

$$= \frac{1}{2}(E + K)u_{i-1,j} + (1 - E)u_{ij} + \frac{1}{2}(E - K)u_{i+1,j} \qquad (5.6)$$

which is *the Crank-Nicolson* scheme with $O[(\Delta x)^2 + (\Delta t)^2]$ accuracy.

The generalized finite-difference equation $(5.2a)$ generates a tridiagonal system of linear equations of the form (2.3) and $(2.4a)$ where

$$c = -(E\theta_2 + K\alpha_3) \qquad a = 1 + E(\theta_1 + \theta_2) - K(\alpha_1 - \alpha_3) \qquad b = -(E\theta_1 - K\alpha_1)$$

$$f_1 = eu_{0j} - cu_{0,j+1} + fu_{1j} + gu_{2j}$$

$$f_i = eu_{i-1,j} + fu_{ij} + gu_{i+1,j} \qquad i = 2, 3, \ldots, m - 1$$

$$f_m = eu_{m-1,j} + fu_{mj} + gu_{m+1,j} - bu_{m+1,j+1}$$

with $e = E\theta_4 + K\alpha_2$, $f = 1 - E(\theta_3 + \theta_4) + K(\alpha_4 - \alpha_2)$, $g = E\theta_3 - K\alpha_4$.

The AGE algorithm is applied to the system $(5.2a)$ or $(2.4a)$ with the usual splitting of the coefficient matrix A resulting in exactly the same equations for computation as was found in Section 3.

6 AGE METHOD TO SOLVE A FIRST ORDER HYPERBOLIC EQUATION

A first order hyperbolic equation takes the general form

$$\frac{\partial U}{\partial t} = -\frac{\partial U}{\partial x} + k(x, t) \qquad 0 \leq x \leq 1, 0 < t \leq T \qquad (6.1)$$

The Dirichlet boundary conditions of $(2.1b)$ may be specified on the above hyperbolic problem. Following the previous arguments, Eq. (6.1) is replaced at the point $(x_i, t_{j+\theta})$ by the difference analogue

$$-\frac{1}{2}\lambda\theta u_{i-1,j+1} + u_{i,j+1} + \frac{1}{2}\lambda\theta u_{i+1,j+1} = \frac{1}{2}\lambda(1 - \theta)u_{i-1,j} + u_{ij}$$

$$-\frac{1}{2}\lambda(1 - \theta)u_{i+1,j} + \Delta t k_{i,j+\theta} \qquad i = 1, 2, \ldots, m \qquad (6.2a)$$

Again, this tridiagonal system of linear equations can be presented in the matrix form [Eqs. (2.3) and (2.4a)] where,

$$c = -\frac{1}{2}\lambda\theta \qquad a = 1 \qquad b = \frac{1}{2}\lambda\theta$$

$$f_1 = u_{1j} + \frac{1}{2}\lambda(1 - \theta)(u_{0j} - u_{2j}) + \frac{1}{2}\lambda\theta u_{0,j+1} + \Delta t k_{1,j+\theta}$$

$$f_i = \frac{1}{2}\lambda(1 - \theta)(u_{i-1,j} - u_{i+1,j}) + u_{ij} + \Delta t k_{i,j+\theta}$$

$$i = 2, 3, \ldots, m - 2, m - 1$$

and $\quad f_m = u_{mj} + \frac{1}{2}\lambda(1 - \theta)(u_{m-1,j} - u_{m+1,j}) - \frac{1}{2}\lambda\theta u_{m+1,j+1} + \Delta t k_{m,j+\theta}$ $\qquad$ (6.2b)

We note that Eq. (6.2a) corresponds to the explicit (classical), fully implicit (centered-in-distance, backward-in-time) and the Crank-Nicolson type (centered-in-distance, centered-in-time) formula when $\theta = 0$, 1, and $1/2$ with accuracies to the order of $O(\Delta x + \Delta t)$, $O[(\Delta x)^2 + \Delta t]$, and $O[(\Delta x)^2 + (\Delta t)^2]$, respectively.

When the AGE algorithm is implemented on the above system, we will arrive at exactly the same form of equations along the lines $(p + 1/2)$ and $(p + 1)$ as those that we derived for the heat equation (2.1a). These explicit equations are given by (3.15a)–(3.16b), (3.18)–(3.19), and (3.23a)–(3.25a).

It must be mentioned, however, that the constituent matrices G_1 and G_2 are not symmetric since from Eq. (6.2a), $c \neq b$. In fact, both matrices (without loss of generality, we take the case m odd only) have one real and $(m - 1)$ complex eigenvalues given by $1/2$, $(1/2)(1 + \lambda\theta i_c)$ [of multiplicity $(1/2)(m - 1)$] and $(1/2)(1 - \lambda\theta i_c)$ [also of multiplicity $(1/2)(m - 1)$] where $i_c = \sqrt{-1}$. The same analysis to prove Theorem 2.1 cannot, therefore, be used to establish the condition (2.15) for convergence. However, an attempt made to employ the inequality (2.16) turned out to be unsuccessful. We have

$$(rI - G_1)(G_1 + rI)^{-1} = \frac{1}{\Delta}$$

$$\times \begin{bmatrix} \Delta r_1/r_2 & & & \\ & (r_1r_2 + bc) & -b(r_1 + r_2) & & \\ & -c(r_1 + r_2) & (r_1r_2 + bc) & & O \\ & & & \ddots & \\ & & O & & (r_1r_2 + bc) & -b(r_1 + r_2) \\ & & & & -c(r_1 + r_2) & (r_1r_2 + bc) \end{bmatrix} \qquad (6.3)$$

and similarly,

$$(rI - G_2)(G_2 + rI)^{-1} = \frac{1}{\Delta}$$

$$\times \begin{bmatrix} \begin{matrix} (r_1r_2 + bc) & -b(r_1 + r_2) \\ -c(r_1 + r_2) & r_1r_2 + bc \end{matrix} & & & \\ & \bigcirc & & \bigcirc \\ & & \begin{matrix} (r_1r_2 + bc) & -b(r_1 + r_2) \\ -c(r_1 + r_2) & (r_1r_2 + bc) \end{matrix} & \\ & & & \Delta r_1/r_2 \end{bmatrix} \tag{6.4}$$

By virtue of Eq. (6.2b), we obtain

$$r_1 = r - \frac{1}{2} \qquad r_2 = r + \frac{1}{2} \qquad r_1r_2 + bc = r^2 - \frac{1}{4}(1 + \lambda^2\theta^2)$$

$$-b(r_1 + r_2) = -\lambda\theta r \qquad -c(r_1 + r_2) = \lambda\theta r$$

$$\Delta = \left(r + \frac{1}{2}\right)^2 + \frac{1}{4}\lambda^2\theta^2 \tag{6.5}$$

Hence by taking the L_∞ norm on inequality (2.16), we find that

$$\left\|(rI - G_1)(G_1 + rI)^{-1}\right\|_\infty = \left\|(rI - G_2)(G_2 + rI)^{-1}\right\|_\infty$$

$$= \max\left[\frac{|r - 1/2|}{(r + 1/2)}, \frac{|r^2 - 1/4(1 + \lambda^2\theta^2)| + \lambda\theta r}{(r + 1/2)^2 + 1/4(\lambda^2\theta^2)}\right]$$

$$\text{for } \lambda, r > 0 \text{ and } 0 \le \theta \le 1. \tag{6.6}$$

If, for example, we prescribe $\lambda = 4$, $r = 1$, and $\theta = 1$, we get

$$\left\|(rI - G_1)(G_1 + rI)^{-1}\right\|_\infty = \left\|(rI - G_2)(G_2 + rI)^{-1}\right\|_\infty$$

$$= 1.16$$

which leads to

$$\|\tilde{M}(r)\|_\infty > 1 \tag{6.7}$$

and this clearly does not satisfy the *sufficient* condition for convergence (i.e., $\|\tilde{M}(r)\| < 1$). The failure of this test does not necessarily imply nonconvergence of the AGE iterative process. It serves to confirm the theoretical difficulty that arises in dealing with unsymmetric matrices. A direct derivation of the eigenvalues of the AGE iteration matrix therefore becomes necessary and this can be very cumbersome if not impossible. An alternative to this analytical approach is to evaluate them numerically by means of, for example, the *power method* to obtain the dominant eigenvalue. This would enable us to show that $\rho[M(r)] < 1$.

7 AGE METHOD TO SOLVE A SECOND ORDER HYPERBOLIC (WAVE) EQUATION

(a) Wave Equation with Dirichlet Boundary Conditions
We seek the solution to the wave equation

$$\frac{\partial^2 U}{\partial x^2} = \frac{\partial^2 U}{\partial t^2} \qquad 0 \le x \le 1, 0 \le t < T \qquad (7.1a)$$

subject to the following auxiliary conditions

$$U(x, 0) = f(x)$$

$$\frac{\partial U}{\partial t}(x, 0) = g(x) \qquad 0 \le x \le 1 \qquad (7.1b)$$

$$\begin{aligned} U(0, t) &= h(t) \\ U(1, t) &= k(t) \end{aligned} \qquad 0 \le t < T \qquad (7.1c)$$

the general three-level implicit formula approximating $(7.1a)$ at the point $(i\Delta x, j\Delta t)$ is

$$\frac{1}{(\Delta t)^2} \delta_t^2 u_{ij} = \frac{1}{(\Delta x)^2} [\alpha \delta_x^2 u_{i,j+1} + (1 - 2\alpha)\delta_x^2 u_{ij} + \alpha \delta_x^2 u_{i,j-1}] \qquad (7.2)$$

where α, a weighting factor, takes the values of $\alpha \ge \frac{1}{4}$ for stability. The truncation error is $O[(\Delta x)^2 + (\Delta t)^2]$ for $\alpha = \frac{1}{4}$ and $\frac{1}{2}$. On expanding Eq. (7.2), we obtain

$$-\alpha \lambda^2 u_{i-1,j+1} + (1 + 2\alpha \lambda^2)u_{i,j+1} - \alpha \lambda^2 u_{i+1,j+1} = (1 - 2\alpha)\lambda^2 u_{i-1,j}$$

$$+ 2[1 - (1 - 2\alpha)\lambda^2]u_{ij} + (1 - 2\alpha)\lambda^2 u_{i+1,j} + \alpha \lambda^2 u_{i-1,j-1}$$

$$- (1 + 2\alpha \lambda^2)u_{i,j-1} + \alpha \lambda^2 u_{i+1,j-1} \qquad i = 1, 2, \ldots, m \qquad (7.3a)$$

which gives a tridiagonal system of equations that can be displayed in the matrix form [Eqs. (2.3) and $(2.4a)$] where

$$c = -\alpha \lambda^2 \qquad a = 1 + 2\alpha \lambda^2 \qquad b = -\alpha \lambda^2 \qquad \lambda = \frac{\Delta t}{\Delta x}$$

$$f_1 = 2[1 - (1 - 2\alpha)\lambda^2]u_{1j} + (1 - 2\alpha)\lambda^2 u_{2j} + [-(1 + 2\alpha \lambda^2)u_{1,j-1}$$

$$+ \alpha \lambda^2 u_{2,j-1}] + \alpha \lambda^2 (u_{0,j+1} + u_{0,j-1}) + (1 - 2\alpha)\lambda^2 u_{0j}$$

$$f_i = (1 - 2\alpha)\lambda^2 u_{i-1,j} + 2[1 - (1 - 2\alpha)\lambda^2]u_{ij} + (1 - 2\alpha)\lambda^2 u_{i+1,j}$$

$$+ \alpha \lambda^2 u_{i-1,j-1} - (1 + 2\alpha \lambda^2)u_{i,j-1} + \alpha \lambda^2 u_{i+1,j-1}$$

$$i = 2, 3, \ldots, m - 1$$

and

$$f_m = [(1 - 2\alpha)\lambda^2 u_{m-1,j} + 2[1 - (1 - 2\alpha)\lambda^2]u_{mj} + \alpha\lambda^2 u_{m-1,j-1}$$
$$- (1 + 2\alpha\lambda^2)u_{m,j-1} + \alpha\lambda^2(u_{m+1,j+1} + u_{m+1,j-1})$$
$$+ (1 - 2\alpha)\lambda^2 u_{m+1,j}$$

The u values on the first time level are given by the initial condition. Values on the second time level are obtained from applying the forward difference approximation to first order, at $t = 0$,

$$\frac{\partial U}{\partial t}(x_i, 0) \approx \frac{u_{i,1} - u_{i,0}}{\Delta t}$$

$$= g_i \qquad \text{[from Eq. (7.1b)]} \qquad (7.3b)$$

giving $u_{i1} = u_{i0} + \Delta t g_i$. Solutions on the third and subsequent time levels are generated iteratively by applying the AGE algorithm on lines $(p + 1/2)$ and $(p + 1)$ and utilizing the same explicit equations used by the first order hyperbolic equation of Eq. (6.1), i.e., the formulas (3.15a)–(3.16a), (3.18)–(3.19) and (3.23a)–(3.25a) for both cases of even and odd number of intervals.

(b) Wave Equation with Derivative Boundary Condition

We will again solve the wave equation (7.1a) together with the initial conditions (7.1b), but now the boundary conditions (7.1c) are replaced by

$$\frac{\partial U}{\partial x}(0, t) = h(t)$$

and $\qquad\qquad U(1, t) = k(t) \qquad\qquad (7.4)$

A central difference approximation is used to represent the boundary condition at $x = 0$, i.e.,

$$\frac{u_{1j} - u_{-1,j}}{2\Delta x} = h_j$$

giving $\qquad\qquad u_{-1,j} = u_{1j} - 2\Delta x h_j \qquad\qquad (7.5a)$

Similarly we have

$$u_{-1,j-1} = u_{1,j-1} - 2\Delta x h_{j-1} \qquad\qquad (7.5b)$$

and $\qquad\qquad u_{-1,j+1} = u_{1,j+1} - 2\Delta x h_{j+1} \qquad\qquad (7.5c)$

Now, returning to Eq. (7.3a) and putting $i = 0$, we get

$$-\alpha\lambda^2 u_{-1,j+1} + (1 - 2\alpha\lambda^2)u_{0,j+1} - \alpha\lambda^2 u_{1,j+1} = (1 - 2\alpha)\lambda^2 u_{-1,j}$$
$$+ 2[1 - (1 - 2\alpha)\lambda^2]u_{0,j} + (1 - 2\alpha)\lambda^2 u_{1,j} + \alpha\lambda^2 u_{-1,j-1}$$
$$- (1 + 2\alpha\lambda^2)u_{0,j-1} + \alpha\lambda^2 u_{1,j-1} \qquad\qquad (7.6)$$

The insertion of the calculated values (7.5a)–(7.5c) into Eq. (7.6) at the fictitious points $(-1, j-1)$, $(-1, j)$, and $(-1, j+1)$ yields

$$(1 + 2\alpha\lambda^2)u_{0,j+1} - 2\alpha\lambda^2 u_{1,j+1} = 2[1 - (1 - 2\alpha)\lambda^2]u_{0j}$$

$$+ 2(1 - 2\alpha)\lambda^2 u_{1j} - (1 + 2\alpha\lambda^2)u_{0,j-1} + 2\alpha\lambda^2 u_{1,j-1}$$

$$- 2(\Delta x)\alpha\lambda^2 h_{j-1} - 2(\Delta x)(1 - 2\alpha)\lambda^2 h_j - 2(\Delta x)\alpha\lambda^2 h_{j+1} \qquad (7.7)$$

Hence when the tridiagonal system of equations incorporating Eqs. (7.7) and (7.3a) (for $i = 2, 3, \ldots, m-1, m$) is written in matrix form, we have

$$\mathbf{Au} = \mathbf{f} \qquad (7.8)$$

where

$$A = \begin{bmatrix} a & b_1 & & & & \\ c & a & b & & & \\ & c & a & b & \bigcirc & \\ & & \ddots & \ddots & \ddots & \\ & \bigcirc & & c & a & b \\ & & & & c & a \end{bmatrix}_{(m+1)\times(m+1)} \qquad (7.9)$$

$$\mathbf{u} = (u_0, u_1, \ldots, u_{m-1}, u_m)^T \text{ [the unknown values at level } (j + 1)]$$

$$\mathbf{f} = (f_0, f_1, \ldots, f_{m-1}, f_m)^T \text{ [the known values at the preceding levels]}$$

and $\quad c = -\alpha\lambda^2 \quad a = 1 + 2\alpha\lambda^2 \quad b = -\alpha\lambda^2 \quad b_1 = -2\alpha\lambda^2$

As in (a), the solutions on the first time level are obtained from the initial condition and the values on the second level from Eq. (7.3b).

To be in conformity with previous convention, we shall take the values of u in the order of $u_1, u_2, \ldots, u_M$ where $M = m + 1$. Hence, u_1 corresponds with the value of u at $x = 0$, u_M with u at $x = (M - 1)\Delta x = m\Delta x$ and u_{M+1} with u at $x = M\Delta x = (m + 1)\Delta x = 1$ on the line segment $0 \le x \le 1$. We then have, from Eq. (7.8),

$$\begin{bmatrix} a & b_1 & & & & \\ c & a & b & & & \\ & c & a & b & \bigcirc & \\ & & \ddots & \ddots & \ddots & \\ & \bigcirc & & c & a & b \\ & & & & c & a \end{bmatrix}_{M\times M} \begin{bmatrix} u_1 \\ u_2 \\ \vdots \\ \vdots \\ u_{M-1} \\ u_M \end{bmatrix} = \begin{bmatrix} f_1 \\ f_2 \\ \vdots \\ \vdots \\ f_{M-1} \\ f_M \end{bmatrix} \qquad (7.10)$$

with

$$f_1 = 2[1 - (1 - 2\alpha)\lambda^2]u_{1j} + 2(1 - 2\alpha)\lambda^2 u_{2j} - (1 + 2\alpha\lambda^2)u_{1,j-1} + 2\alpha\lambda^2 u_{2,j-1}$$
$$- 2(\Delta x)\alpha\lambda^2 h_{j-1} - 2(\Delta x)(1 - 2\alpha)\lambda^2 h_j - 2(\Delta x)\alpha\lambda^2 h_{j+1}$$

$$f_i = (1 - 2\alpha)\lambda^2 u_{i-1,j} + 2[1 - (1 - 2\alpha)\lambda^2]u_{ij} + (1 - 2\alpha)\lambda^2 u_{i+1,j} + \alpha\lambda^2 u_{i-1,j-1}$$
$$- (1 + 2\alpha\lambda^2)u_{i,j-1} + \alpha\lambda^2 u_{i+1,j+1} \qquad i = 2, 3, \ldots, M - 1$$

and $f_M = (1 - 2\alpha)\lambda^2 u_{M-1,j} + 2[1 - (1 - 2\alpha)\lambda^2]u_{Mj} + \alpha\lambda^2 u_{M-1,j-1}$
$$- (1 + 2\alpha\lambda^2)u_{M,j-1} + \alpha\lambda^2(u_{M+1,j+1} + u_{M+1,j-1}) + (1 - 2\alpha)\lambda^2 u_{M+1,j}$$

If we assume M *odd*, then with the usual splitting $A = G_1 + G_2$, G_1 takes exactly the same form (with m replaced by M) as in Eq. (2.6). The other constituent matrix G_2, however, is given by

$$G_2 = \begin{bmatrix} a/2 & b_1 & & & & & \\ c & a/2 & & & & & \\ & & a/2 & b & & & \\ & & c & a/2 & & & \\ & & & & a/2 & b & \\ & & & & c & a/2 & \\ & & & & & & a/2 \end{bmatrix}_{M \times M} \qquad (7.11)$$

Similarly $(G_1 + rI)^{-1}$ is given by Eqs. (3.8)–(3.9) while $(G_2 + rI)^{-1}$ turns out to be

$$(G_2 + rI)^{-1} = \frac{1}{\Delta} \begin{bmatrix} \Delta r_2/\Delta_2 & -\Delta b_1/\Delta_2 & & & & & \\ -\Delta c/\Delta_2 & \Delta r_2/\Delta_2 & & & & & \\ & & r_2 & -b & & & \\ & & -c & r_2 & & & \\ & & & & r_2 & -b & \\ & & & & -c & r_2 & \\ & & & & & & \Delta/r_2 \end{bmatrix}_{M \times M}$$

$$(7.12)$$

where $$\Delta_2 = r_2^2 - b_1 c \qquad (7.13)$$

It is clear that the required explicit equations for the computation of the AGE scheme at level $(p + 1/2)$ remain the same as [Eqs. (3.15a) and (3.15b)] except the first equation which is replaced by $u_1^{(p+1/2)} = (r_1 u_1^{(p)} - b_1 u_2^{(p)} + f_1)/r_2$.

We arrive at the following AGE equations at level $(p + 1)$:

$$u_1^{(p+1)} = \frac{r_2(r_3u_1^{(p)} + b_1u_2^{(p)} + r_4u_1^{(p+1/2)}) - b_1(cu_1^{(p)} + r_3u_2^{(p)} + r_4u_2^{(p+1/2)})}{\Delta_2}$$

$$u_2^{(p+1)} = \frac{-c(r_3u_1^{(p)} + b_1u_2^{(p)} + r_4u_1^{(p+1/2)}) + r_2(cu_1^{(p)} + r_3u_2^{(p)} + r_4u_2^{(p+1/2)})}{\Delta_2}$$

$$u_i^{(p+1)} = \frac{Pu_i^{(p+1/2)} + Qu_{i+1}^{(p+1/2)} + Ru_i^{(p)} + Su_{i+1}^{(p)}}{\Delta}$$

$$u_{i+1}^{(p+1)} = \frac{\tilde{P}u_i^{(p+1/2)} + \tilde{Q}u_{i+1}^{(p+1/2)} + \tilde{R}u_i^{(p)} + \tilde{S}u_{i+1}^{(p)}}{\Delta}$$

$$i = 3, 4, \ldots, M - 2$$

$\left.\begin{array}{c} \\ \\ \\ \\ \\ \\ \\ \end{array}\right\}$ (7.14)

and

$$u_M^{(p+1)} = \frac{r_3u_M^{(p)} + r_4u_M^{(p+1/2)}}{r_2}$$

where P, $\tilde{P}$, Q, $\tilde{Q}$, R, $\tilde{R}$, S, and $\tilde{S}$ are given by Eq. (3.19).

If M is even, we find that

$$G_1 = \begin{bmatrix} a/2 & b_1 & & & & & & \\ c & a/2 & & & & & & \\ & & a/2 & b & & & & \\ & & c & a/2 & & & & \\ & & & & \bigcirc & & & \\ & & & & & a/2 & b & \\ & & & \bigcirc & & c & a/2 & \\ & & & & & & & a/2 & b \\ & & & & & & & c & a/2 \end{bmatrix}_{(M \times M)}$$ (7.15)

with G_2 as in Eq. (2.36),

$$(G_1 + rI)^{-1} = \frac{1}{\Delta} \begin{bmatrix} \Delta r_2/\Delta_2 & -\Delta b_1/\Delta_2 & & & & & \\ -\Delta c/\Delta_2 & r_2/\Delta_2 & & & & & \\ & & r_2 & -b & & & \\ & & -c & r_2 & \bigcirc & & \\ & & & & & & \\ & & \bigcirc & & r_2 & -b & \\ & & & & -c & r_2 & \\ & & & & & & r_2 & -b \\ & & & & & & -c & r_2 \end{bmatrix}$$ (7.16)

and $(G_2 + rI)^{-1}$ as in Eq. (3.22). Hence, the explicit expressions for the generated AGE scheme at level $(p + 1/2)$ are given by

$$
u_1^{(p+1/2)} = \frac{r_2(r_1 u_1^{(p)} + f_1) - b_1(r_1 u_2^{(p)} - b u_3^{(p)} + f_2)}{\Delta_2}
$$

$$
u_2^{(p+1/2)} = \frac{-c(r_1 u_1^{(p)} + f_1) + r_2(r_1 u_2^{(p)} - b u_3^{(p)} + f_2)}{\Delta_2}
$$

$$
u_i^{(p+1/2)} = \frac{A u_{i-1}^{(p)} + B u_i^{(p)} + C u_{i+1}^{(p)} + D u_{i+2}^{(p)} + E_i}{\Delta}
$$

and

$$
u_{i+1}^{(p+1/2)} = \frac{\tilde{A} u_{i-1}^{(p)} + \tilde{B} u_i^{(p)} + \tilde{C} u_{i+1}^{(p)} + \tilde{D} u_{i+2}^{(p)} + \tilde{E}_i}{\Delta}
$$

$$
i = 3, 5, \ldots, M - 1
$$

(7.17)

where A, $\tilde{A}$, B, $\tilde{B}$, C, $\tilde{C}$, D, $\tilde{D}$, E_i, and $\tilde{E}_i$ are obtained from Eq. (3.23b). The same equations in (3.24a) remain valid for use at level $(p + 1)$.

8 AGE METHOD TO SOLVE NONLINEAR PARABOLIC AND HYPERBOLIC EQUATIONS

The concept of the AGE method is now extended to a variety of *nonlinear* problems.

(i) *Solving the equation:*

$$
\frac{\partial U}{\partial t} = \frac{\partial^2 U^n}{\partial x^2} \qquad n \geq 2
$$

We shall now consider implementing the AGE algorithm on the following nonlinear parabolic problem of the form

$$
\frac{\partial U}{\partial t} = \frac{\partial^2 U^n}{\partial x^2} \qquad n \geq 2 \tag{8.1a}
$$

given the initial condition

$$
U(x, 0) = f(x) \qquad 0 < x < 1
$$

and the boundary conditions

$$
U(0, t) = g(t)
$$

$$
0 < t \leq T \tag{8.1b}
$$

and

$$
U(1, t) = h(t)
$$

Equation (8.1a) is approximated at the grid points by finite-difference schemes and we shall adopt the approach of Richtmyer and Lees to linearize them, which results in tridiagonal systems of equations as before.

(a) Richtmyer's linearization

The nonlinear equation (8.1a) is approximated by the implicit weighted average difference formula

$$\frac{1}{\Delta t}(u_{i,j+1} - u_{ij}) = \frac{1}{(\Delta x)^2}[\theta\delta_x^2(u_{i,j+1}^n)$$

$$+ (1 - \theta)\delta_x^2(u_{ij}^n)] \qquad \text{with } i = 1, 2, \ldots, m \qquad (8.2)$$

As in the linear case $n = 1$, the above formula corresponds to the fully implicit, the Crank-Nicolson and the Douglas schemes when $\theta = 1$, $1/2$, and $(6\lambda - 1)/12\lambda$, respectively, with $\lambda = \Delta t/(\Delta x)^2$. By resorting to the Taylor series expansion of $u_{i,j+1}^n$ about the point (x_i, t_j) we have

$$u_{i,j+1}^n = u_{ij}^n + (\Delta t)\frac{\partial u_{ij}^n}{\partial t} + \cdots$$

$$= u_{ij}^n + (\Delta t)\frac{\partial u_{ij}^n}{\partial u_{ij}}\frac{\partial u_{ij}}{\partial t} + \cdots$$

Hence to terms of order n, the approximation

$$u_{i,j+1}^n = u_{ij}^n + nu_{ij}^{n-1}(u_{i,j+1} - u_{ij}) \qquad (8.3)$$

replaces the nonlinear unknown $u_{i,j+1}^n$ by an approximation which is linear in $u_{i,j+1}$. Now, if we let

$$w_i = u_{i,j+1} - u_{ij} \qquad (8.4)$$

then, using Eqs. (8.2) and (8.3), we obtain

$$\frac{1}{\Delta t}w_i = \frac{1}{(\Delta x)^2}[\theta\delta_x^2(u_{ij}^n + nu_{ij}^{n-1}w_i) + (1 - \theta)\delta_x^2 u_{ij}^n]$$

$$= \frac{1}{(\Delta x)^2}(n\theta\delta_x^2 u_{ij}^{n-1}w_i + \delta_x^2 u_{ij}^n)$$

$$= \frac{1}{(\Delta x)^2}[n\theta(u_{i-1,j}^{n-1}w_{i-1} - 2u_{ij}^{n-1}w_i + u_{i+1,j}^{n-1}w_{i+1})$$

$$+ (u_{i-1,j}^n - 2u_{ij}^n + u_{i+1,j}^n)] \qquad (8.5)$$

which gives a set of linear equations for the w_i. Now the system of Eq. (8.5) can be written in the more compact matrix form as (*with* $n = 2$),

$$A\mathbf{w} = \mathbf{f} \qquad (8.6)$$

which is solved for $\mathbf{w}$ and by means of Eq. (8.4), the solution at time level $(j + 1)$ is given by

$$\mathbf{u}_{j+1} = \mathbf{w} + \mathbf{u}_j \qquad (8.7)$$

The coefficient matrix A takes the following tridiagonal form

$$A = \begin{bmatrix} a_{1j} & b_{1j} & & & & \\ c_{2j} & a_{2j} & b_{2j} & & \bigcirc & \\ & & \ddots & \ddots & & \\ & \bigcirc & & \ddots & \ddots & \\ & & c_{m-1,j} & a_{m-1,j} & b_{m-1,j} \\ & & & c_{mj} & a_{mj} \end{bmatrix}_{(m \times m)} \qquad (8.8)$$

where

$$a_{ij} = 1 + 4\lambda\theta u_{ij} \qquad i = 1, 2, \ldots, m$$

$$b_{ij} = -2\lambda\theta u_{i+1,j} \qquad i = 1, 2, \ldots, m - 1$$

and $$c_{ij} = -2\lambda\theta u_{i-1,j} \qquad i = 2, 3, \ldots, m$$

The entries of the right-hand side vector of Eq. (8.6) are given by

$$f_1 = (-2\lambda u_{1j})u_{1j} + (\lambda u_{2j})u_{2j} + \lambda u_{0j}[u_{0j} + 2\theta(u_{0,j+1} - u_{0j})]$$

$$f_i = (\lambda u_{i-1,j})u_{i-1,j} + (-2\lambda u_{ij})u_{ij} + (\lambda u_{i+1,j})u_{i+1,j} \qquad i = 2, 3, \ldots, m - 1$$

and

$$f_m = (\lambda u_{m-1,j})u_{m-1,j} + (-2\lambda u_{mj})u_{mj} + \lambda u_{m+1,j}[u_{m+1,j} + 2\theta(u_{m+1,j+1} - u_{m+1,j})]$$

If we assume m *odd* then when A is split into the sum of the matrices G_1 and G_2, these constituent matrices take the form

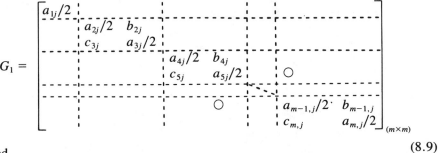

$$(8.9)$$

and

$$(8.10)$$

If we define

$$G^{(i)} = \begin{bmatrix} (1/2)a_{2i,j} + r & b_{2i,j} \\ c_{2i+1,j} & (1/2)a_{2i+1,j} + r \end{bmatrix} \qquad (8.11)$$

then

$$\alpha_i = |G^{(i)}|$$

$$= \left(\frac{1}{2}a_{2i,j} + r\right)\left(\frac{1}{2}a_{2i+1,j} + r\right) - b_{2i,j}c_{2i+1,j} \qquad (8.12)$$

and

$$(G_1 + rI)^{-1} = \qquad (8.13)$$

where

$$\hat{G}^{(i)} = (G^{(i)})^{-1}$$

$$= \frac{1}{\alpha_i}\begin{bmatrix} (1/2)a_{2i+1,j} + r & -b_{2i,j} \\ -c_{2i+1,j} & (1/2)a_{2i,j} + r \end{bmatrix} \qquad i = 1, 2, \ldots, \frac{1}{2}(m-1) \quad (8.14)$$

Similarly, we have

$$(G_2 + rI)^{-1} = \qquad (8.15)$$

where

$$\hat{G}{}^{(i)} = \frac{1}{\hat{\alpha}_i}\begin{bmatrix} (1/2)a_{2i,j} + r & -b_{2i-1,j} \\ -c_{2i,j} & (1/2)a_{2i-1,j} + r \end{bmatrix} \qquad (8.16)$$

with

$$\hat{\alpha}_i = \left(\frac{1}{2}a_{2i,j} + r\right)\left(\frac{1}{2}a_{2i-1,j} + r\right) - b_{2i-1,j}c_{2i,j} \qquad i = 1, 2, \ldots, \frac{1}{2}(m-1) \quad (8.17)$$

We, therefore, arrive at the following equations for the computation of the solution of the nonlinear problem using the generalized AGE scheme.

(1) at level (iterate) $(p + 1/2)$

$$w_1^{(p+1/2)} = \frac{\bar{s}_1 w_1^{(p)} - b_1 w_2^{(p)} + f_1}{\bar{r}_1}$$

$$w_i^{(p+1/2)} = \frac{A_i w_{i-1}^{(p)} + B_i w_i^{(p)} + C_i w_{i+1}^{(p)} + D_i w_{i+2}^{(p)} + E_i}{\alpha_{i/2}} \qquad (8.18a)$$

$$i = 2, 4, \ldots, m - 1$$

$$w_{i+1}^{(p+1/2)} = \frac{(\tilde{A}_i w_{i-1}^{(p)} + \tilde{B}_i w_i^{(p)} + \tilde{C}_i w_{i+1}^{(p)} + \tilde{D}_i w_{i+2}^{(p)} + \tilde{E}_i)}{\alpha_{i/2}}$$

where

$$A_i = -c_i \bar{r}_{i+1} \qquad B_i = \bar{r}_{i+1} \bar{s}_i \qquad C_i = -b_i \bar{s}_{i+1}$$

$$D_i = \begin{cases} 0 & \text{for } i = m - 1 \\ b_i b_{i+1} & \text{otherwise} \end{cases} \qquad E_i = \bar{r}_{i+1} f_i - b_i f_{i+1}$$

$$\tilde{A}_i = c_i c_{i+1} \qquad \tilde{B}_i = -c_{i+1} \bar{s}_i \qquad \tilde{C}_i = \bar{r}_i \bar{s}_{i+1} \qquad (8.18b)$$

$$\tilde{D}_i = \begin{cases} 0 & \text{for } i = m - 1 \\ -b_{i+1} \bar{r}_i & \text{otherwise} \end{cases} \qquad \tilde{E}_i = -c_{i+1} f_i + \bar{r}_i f_{i+1}$$

with $\qquad \bar{r}_i = r + \dfrac{1}{2} a_i \quad$ and $\quad \bar{s}_i = r - \dfrac{1}{2} a_i \qquad i = 1, 2, \ldots, m$

(2) at level (iterate) $(p + 1)$

$$w_i^{(p+1)} = \frac{P_i w_i^{(p)} + Q_i w_{i+1}^{(p)} + R_i w_i^{(p+1/2)} + S_i w_{i+1}^{(p+1/2)}}{\hat{\alpha}_{(i+1)/2}}$$

$$i = 1, 3, \ldots, m - 2$$

$$w_{i+1}^{(p+1)} = \frac{\tilde{P}_i w_i^{(p)} + \tilde{Q}_i w_{i+1}^{(p)} + \tilde{R}_i w_i^{(p+1/2)} + \tilde{S}_i w_{i+1}^{(p+1/2)}}{\hat{\alpha}_{(i+1)/2}} \qquad (8.19a)$$

$$w_m^{(p+1)} = \frac{\bar{q}_m w_m^{(p)} + d w_m^{(p+1/2)}}{\bar{r}_m}$$

where

$$P_i = \bar{r}_{i+1}\bar{q}_i - b_i c_{i+1} \qquad Q_i = b_i(\bar{r}_{i+1} - \bar{q}_{i+1})$$

$$R_i = \bar{r}_{i+1}d \qquad S_i = -b_i d$$

and

$$\tilde{P}_i = c_{i+1}(\bar{r}_i - \bar{q}_i) \qquad \tilde{Q}_i = -b_i c_{i+1} + \bar{r}_i \bar{q}_{i+1}$$

$$\tilde{R}_i = -c_{i+1}d \qquad \tilde{S}_i = \bar{r}_i d$$

$$(8.19b)$$

with

$$\bar{r}_i = r + \frac{1}{2} a_i \qquad \bar{q}_i = \frac{1}{2} a_i - (1 - \omega)r \qquad d = (2 - \omega)r$$

$$i = 1, 2, \ldots, m$$

If, on the other hand, we assume that m is *even* then we obtain the following forms of the constituent matrices:

$$G_1 = \begin{bmatrix} a_{1j}/2 & b_{1j} & & & & & \\ c_{2j} & a_{2j}/2 & & & & & \\ & & a_{3j}/2 & b_{3j} & & \bigcirc & \\ & & c_{4j} & a_{4j}/2 & & & \\ & & & & & & \\ & \bigcirc & & & & a_{m-1,j}/2 & b_{m-1,j} \\ & & & & & c_{m,j} & a_{mj}/2 \end{bmatrix}_{(m \times m)} \qquad (8.20)$$

$$G_2 = \begin{bmatrix} a_{1j}/2 & & & & & \\ & a_{2j}/2 & b_{2j} & & \bigcirc & \\ & c_{3j} & a_{3j}/2 & & & \\ & & & \bigcirc & a_{m-2,j}/2 & b_{m-2,j} \\ & & & & c_{m-1,j} & a_{m-1,j}/2 \\ & & & & & a_{mj}/2 \end{bmatrix}_{(m \times m)} \qquad (8.21)$$

$$(G_1 + rI)^{-1} = \begin{bmatrix} \hat{G}^{(1)} & & \\ & \hat{G}^{(2)} & \bigcirc \\ & \bigcirc & \hat{G}^{(m/2)} \end{bmatrix}_{(m \times m)} \qquad (8.22)$$

and

$$(G_2 + rI)^{-1} = \begin{bmatrix} \dfrac{1}{(r + (1/2)a_{1j})} & & & \\ & \hat{G}^{(1)} & & \bigcirc \\ & & & \\ & \bigcirc & \hat{G}^{(m-2/2)} & \\ & & & \dfrac{1}{(r + (1/2)a_{mj})} \end{bmatrix}_{(m \times m)}$$

$$(8.23)$$

where $\hat{G}^{(i)}$ and $\hat{G}^{(i)}$ are given by Eqs. (8.14) and (8.16), respectively. The AGE equations can be derived along similar lines as before and are given by

(1) at level $(p + 1/2)$

$$w_i^{(p+1/2)} = \frac{A_i w_{i-1}^{(p)} + B_i w_i^{(p)} + C_i w_{i+1}^{(p)} + D_i w_{i+2}^{(p)} + E_i}{\hat{\alpha}_{(i+1)/2}}$$

$$w_{i+1}^{(p+1/2)} = \frac{\tilde{A}_i w_{i-1}^{(p)} + \tilde{B}_i w_i^{(p)} + \tilde{C}_i w_{i+1}^{(p)} + \tilde{D}_i w_{i+2}^{(p)} + \tilde{E}_i}{\hat{\alpha}_{(i+1)/2}} \qquad (8.24a)$$

$$i = 1, 3, \ldots, m - 1$$

where

$$A_i = \begin{cases} 0 & \text{for } i = 1 \\ -c_i \bar{r}_{i+1} & \text{otherwise} \end{cases} \qquad B_i = \bar{r}_{i+1}\bar{s}_i \qquad C_i = -b_i \bar{s}_{i+1}$$

$$D_i = \begin{cases} 0 & \text{for } i = m - 1 \\ b_i b_{i+1} & \text{otherwise} \end{cases} \qquad E_i = \bar{r}_{i+1}f_i - b_i f_{i+1}$$

$$\tilde{A}_i = \begin{cases} 0 & \text{for } i = 1 \\ c_i c_{i+1} & \text{otherwise} \end{cases} \qquad \tilde{B}_i = -c_{i+1}\bar{s}_i \qquad \tilde{C}_i = \bar{r}_i\bar{s}_{i+1}$$

$$\tilde{D}_i = \begin{cases} 0 & \text{for } i = m - 1 \\ -\bar{r}_i b_{i+1} & \text{otherwise} \end{cases} \qquad \tilde{E}_i = -c_{i+1}f_i + \bar{r}_i f_{i+1}$$

$$\left.\vphantom{\begin{cases}0\\0\\0\\0\\0\\0\\0\\0\end{cases}}\right\} \quad (8.24b)$$

(2) at level $(p + 1)$

$$w_1^{(p+1)} = \frac{\bar{q}_1 w_1^{(p)} + dw_1^{(p+1/2)}}{\bar{r}_1}$$

$$w_i^{(p+1)} = \frac{P_i w_i^{(p)} + Q_i w_{i+1}^{(p)} + R_i w_i^{(p+1/2)} + S_i w_{i+1}^{(p+1/2)}}{\alpha_{i/2}}$$

$$i = 2, 4, \ldots, m - 2 \qquad (8.25a)$$

$$w_{i+1}^{(p+1)} = \frac{\tilde{P}_i w_i^{(p)} + \tilde{Q}_i w_{i+1}^{(p)} + \tilde{R}_i w_i^{(p+1/2)} + \tilde{S}_i w_{i+1}^{(p+1/2)}}{\alpha_{i/2}}$$

$$w_m^{(p+1)} = \frac{\bar{q}_m w_m^{(p)} + dw_m^{(p+1/2)}}{\bar{r}_m}$$

where

$$\begin{aligned} P_i &= (\bar{r}_{i+1}\bar{q}_i - b_i c_{i+1}) & Q_i &= b_i(\bar{r}_{i+1} - \bar{q}_{i+1}) \\ R_i &= d\bar{r}_{i+1} & S_i &= -db_i \\ \tilde{P}_i &= c_{i+1}(\bar{r}_i - \bar{q}_i) & \tilde{Q}_i &= \bar{r}_i\bar{q}_{i+1} - c_{i+1}b_i \\ \tilde{R}_i &= -dc_{i+1} & \tilde{S}_i &= d\bar{r}_i \end{aligned} \qquad (8.25b)$$

The iterative process is continued until the convergence requirement is met.

(b) Lees' three-level linearization

Lees (1966) considered the nonlinear equation

$$b(U) \frac{\partial U}{\partial t} = \frac{\partial}{\partial x} \left[a(U) \frac{\partial U}{\partial x} \right] \qquad a(U) > 0, \, b(U) > 0 \qquad (8.26)$$

and investigated a difference scheme that achieved *linearity* in the unknowns $u_{i,j+1}$ by evaluating all coefficients of $u_{i,j+1}$ at a time level of known solution values; preserved *stability* by averaging u_{ij} over three time levels; and maintained *accuracy* by using central-difference approximations. Since

$$\left(\frac{\partial U}{\partial x} \right)_{i,j} \approx \frac{1}{\Delta x} (U_{i+1/2,j} - U_{i-1/2,j})$$

$$= \frac{1}{\Delta x} \delta_x U_{ij}$$

then a central-difference approximation to Eq. (8.26) is given by

$$b(u_{i,j}) \frac{1}{2\Delta t} (u_{i,j+1} - u_{i,j-1}) = \frac{1}{\Delta x} \delta_x \left[a(u_{ij}) \frac{1}{\Delta x} \delta_x u_{ij} \right]$$

$$= \frac{1}{(\Delta x)^2} \delta_x [a(u_{ij}) \delta_x] u_{ij} \qquad (8.27)$$

which reduces to the Richardson formula when $a(u) = b(u) = 1$ and is therefore unconditionally unstable. However, in *the linear constant coefficient case* (see Mitchell and Griffiths (1980), pages 89–92), unconditional stability is obtained by replacing $\delta_x^2 u_{i,j}$ by $\frac{1}{3}\delta_x^2(u_{i,j+1} + u_{ij} + u_{i,j-1})$. Following this procedure, Eq. (8.27) is rewritten as

$$b(u_{ij})(u_{i,j+1} - u_{i,j-1}) = 2\lambda [a(u_{i+1/2,j})(u_{i+1,j} - u_{ij})$$

$$- a(u_{i-1/2,j})(u_{ij} - u_{i-1,j})]$$

and then $u_{i+1,j}$, u_{ij} and $u_{i-1,j}$ are replaced by

$$\frac{1}{3} (u_{i+1,j+1} + u_{i+1,j} + u_{i+1,j-1}), \frac{1}{3} (u_{i,j+1} + u_{ij} + u_{i,j-1})$$

$$\text{and} \quad \frac{1}{3} (u_{i-1,j+1} + u_{i-1,j} + u_{i-1,j-1})$$

respectively. Furthermore, since $u_{i\pm1/2,j}$ do not fall on the grid points, we replace $a(u_{i+1/2,j})$ and $a(u_{i-1/2,j})$ by $a\,[(u_{i+1,j} + u_{ij})/2]$ and $a[(u_{ij} + u_{i-1,j})/2]$ respectively.

This leads to *the linearized three-level formula*

$$b(u_{ij})(u_{i,j+1} - u_{i,j-1}) = \frac{2}{3}\lambda\{\beta^+[(u_{i+1,j+1} - u_{i,j+1})$$

$$+ (u_{i+1,j} - u_{ij}) + (u_{i+1,j-1} - u_{i,j-1})]$$

$$- \beta^-[(u_{i,j+1} - u_{i-1,j+1}) + (u_{ij} - u_{i-1,j})$$

$$+ (u_{i,j-1} - u_{i-1,j-1})]\} \tag{8.28}$$

where

$$\beta^+ = a\frac{u_{i+1,j} + u_{ij}}{2} \quad \text{and} \quad \beta^- = a\frac{u_{ij} + u_{i-1,j}}{2} \tag{8.29}$$

Lees (1966) proved the convergence result for Eq. (8.28) by showing that for sufficiently small values of Δx and Δt,

$$\max_{i,j} |U_{ij} - u_{ij}| \le K[(\Delta x)^2 + (\Delta t)^2]$$

where K is a constant. For this method to be applied to Eq. (8.1a), it is necessary to write the equation in the *self-adjoint form* as

$$\frac{\partial U}{\partial t} = \frac{\partial^2 U^n}{\partial x^2} \quad n \ge 2$$

$$= \frac{\partial}{\partial x}\left(nU^{n-1}\frac{\partial U}{\partial x}\right)$$

On comparing this equation with Eq. (8.26), we find that for the particular value of $n = 2$, $a(U) = 2U$ and $b(U) = 1$ and from Eq. (8.29),

$$\beta^+ = u_{i+1,j} + u_{ij} \qquad \beta^- = u_{ij} + u_{i-1,j} \tag{8.30}$$

Hence, the formula (8.28) becomes

$$-\frac{2}{3}\lambda\beta^- u_{i-1,j+1} + \left[1 + \frac{2}{3}\lambda(\beta^+ + \beta^-)\right]u_{i,j+1} - \frac{2}{3}\lambda\beta^+ u_{i+1,j+1}$$

$$= \frac{2}{3}\lambda\beta^- u_{i-1,j} - \frac{2}{3}\lambda(\beta^+ + \beta^-)u_{ij} + \frac{2}{3}\lambda\beta^+ u_{i+1,j}$$

$$+ \frac{2}{3}\lambda\beta^- u_{i-1,j-1} + \left[1 - \frac{2}{3}\lambda(\beta^+ + \beta^-)\right]u_{i,j-1}$$

$$+ \frac{2}{3}\lambda\beta^+ u_{i+1,j-1} \qquad \text{for } i = 1, 2, \ldots, m \tag{8.31}$$

which is a tridiagonal system of equations that can be written in the matrix form (8.6) (with **u** replacing **w**) where A takes the form (8.8) and

$$a_{ij} = 1 + \frac{2}{3}\lambda(\beta^+ + \beta^-)$$

$$= 1 + \frac{2}{3}\lambda(u_{i-1,j} + 2u_{ij} + u_{i+1,j}) \qquad i = 1, 2, \ldots, m$$

$$b_{ij} = -\frac{2}{3}\lambda\beta^+$$

$$= -\frac{2}{3}\lambda(u_{i+1,j} + u_{ij}) \qquad i = 1, 2, \ldots, m-1$$

and
$$c_{ij} = -\frac{2}{3}\lambda\beta^- u_{i-1,j+1}$$

$$= -\frac{2}{3}\lambda(u_{ij} + u_{i-1,j}) \qquad i = 2, 3, \ldots, m$$

The components of the right-hand side vector **f** are given by

$$f_1 = \frac{2}{3}\lambda[(u_{1j} + u_{0j})(u_{0,j-1} + u_{0,j} + u_{0,j+1}) + (u_{2j} + u_{1j})(u_{2,j-1} + u_{2j})$$

$$- (u_{0j} + 2u_{1j} + u_{2j})u_{1j}] + \left[1 - \frac{2}{3}\lambda(u_{0j} + 2u_{1j} + u_{2j})\right]u_{1,j-1}$$

$$f_i = \frac{2}{3}\lambda[\beta^+(u_{i+1,j} + u_{i+1,j-1}) + \beta^-(u_{i-1,j} + u_{i-1,j-1}) - (\beta^+ + \beta^-)u_{ij}]$$

$$+ \left[1 - \frac{2}{3}\lambda(\beta^+ + \beta^-)\right]u_{i,j-1} \qquad \text{for } i = 2, 3, \ldots, m-1$$

and

$$f_m = \frac{2}{3}\lambda[(u_{m+1,j} + u_{mj})(u_{m+1,j-1} + u_{m+1,j} + u_{m+1,j+1})$$

$$+ (u_{mj} + u_{m-1,j})(u_{m-1,j-1} + u_{m-1,j}) - (u_{m+1,j} + 2u_{mj} + u_{m-1,j})u_{mj}]$$

$$+ \left[1 - \frac{2}{3}\lambda(u_{m+1,j} + 2u_{mj} + u_{m-1,j})\right]u_{m,j-1}$$

When the AGE procedure is implemented on the above tridiagonal system of equations, we will arrive at the same computational formulas (with **w** replaced by **u**) at the $(p + 1/2)$ and $(p + 1)$ iterates as that for Richtmyer's linearization. This implies that Eqs. (8.18a)–(8.19a) (for the case m odd) and Eqs. (8.24a)–(8.25a) (when m is even) will be used for our iterative process.

(ii) Solving Burger's equation:

$$\epsilon \frac{\partial^2 U}{\partial x^2} = \frac{\partial U}{\partial t} + U \frac{\partial U}{\partial x} \qquad \epsilon > 0$$

The *general nonlinear parabolic equation* for initial boundary value problems is given by

$$\frac{\partial U}{\partial t} = \phi\left(x, t, U, \frac{\partial U}{\partial x}, \frac{\partial^2 U}{\partial x^2}\right) \qquad 0 < x < 1, 0 < t \leq T \qquad (8.32)$$

subject to smooth initial and boundary conditions. This problem is well posed in the region (see, for example, Friedman, 1964) if

$$\frac{\partial \phi}{\partial U_{xx}} \geq a > 0 \qquad (8.33)$$

If this holds, then the implicit relation (8.32) may be solved for $\partial^2 U / \partial x^2$. Thus, we assume the partial differential equation to have the form

$$\frac{\partial^2 U}{\partial x^2} = \psi\left(x, t, U, \frac{\partial U}{\partial x}, \frac{\partial U}{\partial t}\right) \qquad (8.34)$$

where the properly posed requirement is

$$\frac{\partial \psi}{\partial U_t} \geq a > 0 \qquad (8.35)$$

In some instances, Eq. (8.34) may be written in the quasilinear form

$$\frac{\partial^2 U}{\partial x^2} + f(x, t, U) \frac{\partial U}{\partial x} + g(x, t, U) = p(x, t, U) \frac{\partial U}{\partial t} \qquad (8.36)$$

(Specifically Burger's equation

$$\epsilon \frac{\partial^2 U}{\partial x^2} = \frac{\partial U}{\partial t} + U \frac{\partial U}{\partial x}$$

is of this form).

(a) The fully implicit form

At the point $(i, j + 1)$, Eq. (8.36) can be approximated by the formula

$$\frac{1}{(\Delta x)^2} \delta_x^2 u_{i,j+1} + \frac{1}{2\Delta x} f[i\Delta x, (j + 1)\Delta t, u_{ij}] \mu \delta_x u_{i,j+1}$$

$$+ g[i\Delta x, (j + 1)\Delta t, u_{ij}]$$

$$= p[i\Delta x, (j + 1)\Delta t, u_{ij}] \frac{u_{i,j+1} - u_{ij}}{\Delta t} \qquad (8.37)$$

which contains $u_{i,j+1}$ only linearly and where the difference operators δ and μ are defined by

$$\delta y_n = y_{n+1/2} - y_{n-1/2} \qquad \text{(central)}$$

and

$$\mu y_n = \frac{1}{2} [y_{n+1/2} + y_{n-1/2}] \quad \text{(averaging)}$$

Thus, the algebraic problem is *linear and tridiagonal* at each time step. For Burger's equation, we have the analogue,

$$\frac{\epsilon}{(\Delta x)^2} \delta_x^2 u_{i,j+1} + \frac{1}{2\Delta x} (-u_{ij}) \mu \delta_x u_{i,j+1} = \frac{u_{i,j+1} - u_{ij}}{\Delta t}$$

which leads to

$$-\lambda \left(\epsilon + \frac{\Delta x}{4} u_{ij} \right) u_{i-1,j+1} + (1 + 2\epsilon\lambda) u_{i,j+1}$$

$$- \lambda \left(\epsilon - \frac{\Delta x}{4} u_{ij} \right) u_{i+1,j+1} = u_{ij} \qquad i = 1, \ldots, m$$

(8.38)

(b) The Crank-Nicolson form

The application of the Crank-Nicolson concept to Eq. (8.36) gives

$$\frac{1}{2(\Delta x)^2} \delta_x^2 (u_{i,j+1} + u_{ij})$$

$$+ \frac{1}{2\Delta x} f \left[i\Delta x, \left(j + \frac{1}{2} \right) \Delta t, \frac{1}{2} (u_{i,j+1} + u_{ij}) \mu \delta_x (u_{i,j+1} + u_{ij}) \right]$$

$$+ g \left[i\Delta x, \left(j + \frac{1}{2} \right) \Delta t, \frac{1}{2} (u_{i,j+1} + u_{ij}) \right]$$

$$= p \left[i\Delta x, \left(j + \frac{1}{2} \right) \Delta t, \frac{1}{2} (u_{i,j+1} + u_{ij}) \right] \frac{(u_{i,j+1} - u_{ij})}{\Delta t}$$

(8.39)

which are *nonlinear*. For Burger's equation, these simplify to

$$u_{i,j+1} - u_{ij} = \frac{\epsilon\lambda}{2} [(u_{i-1,j} - 2u_{ij} + u_{i+1,j})$$

$$+ (u_{i-1,j+1} - 2u_{i,j+1} + u_{i+1,j+1})]$$

$$- \frac{\lambda \Delta x}{4} [(u_{i+1,j} - u_{i-1,j}) \alpha_{ij}$$

$$+ (u_{i+1,j+1} - u_{i-1,j+1}) \beta_{ij}]$$

(8.40)

where $\alpha_{ij} = \beta_{ij} = (u_{i,j+1} + u_{ij})/2$. This equation, however, can be *linearized* if we replace α_{ij} by $u_{i,j+1}$ and β_{ij} by u_{ij}. Thus, Eq. (8.40) becomes

$$-\left(\frac{\epsilon\lambda}{2} + \frac{\Delta x}{4}\lambda u_{ij}\right)u_{i-1,j+1} + \left[1 + \epsilon\lambda + \frac{\Delta x}{4}\lambda(u_{i+1,j} - u_{i-1,j})\right]u_{i,j+1}$$

$$-\left(\frac{\epsilon\lambda}{2} - \frac{\Delta x}{4}\lambda u_{ij}\right)u_{i+1,j+1} = \frac{\epsilon\lambda}{2}u_{i-1,j}$$

$$+ (1 - \epsilon\lambda)u_{ij} + \frac{\epsilon\lambda}{2}u_{i+1,j} \qquad i = 1, 2, \ldots, m \tag{8.41}$$

(c) The Predictor-Corrector form

Nonlinear algebraic equations can be avoided if two-step methods called the predictor-corrector methods are used. The "predictor" gives a first approximation to the solution and the "corrector" is used repeatedly, if necessary, to provide the final result. If ψ of Eq. (8.34) assumes the form

$$\psi = f_1(x, t, U)\frac{\partial U}{\partial t} + f_2(x, t, U)\frac{\partial U}{\partial x} + f_3(x, t, U) \tag{8.42a}$$

or

$$\psi = g_1\left(x, t, U, \frac{\partial U}{\partial x}\right)\frac{\partial U}{\partial t} + g_2\left(x, t, U, \frac{\partial U}{\partial x}\right) \tag{8.42b}$$

then a predictor-corrector modification of the Crank-Nicolson procedure is possible so that the resulting algebraic problem is linear. The class of Eq. (8.42a) includes Burger's equation and if ψ is of the form (8.42a), then one possibility of the predictor is

$$\frac{1}{(\Delta x)^2}\delta_x^2 u_{i,j+1/2} = \psi\left[i\Delta x, \left(j + \frac{1}{2}\right)\Delta t, u_{ij}, \frac{1}{\Delta x}\mu\delta_x u_{ij}, \frac{2}{\Delta t}(u_{i,j+1/2} - u_{ij})\right] \tag{8.43}$$

for $i = 1, 2, \ldots, m$ followed by the corrector,

$$\frac{1}{2(\Delta x)^2}\delta_x^2[u_{i,j+1} + u_{ij}] = \psi\left[i\Delta x, \left(j + \frac{1}{2}\right)\Delta t, u_{i,j+1/2}, \frac{1}{2\Delta x}\mu\delta_x(u_{i,j+1}\right.$$

$$\left. + u_{ij}), \frac{1}{\Delta t}(u_{i,j+1} - u_{ij})\right] \tag{8.44}$$

For Burgers' equation, the corresponding predictor-corrector (P-C) pair are given by

$$P: \epsilon\lambda u_{i-1,j+1/2} - 2(1 + \epsilon\lambda)u_{i,j+1/2} + \epsilon\lambda u_{i+1,j+1/2}$$

$$= \left[\frac{1}{2}\lambda(\Delta x)(u_{i+1,j} - u_{i-1,j}) - 2\right]u_{ij} \tag{8.45}$$

and

$$C: -\frac{\lambda}{2}\left(\epsilon + \frac{\Delta x}{2}u_{i,j+1/2}\right)u_{i-1,j+1} + (1 + \epsilon\lambda)u_{i,j+1}$$

$$-\frac{\lambda}{2}\left(\epsilon - \frac{\Delta x}{2}u_{i,j+1/2}\right)u_{i+1,j+1} = \frac{\epsilon\lambda}{2}(u_{i-1,j} - 2u_{ij} + u_{i+1,j}) + u_{ij}$$

$$-\frac{\lambda\Delta x}{4}u_{i,j+1/2}(u_{i+1,j} - u_{i-1,j}) \tag{8.46a}$$

The above predictor-corrector formulas are known to have second-order accuracy in both space and time (Douglas and Jones, 1963). Note that Eq. (8.43) is a backward difference equation. One may also use the following modified Crank-Nicolson predictor,

$$\frac{1}{2(\Delta x)^2}\delta_x^2(u_{i,j+1/2} + u_{ij})$$

$$= \psi\left[i\Delta x, (j + 1/2)\Delta t, u_{ij}, \frac{1}{\Delta x}\mu\delta_x u_{ij}, \frac{2}{\Delta t}(u_{i,j+1/2} - u_{ij})\right] \tag{8.46b}$$

While the procedure (8.43) and (8.44) leads to a set of linear algebraic equations to solve for ψ when it is of the form (8.42a), it does not for ψ of the form (8.42b). If Eq. (8.44) is replaced by

$$\frac{1}{2(\Delta x)^2}\delta_x^2(u_{i,j+1} + u_{ij})$$

$$= \psi\left[i\Delta x, \left(j + \frac{1}{2}\right)\Delta t, u_{i,j+1/2}, \frac{1}{\Delta x}\mu\delta_x u_{i,j+1/2}, \frac{1}{\Delta t}(u_{i,j+1} - u_{ij})\right] \tag{8.47a}$$

the system (8.43) and (8.47) does produce the desired linear algebraic equations with a local truncation error $O[(\Delta x)^2] + O[(\Delta t)^{3/2}]$.

All of the Eqs. (8.38), (8.41), and (8.46) generate, as before, tridiagonal systems of the form

$$A\mathbf{u} = \mathbf{f} \tag{8.47b}$$

where A takes the same form as Eq. (8.8), $\mathbf{u} = (u_{1,j+1}, u_{2,j+1}, \ldots, u_{m,j+1})^T$ and $\mathbf{f} = (f_1, f_2, \ldots, f_m)^T$. Hence, we have

(1) for the implicit form (8.38)

$$a_{i,j} = 1 + 2\epsilon\lambda \qquad i = 1, 2, \ldots, m$$

$$b_{i,j} = -\lambda\left(\epsilon - \frac{\Delta x}{4}u_{ij}\right) \qquad i = 1, 2, \ldots, m - 1$$

$$c_{i,j} = -\lambda\left(\epsilon + \frac{\Delta x}{4}u_{i,j}\right) \qquad i = 2, 3, \ldots, m$$

$$f_1 = \lambda\left(\epsilon + \frac{\Delta x}{4}u_{1j}\right)u_{0,j+1} + u_{1j}$$

$$f_i = u_{ij} \qquad i = 2, 3, \ldots, m - 1$$

and
$$f_m = \lambda\left(\epsilon - \frac{\Delta x}{4}u_{mj}\right)u_{m+1,j+1} + u_{mj}$$

(2) for the Crank-Nicolson form (8.41)

$$a_{i,j} = 1 + \lambda\left[\epsilon + \frac{\Delta x}{4}(u_{i+1,j} - u_{i-1,j})\right] \qquad i = 1, 2, \ldots, m$$

$$b_{i,j} = -\frac{\lambda}{2}\left(\epsilon - \frac{\Delta x}{2}u_{ij}\right) \qquad i = 1, 2, \ldots, m - 1$$

$$c_{i,j} = -\frac{\lambda}{2}\left(\epsilon + \frac{\Delta x}{2}u_{ij}\right) \qquad i = 2, 3, \ldots, m$$

$$f_1 = (1 - \epsilon\lambda)u_{1j} + \frac{\lambda}{2}\left[\epsilon(u_{0j} + u_{2j}) + \left(\epsilon + \frac{\Delta x}{2}u_{1j}\right)u_{0,j+1}\right]$$

$$f_i = \frac{\epsilon\lambda}{2}(u_{i-1,j} + u_{i+1,j}) + (1 - \epsilon\lambda)u_{ij} \qquad i = 2, 3, \ldots, m - 1$$

and

$$f_m = (1 - \epsilon\lambda)u_{mj} + \frac{\lambda}{2}\left[\epsilon(u_{m-1,j} + u_{m+1,j}) + \left(\epsilon - \frac{\Delta x}{2}u_{mj}\right)u_{m+1,j+1}\right]$$

(3) for the predictor-corrector form (8.46a)

$$c_{i,j} = -\frac{\lambda}{2}\left(\epsilon + \frac{\Delta x}{2}u_{i,j+1/2}\right) \qquad i = 1, 2, \ldots, m$$

$$a_{i,j} = 1 + \epsilon\lambda \qquad i = 1, 2, \ldots, m - 1$$

$$b_{i,j} = -\frac{\lambda}{2}\left(\epsilon - \frac{\Delta x}{2}u_{i,j+1/2}\right) \qquad i = 2, 3, \ldots, m$$

$$f_1 = \frac{\lambda}{2}\left[\epsilon(u_{0j} - 2u_{1j} + u_{2j}) + \left(\epsilon + \frac{\Delta x}{2}u_{1,j+1/2}\right)u_{0,j+1}\right.$$
$$\left. - \frac{\Delta x}{2}u_{1,j+1/2}(u_{2j} - u_{0j})\right] + u_{1j}$$

$$f_i = \frac{\lambda}{2}\left[\epsilon(u_{i-1,j} - 2u_{ij} + u_{i+1,j}) - \frac{\Delta x}{2}u_{i,j+1/2}(u_{i+1,j} - u_{i-1,j})\right]$$

$$+ u_{ij} \qquad i = 2, 3, \ldots, m-1$$

and

$$f_m = \frac{\lambda}{2}\left[\epsilon(u_{m-1,j} - 2u_{mj} + u_{m+1,j}) + \left(\epsilon - \frac{\Delta x}{2}u_{m,j+1/2}\right)u_{m+1,j+1}\right.$$

$$\left. - \frac{\Delta x}{2}u_{m,j+1/2}(u_{m+1,j} - u_{m-1,j})\right] + u_{mj}$$

When the AGE algorithm is implemented, our iterative process will require the same equations for computation as in Eqs. (8.18a) and (8.19a) (for the case m odd) and (8.24a) and (8.25a) (for the case m even) with **w** replaced by **u**. We note, however, that for the predictor-corrector form, the solutions at the *predictor* stage are obtained using the Thomas elimination algorithm.

(iii) A Nonlinear Example for the Reaction-Diffusion Equation

We shall now consider the following one-dimensional *reaction-diffusion equation* taken from Ramos (1985),

$$\frac{\partial U}{\partial t} = \frac{\partial^2 U}{\partial x^2} + S \tag{8.48}$$

where $S = U^2(1 - U)$ and $-\infty < x < \infty$ and $t \geq 0$. This equation has an exact traveling wave solution given by

$$U(x, t) = \frac{1}{1 + \exp[V(x - Vt)]} \tag{8.49}$$

where

$$U(-\infty, t) = 1, U(\infty, t) = 0 \tag{8.50}$$

and V is the *steady-state wave speed* which is equal to $1/\sqrt{2}$. In our numerical experiments, however, Eq. (8.48) was solved in a truncated domain $-50 \leq x \leq 400$, where the locations of the boundaries were selected so that they did not influence the wave propagation. In the truncated domain, the following initial condition was used,

$$U(x, O) = \frac{1}{1 + \exp(Vx)} \tag{8.51}$$

i.e., the initial condition corresponds to the exact solution.

Various schemes in the GE and AGE class of methods are now developed to solve Eq. (8.48).

(a) GE schemes involving an explicit evaluation of the source term: GE-EXP

Following the same argument as in Section 2, a generalized approximation to Eq. (8.48) at the point $(x_i, t_{j+1/2})$ is

$$\frac{u_{i,j+1} - u_{ij}}{\Delta t} = \frac{1}{(\Delta x)^2} (\theta_1 \delta_x u_{i+1/2,j+1} - \theta_2 \delta_x u_{i-1/2,j+1}$$

$$+ \theta_3 \delta_x u_{i+1/2,j} - \theta_4 \delta_x u_{i-1/2,j}) + s_{ij} \qquad (8.52)$$

By letting $\theta_1 = \theta_4 = 1$ and $\theta_2 = \theta_3 = 0$ in Eq. (8.52), we obtain the following asymmetric LR approximation:

$$-\lambda u_{i+1,j+1} + (1 + \lambda)u_{i,j+1} = \lambda u_{i-1,j} + (1 - \lambda)u_{ij} + \Delta t s_{ij} \qquad (8.53)$$

where $\lambda = \Delta t/(\Delta x)^2$. If we choose $\theta_2 = \theta_3 = 1$ and $\theta_1 = \theta_4 = 0$ we arrive at the following RL formula:

$$(1 + \lambda)u_{i,j+1} - \lambda u_{i-1,j+1} = \lambda u_{i+1,j} + (1 - \lambda)u_{ij} + \Delta t s_{ij}$$

or equivalently, at the point $(i + 1, j + 1/2)$,

$$-\lambda u_{i,j+1} + (1 + \lambda)u_{i+1,j+1} = (1 - \lambda)u_{i+1,j} + \lambda u_{i+2,j} + \Delta t s_{i+1,j} \qquad (8.54)$$

When we couple Eqs. (8.53) and (8.54), we obtain

$$\begin{bmatrix} (1 + \lambda) & -\lambda \\ -\lambda & (1 + \lambda) \end{bmatrix} \begin{bmatrix} u_{i,j+1} \\ u_{i+1,j+1} \end{bmatrix} = \begin{bmatrix} 1 - \lambda & 0 \\ 0 & 1 - \lambda \end{bmatrix} \begin{bmatrix} u_{ij} \\ u_{i+1,j} \end{bmatrix}$$

$$+ \begin{bmatrix} \lambda u_{i-1,j} + \Delta t s_{ij} \\ \lambda u_{i+2,j} + \Delta t s_{i+1,j} \end{bmatrix}$$

i.e.,

$$A\mathbf{u}_{j+1} = B\mathbf{u}_j + \hat{\mathbf{u}}_j$$

or

$$\mathbf{u}_{j+1} = A^{-1}(B\mathbf{u}_j + \hat{\mathbf{u}}_j)$$

giving

$$\begin{bmatrix} u_{i,j+1} \\ u_{i+1,j+1} \end{bmatrix} = \frac{1}{(1 + 2\lambda)} \begin{bmatrix} \lambda(1 + \lambda)u_{i-1,j} + (1 - \lambda^2)u_{ij} + \lambda(1 - \lambda)u_{i+1,j} \\ + \lambda^2 u_{i+2,j} + \Delta t[(1 + \lambda)s_{ij} + \lambda s_{i+1,j}] \\ \lambda^2 u_{i-1,j} + \lambda(1 - \lambda)u_{ij} + (1 - \lambda^2)u_{i+1,j} \\ + \lambda(1 + \lambda)u_{i+2,j} + \Delta t[\lambda s_{ij} + (1 + \lambda)s_{i+1,j}] \end{bmatrix}$$

Letting

$$a_1 = \frac{\lambda(1 + \lambda)}{1 + 2\lambda} \qquad a_2 = \frac{(1 - \lambda^2)}{1 + 2\lambda} \qquad a_3 = \frac{\lambda(1 - \lambda)}{1 + 2\lambda}$$

$$a_4 = \frac{\lambda^2}{1 + 2\lambda} \qquad a_5 = \frac{\Delta t(1 + \lambda)}{1 + 2\lambda} \qquad a_6 = \frac{\Delta t\lambda}{1 + 2\lambda}$$

we have the following set of explicit equations defining the GE schemes at the general points,

$$u_{i,j+1} = a_1 u_{i-1,j} + a_2 u_{ij} + a_3 u_{i+1,j} + a_4 u_{i+2,j} + a_5 s_{ij} + a_6 s_{i+1,j} \quad (8.55)$$

and

$$u_{i+1,j+1} = a_4 u_{i-1,j} + a_3 u_{ij} + a_2 u_{i+1,j} + a_1 u_{i+2,j} + a_6 s_{ij} + a_5 s_{i+1,j} \quad (8.56)$$

For the ungrouped point at the right end boundary, we have, using Eq. (8.53),

$$u_{m-1,j+1} = \frac{1}{1+\lambda} [\lambda(u_{m-2,j} + u_{m,j+1}) + (1-\lambda)u_{m-1,j} + \Delta t s_{m-1,j}]$$

i.e.,

$$u_{m-1,j+1} = a_7(u_{m-2,j} + u_{m,j+1}) + a_8 u_{m-1,j} + a_9 s_{m-1,j} \quad (8.57)$$

where

$$a_7 = \frac{\lambda}{1+\lambda} \qquad a_8 = \frac{1-\lambda}{1+\lambda} \qquad a_9 = \frac{\Delta t}{1+\lambda}$$

Similarly, using Eq. (8.54), the ungrouped point at the left end boundary is given by

$$u_{1,j+1} = a_8 u_{1j} + a_7(u_{2j} + u_{0,j+1}) + a_9 s_{1j} \quad (8.58)$$

With Eqs. (8.55)–(8.58) we can then construct the alternating schemes of (S)AGE/(D)AGE-EXP schemes based on their constituent GER and GEL formulas.

(b) GE methods employing the predictor-corrector technique: GE-PC

The formulation is the same as in iii(a) except that the source term S is now approximated by $(s_{i,j} + \bar{s}_i)/2$ where $\bar{s}_i = S(\bar{u}_i)$ and the value $\bar{u}_i$ in the predictor step is determined by the solution of the system of differential equations (the explicit method of lines)

$$\frac{du_i}{dt} = \frac{1}{(\Delta x)^2} (u_{i+1,j} - 2u_{i,j} + u_{i-1,j}) + s_i \qquad i = 1, 2, \ldots, m-1 \quad (8.59)$$

[The system (8.59) is derived by discretizing the spatial derivatives in Eq. (8.48) and keeping the time as a continuous variable. The diffusion terms are evaluated at the previous time step.] These equations are solved by means of *an explicit, fourth order accurate Runge-Kutta method*. The solution is then employed in the corrector step by the appropriate GE schemes whose set of explicit equations could be similarly derived as in iii(a) and given by

$$u_{i,j+1} = a_1 u_{i-1,j} + a_2 u_{ij} + a_3 u_{i+1,j} + a_4 u_{i+2,j} + \frac{a_5(s_{ij} + \bar{s}_i)}{2}$$

$$+ \frac{a_6(s_{i+1,j} + \bar{s}_{i+1})}{2} \quad (8.60)$$

and

$$u_{i+1,j+1} = a_4 u_{i-1,j} + a_3 u_{ij} + a_2 u_{i+1,j} + a_1 u_{i+2,j}$$

$$+ \frac{a_6(s_{ij} + \bar{s}_i)}{2} + \frac{a_5(s_{i+1,j} + \bar{s}_{i+1})}{2} \qquad (8.61)$$

In the same way, the ungrouped point at the right end boundary is given as

$$u_{m-1,j+1} = a_7(u_{m-2,j} + u_{m,j+1}) + a_8 u_{m-1,j} + \frac{a_9(s_{m-1,j} + \bar{s}_{m-1})}{2} \qquad (8.62)$$

and at the left end boundary by

$$u_{1,j+1} = a_8 u_{1j} + a_7(u_{2j} + u_{0,j+1}) + \frac{a_9(s_{1j} + \bar{s}_1)}{2} \qquad (8.63)$$

Having found the basic equations for our GE schemes given by Eqs. (8.60)–(8.63) the (S)AGE/(D)AGE-PC methods can be constructed.

(c) AGE methods employing the predictor-corrector technique on implicit approximation: AGE-PC

As in iii(b), before applying the AGE algorithm on the corrector the explicit method of lines is again employed to determine the values of u_i from the system of ordinary differential equations,

$$\frac{du_i}{dt} = \frac{1}{(\Delta x)^2}(u_{i+1,j} - 2u_{ij} + u_{i-1,j}) + s_i \qquad i = 1, 2, \ldots, m$$

which is solved by means of an explicit, fourth-order accurate, Runge-Kutta method. The solution $\bar{u}_i$ say, is then employed in the following implicit approximation to Eq. (8.48) in the corrector step,

$$\frac{u_{i,j+1} - u_{ij}}{\Delta t} = \frac{1}{2(\Delta x)^2}[(u_{i+1,j} - 2u_{ij} + u_{i-1,j})$$

$$+ (u_{i+1,j+1} - 2u_{i,j+1} + u_{i-1,j+1})] + \frac{s_{ij} + \bar{s}_i}{2}$$

where $\bar{s}_i = S(\bar{u}_i)$. This CN-like scheme can be written in tridiagonal matrix form as

$$-\lambda u_{i-1,j+1} + (2 + 2\lambda)u_{i,j+1} - \lambda u_{i+1,j+1}$$

$$= 2u_{ij} + \lambda(u_{i+1,j} - 2u_{ij} + u_{i-1,j}) + \Delta t(s_{ij} + \bar{s}_i) \qquad (8.64)$$

or

$$\mathbf{Au} = \mathbf{f}$$

where A takes the same form as in Eq. (2.3), $\mathbf{u} = (u_{1,j+1}, u_{2,j+1}, \ldots, u_{m,j+1})^T$ and $\mathbf{f} = (f_1, f_2, \ldots, f_m)^T$

with $c = b = -\lambda$ $\quad a = 2(1 + \lambda)$

$$f_1 = 2u_{1j} + \lambda(u_{0j} + u_{0,j+1} + u_{2j} - 2u_{1j}) + \Delta t(s_{1j} + \bar{s}_1)$$

$$f_i = 2u_{ij} + \lambda(u_{i+1,j} - 2u_{ij} + u_{i-1,j})$$

$$+ \Delta t(s_{ij} + \bar{s}_i) \qquad i = 2, 3, \ldots, m - 1$$

and $\qquad f_m = 2u_{mj} + \lambda(u_{m+1,j} + u_{m+1,j+1} - 2u_{mj} + u_{m-1,j})$

$$+ \Delta t(s_{mj} + \bar{s}_m)$$

The AGE equations required to solve the system (8.64) are given by Eqs. (3.15a) and (3.18) for the case m odd and (3.23a) and (3.25a) for the case m even.

(d) AGE methods employing time linearization techniques on implicit approximation: AGE-TL

Four different time linearization schemes can be derived as approximations to the differential equation (8.48).

(1) First time linearization scheme (1TL)

The familiar fully implicit approximation to Eq. (8.48) is given by

$$u_{i,j+1} = u_{ij} + \lambda(u_{i-1,j+1} - 2u_{i,j+1} + u_{i+1,j+1})$$

$$+ \Delta t s_{i,j+1} \qquad i = 1, 2, \ldots, m \tag{8.65}$$

The source term $s_{i,j+1}$ can be linearized around the previous time step by means of Taylor's series as

$$s_{i,j+1} = s_{ij} + \left(\frac{\partial S}{\partial U}\right)_{ij} (u_{i,j+1} - u_{ij}) \tag{8.66}$$

The substitution of the expression in Eq. (8.66) into Eq. (8.65) yields

$$-\lambda u_{i-1,j+1} + \left[1 + 2\lambda - \Delta t \left(\frac{\partial S}{\partial U}\right)_{ij}\right] u_{i,j+1} - \lambda u_{i+1,j+1}$$

$$= u_{ij} + \Delta t \left[s_{ij} - \left(\frac{\partial S}{\partial U}\right)_{ij} u_{ij}\right] \tag{8.67}$$

(2) Second time linearization scheme (2TL)

By applying the Crank-Nicolson concept to the differential equation (8.48), we obtain the following aproximation:

$$u_{i,j+1} = u_{ij} + \lambda \frac{u_{i+1,j+1} - 2u_{i,j+1} + u_{i-1,j+1}}{2}$$

$$+ \lambda \frac{u_{i+1,j} - 2u_{ij} + u_{i-1,j}}{2} + \Delta t \frac{s_{ij} + s_{i,j+1}}{2} \qquad \text{for } i = 1, 2, \ldots, m \tag{8.68}$$

The substitution of Eq. (8.66) into Eq. (8.68) leads to

$$-\lambda u_{i-1,j+1} + \left[2 + 2\lambda - \Delta t\left(\frac{\partial S}{\partial U}\right)_{ij}\right]u_{i,j+1} - \lambda u_{i+1,j+1}$$

$$= 2u_{ij} + \lambda(u_{i+1,j} - 2u_{ij} + u_{i-1,j}) + \Delta t\left[2s_{ij} - \left(\frac{\partial S}{\partial U}\right)_{ij}u_{ij}\right] \quad (8.69)$$

(3) Third time linearization scheme (3TL)
  This technique employs a first-order accurate time discretization and a three-point compact formula for the diffusion terms (Kopal, 1961) and can be written as

$$u_{i,j+1} = u_{ij} + \lambda\frac{\delta_x^2}{1 + \delta_x^2/12}u_{i,j+1} + \Delta t s_{i,j+1} \quad i = 1, 2, \ldots, m$$

or

$$\left(1 + \frac{\delta_x^2}{12}\right)u_{i,j+1} = \left(1 + \frac{\delta_x^2}{12}\right)u_{ij} + \lambda\delta_x^2 u_{i,j+1} + \Delta t\left(1 + \frac{\delta_x^2}{12}\right)s_{i,j+1} \quad (8.70)$$

where δ denotes the usual central difference operator. By substituting Eq. (8.66) into Eq. (8.70), we obtain

$$\left\{-\lambda + \frac{1}{12}\left[1 - \Delta t\left(\frac{\partial S}{\partial U}\right)_{i-1,j}\right]\right\}u_{i-1,j+1}$$

$$+ \left\{2\lambda + \frac{5}{6}\left[1 - \Delta t\left(\frac{\partial S}{\partial U}\right)_{i,j}\right]\right\}u_{i,j+1}$$

$$+ \left\{-\lambda + \frac{1}{12}\left[1 - \Delta t\left(\frac{\partial S}{\partial U}\right)_{i+1,j}\right]\right\}u_{i+1,j+1}$$

$$= \frac{1}{12}\left[1 - \Delta t\left(\frac{\partial S}{\partial U}\right)_{i-1,j}\right]u_{i-1,j} + \frac{5}{6}\left[1 - \Delta t\left(\frac{\partial S}{\partial U}\right)_{ij}\right]u_{ij}$$

$$+ \frac{1}{12}\left[1 - \Delta t\left(\frac{\partial S}{\partial U}\right)_{i+1,j}\right]u_{i+1,j}$$

$$+ \frac{1}{6}\Delta t\left[5s_{ij} + \frac{1}{2}(s_{i-1,j} + s_{i+1,j})\right] \quad i = 1, 2, \ldots, m \quad (8.71)$$

(4) Fourth time linearization scheme (4TL)
  This scheme employs a second-order accurate time discretization and a three-point compact formula for the diffusion terms (Kopal, 1961) which is fourth-order accurate in space. The finite-difference form of the 4TL method can be written

as

$$u_{i,j+1} = u_{ij} + \lambda \left(\frac{\delta_x^2}{1 + \delta_x^2/12} \right) \left(\frac{u_{ij} + u_{i,j+1}}{2 + \Delta t} \right) \left(\frac{s_{ij} + s_{i,j+1}}{2} \right) \qquad i = 1, 2, \ldots, m$$

or

$$\left(2 + \frac{\delta_x^2}{6} \right) u_{i,j+1} = \left(2 + \frac{\delta_x^2}{6} \right) u_{ij} + \lambda \delta_x^2 (u_{ij} + u_{i,j+1})$$

$$+ \Delta t \left(1 + \frac{\delta_x^2}{12} \right) (s_{ij} + s_{i,j+1}) \qquad (8.72)$$

The substitution of Eq. (8.66) into Eq. (8.72) yields

$$\left\{ \frac{1}{6} \left[1 - \frac{\Delta t}{2} \left(\frac{\partial S}{\partial U} \right)_{i-1,j} \right] - \lambda \right\} u_{i-1,j+1}$$

$$+ \left\{ \frac{5}{6} \left[2 - \Delta t \left(\frac{\partial S}{\partial U} \right)_{i,j} \right] + 2\lambda \right\} u_{i,j+1}$$

$$+ \left\{ \frac{1}{6} \left[1 - \frac{\Delta t}{2} \left(\frac{\partial S}{\partial U} \right)_{i+1,j} \right] - \lambda \right\} u_{i+1,j+1}$$

$$= \left\{ \frac{1}{6} \left[1 - \frac{\Delta t}{2} \left(\frac{\partial S}{\partial U} \right)_{i-1,j} \right] + \lambda \right\} u_{i-1,j}$$

$$+ \left\{ \frac{5}{6} \left[2 - \Delta t \left(\frac{\partial S}{\partial U} \right)_{ij} \right] - 2\lambda \right\} u_{ij}$$

$$+ \left\{ \frac{1}{6} \left[1 - \frac{\Delta t}{2} \left(\frac{\partial S}{\partial U} \right)_{i+1,j} \right] + \lambda \right\} u_{i+1,j}$$

$$+ \frac{\Delta t}{6} (S_{i-1,j} + 10 S_{ij} + S_{i+1,j}) \qquad (8.73)$$

We note that Eqs. (8.67), (8.69), (8.71), and (8.73) generate tridiagonal systems of the form (8.47b). Hence, we have

(1) *for 1TL (8.67)*

$$c_{ij} = -\lambda \qquad i = 2, 3, \ldots, m$$

$$a_{ij} = 1 + 2\lambda - \Delta t \left(\frac{\partial S}{\partial U} \right)_{ij} \qquad i = 1, 2, \ldots, m$$

$$b_{ij} = -\lambda \qquad i = 1, 2, \ldots, m - 1$$

$$f_1 = u_{1j} + \Delta t\left[s_{1j} - \left(\frac{\partial S}{\partial U}\right)_{1,j} u_{1j}\right] + \lambda u_{0,j+1}$$

$$f_i = u_{ij} + \Delta t\left[s_{ij} - \left(\frac{\partial S}{\partial U}\right)_{ij} u_{ij}\right] \qquad i = 2, 3, \ldots, m-1$$

and

$$f_m = u_{mj} + \Delta t\left[s_{mj} - \left(\frac{\partial S}{\partial U}\right)_{mj} u_{mj}\right] + \lambda u_{m+1,j+1}$$

(2) for 2TL (8.69)

$$c_{ij} = -\lambda \qquad i = 2, 3, \ldots, m$$

$$a_{ij} = 2 + 2\lambda - \Delta t\left(\frac{\partial S}{\partial U}\right)_{i,j} \qquad i = 1, 2, \ldots, m$$

$$b_{ij} = -\lambda \qquad i = 1, 2, \ldots, m-1$$

$$f_1 = 2u_{1j} + \lambda(u_{2j} - 2u_{1j}) + \Delta t\left[2s_{1j} - \left(\frac{\partial S}{\partial U}\right)_{1j} u_{1j}\right] + \lambda(u_{0j} + u_{0,j+1})$$

$$f_i = 2u_{ij} + \lambda(u_{i-1,j} - 2u_{ij} + u_{i+1,j})$$

$$+ \Delta t\left(2s_{ij} - \left(\frac{\partial S}{\partial U}\right)_{ij} u_{ij}\right) \qquad i = 2, \ldots, m-1$$

and

$$f_m = 2u_{mj} + \lambda(u_{m-1,j} - 2u_{mj}) + \Delta t\left[2s_{mj} - \left(\frac{\partial S}{\partial U}\right)_{mj} u_{mj}\right]$$

$$+ \lambda(u_{m+1,j} + u_{m+1,j+1})$$

(3) for 3TL (8.71)

$$c_{ij} = -\lambda + \frac{1}{12}\left[1 - \Delta t\left(\frac{\partial S}{\partial U}\right)_{i-1,j}\right] \qquad i = 2, 3, \ldots, m$$

$$a_{ij} = 2\lambda + \frac{5}{6}\left[1 - \Delta t\left(\frac{\partial S}{\partial U}\right)_{ij}\right] \qquad i = 1, 2, \ldots, m$$

$$b_{ij} = -\lambda + \frac{1}{12}\left[1 - \Delta t\left(\frac{\partial S}{\partial U}\right)_{i+1,j}\right] \qquad i = 1, 2, \ldots, m-1$$

$$f_1 = \frac{1}{12}\left[1 - \Delta t\left(\frac{\partial S}{\partial U}\right)_{0,j}\right](u_{0j} - u_{0,j+1}) + \lambda u_{0,j+1}$$

$$+ \frac{5}{6}\left[1 - \Delta t\left(\frac{\partial S}{\partial U}\right)_{1,j}\right]u_{1j} + \frac{1}{6}\Delta t\left[5s_{1j} + \frac{1}{2}(s_{0j} + s_{2j})\right]$$

$$f_i = \frac{1}{12}\left[1 - \Delta t\left(\frac{\partial S}{\partial U}\right)_{i-1,j}\right]u_{i-1,j} + \frac{5}{6}\left[1 - \Delta t\left(\frac{\partial S}{\partial U}\right)_{ij}\right]$$

$$+ \frac{1}{12}\left[1 - \Delta t\left(\frac{\partial S}{\partial U}\right)_{i+1,j}\right]u_{i+1,j}$$

$$+ \frac{1}{6}\Delta t\left[5s_{ij} + \frac{1}{2}(s_{i-1,j} + s_{i+1,j})\right]$$

$$i = 2, \ldots, m - 1$$

and

$$f_m = \frac{1}{12}\left[1 - \Delta t\left(\frac{\partial S}{\partial U}\right)_{m+1,j}\right][u_{m+1,j} - u_{m+1,j+1}]$$

$$+ \lambda u_{m+1,j+1} + \frac{1}{12}\left[1 - \Delta t\left(\frac{\partial S}{\partial U}\right)_{m-1,j}\right]u_{m-1,j}$$

$$+ \frac{1}{6}\left\{5\left[1 - \Delta t\left(\frac{\partial S}{\partial U}\right)_{mj}\right]u_{mj}\right.$$

$$+ \left.\Delta t\left[5s_{mj} + \frac{1}{2}(s_{m-1,j} + s_{m+1,j})\right]\right\};$$

(4) for 4TL (8.73)

$$c_{ij} = \frac{1}{6}\left[1 - \frac{\Delta t}{2}\left(\frac{\partial S}{\partial U}\right)_{i-1,j}\right] - \lambda \qquad i = 2, 3, \ldots, m$$

$$a_{ij} = \frac{5}{6}\left[2 - \Delta t\left(\frac{\partial S}{\partial U}\right)_{ij}\right] + 2\lambda \qquad i = 1, 2, \ldots, m$$

$$b_{ij} = \frac{1}{6}\left[1 - \frac{\Delta t}{2}\left(\frac{\partial S}{\partial U}\right)_{i+1,j}\right] - \lambda \qquad i = 1, 2, \ldots, m - 1$$

$$f_1 = \frac{1}{6}\left[1 - \frac{\Delta t}{2}\left(\frac{\partial S}{\partial U}\right)_{0,j}\right](u_{0j} - u_{0,j+1}) + \lambda(u_{0j} + u_{0,j+1})$$

$$+ \left\{\frac{5}{6}\left[2 - \Delta t\left(\frac{\partial S}{\partial U}\right)_{1j}\right] - 2\lambda\right\}u_{1j} + \left\{\frac{1}{6}\left[1 - \frac{\Delta t}{2}\left(\frac{\partial S}{\partial U}\right)_{2,j}\right] + \lambda\right\}u_{2j}$$

$$+ \frac{\Delta t}{6}(s_{0j} + 10s_{1j} + s_{2j})$$

$$f_i = \left\{\frac{1}{6}\left[1 - \frac{\Delta t}{2}\left(\frac{\partial S}{\partial U}\right)_{i-1,j}\right] + \lambda\right\}u_{i-1,j} + \left\{\frac{5}{6}\left[2 - \Delta t\left(\frac{\partial S}{\partial U}\right)_{ij}\right] - 2\lambda\right\}u_{ij}$$

$$+ \left\{\frac{1}{6}\left[1 - \frac{\Delta t}{2}\left(\frac{\partial S}{\partial U}\right)_{i+1,j}\right] + \lambda\right\}u_{i+1,j}$$

$$+ \frac{\Delta t}{6}(s_{i-1,j} + 10s_{ij} + s_{i+1,j}) \qquad i = 2, 3, \ldots, m - 1$$

and

$$f_m = \frac{1}{6}\left[1 - \frac{\Delta t}{2}\left(\frac{\partial S}{\partial U}\right)_{m+1,j}\right](u_{m+1,j} - u_{m+1,j+1}) + \lambda(u_{m+1,j} + u_{m+1,j+1})$$

$$+ \left\{\frac{1}{6}\left[1 - \frac{\Delta t}{2}\left(\frac{\partial S}{\partial U}\right)_{m-1,j}\right] + \lambda\right\}u_{m-1,j}$$

$$+ \left\{\frac{5}{6}\left[2 - \Delta t\left(\frac{\partial S}{\partial U}\right)_{m,j}\right] - 2\lambda\right\}u_{mj}$$

$$+ \frac{\Delta t}{6}(s_{m-1,j} + 10s_{mj} + s_{m+1,j})$$

Again we find that the equations governing the convergence of the AGE iterative process that utilizes each of the above time linearization schemes are given by Eqs. (8.18a) and (8.19a) (for the case m odd) and (8.24a) and (8.25a) (for the case m even) with $\mathbf{w}$ replaced by $\mathbf{u}$.

(iv) Solving the Nonlinear First Order Hyperbolic (Convection) Equation
A nonlinear first order hyperbolic equation may take the following form:

$$\frac{\partial U}{\partial t} = -U\frac{\partial U}{\partial x} \tag{8.74}$$

The generalized difference formula developed as an approximation to Eq. (8.74) at the point $(i, j + \theta)$ is given by

$$\frac{u_{i,j+1} - u_{ij}}{\Delta t} = -u_{i,j+\theta}\left\{\frac{1}{\Delta x}[(1 - w)u_{i+1,j+1} + (2w - 1)u_{i,j+1}\right.$$

$$- wu_{i-1,j+1}] + (1 - \theta)[(1 - w)u_{i+1,j}$$

$$\left. + (2w - 1)u_{ij} - wu_{i-1,j}]\right\} \qquad \text{for } i = 1, 2, \ldots, m \tag{8.75}$$

where $0 \le w, \theta \le 1$.

(a) Implicit, centered-in-distance, backward-in-time scheme (CDBT)
If we choose $w = 1/2$ and $\theta = 1$ in Eq. (8.75), we obtain the following formula:

$$\frac{u_{i,j+1} - u_{ij}}{\Delta t} = -\frac{\alpha_{i,j+1}}{2\Delta x}(u_{i+1,j+1} - u_{i-1,j+1}) \tag{8.76}$$

where $\alpha_{i,j+1} = u_{i,j+1}$. Equation (8.76) is linearized by replacing $\alpha_{i,j+1}$ by $u_{i,j+1}^{(p)}$ (the old value in our iterative process) leading to the tridiagonal system of equations,

$$-\frac{\lambda}{2}u_{i,j+1}^{(p)}u_{i-1,j+1} + u_{i,j+1} + \frac{\lambda}{2}u_{i,j+1}^{(p)}u_{i+1,j+1} = u_{ij} \qquad i = 1, 2, \ldots, m \quad (8.77)$$

where $\lambda = \Delta t/\Delta x$, the mesh ratio.

(b) Centered-in-distance, centered-in-time scheme (CDCT)

The choice of $w = 1/2$ and $\theta = 1/2$ in Eq. (8.75) gives the difference analogue,

$$\frac{-u_{i,j+1} - u_{ij}}{\Delta t} = \frac{u_{i,j+1/2}}{4\Delta x}(u_{i+1,j+1} - u_{i-1,j+1} + u_{i+1,j} - u_{i-1,j}) \qquad (8.78)$$

Since $u_{i,j+1/2}$ does not fall on the grid point, it can be replaced by $(u_{i,j} + u_{i,j+1})/2$ and Eq. (8.78) becomes

$$\frac{u_{i,j+1} - u_{ij}}{\Delta t} = -\frac{1}{4\Delta x}[\alpha_{ij}(u_{i+1,j+1} - u_{i-1,j+1}) + \beta_{ij}(u_{i+1,j} - u_{i-1,j})]$$

which is nonlinear with $\alpha_{ij} = \beta_{ij} = (u_{ij} + u_{i,j+1})/2$. It can, however, be linearized by letting instead $\alpha_{ij} = u_{ij}$ and $\beta_{ij} = u_{i,j+1}$, and consequently we get the tridiagonal system of equations

$$-\frac{\lambda}{4}u_{ij}u_{i-1,j+1} + \left[1 + \frac{\lambda}{4}(u_{i+1,j} - u_{i-1,j})\right]u_{i,j+1} + \frac{\lambda}{4}u_{ij}u_{i+1,j+1} = u_{ij} \quad (8.79)$$

As previously, the tridiagonal systems (8.77) and (8.79) are of the form (8.47b) and therefore we have

(1) for CDBT (8.77)

$$c_{ij} = -\frac{\lambda}{2}u_{i,j+1}^{(p)} \qquad i = 2, 3, \ldots, m$$

$$a_{ij} = 1 \qquad i = 1, 2, \ldots, m$$

$$b_{ij} = \frac{\lambda}{2}u_{i,j+1}^{(p)} \qquad i = 1, 2, \ldots, m - 1$$

$$f_1 = u_{1j} + \left(\frac{\lambda}{2}u_{1,j+1}^{(p)}\right)u_{0,j+1}$$

$$f_i = u_{ij} \qquad i = 2, 3, \ldots, m - 1$$

$$f_m = u_{mj} - \left(\frac{\lambda}{2}u_{m,j+1}^{(p)}\right)u_{m+1,j+1}$$

(2) for CDCT (8.79)

$$c_{ij} = -\frac{\lambda}{4} u_{ij} \qquad i = 2, 3, \ldots, m$$

$$a_{ij} = 1 + \frac{\lambda}{4}(u_{i+1,j} - u_{i-1,j}) \qquad i = 1, 2, \ldots, m$$

$$b_{ij} = \frac{\lambda}{4} u_{ij} \qquad i = 1, 2, \ldots, m - 1$$

$$f_1 = u_{1j}\left(1 + \frac{\lambda}{4} u_{0,j+1}\right)$$

$$f_i = u_{ij} \qquad i = 2, \ldots, m - 1$$

and

$$f_m = u_{mj}\left(1 - \frac{\lambda}{4} u_{m+1,j+1}\right)$$

The AGE iterative process is completed when convergence is reached using Eqs. (8.18*a*) and (8.19*b*) (if *m* odd) and (8.24*a*) and (8.25*a*) (if *m* even) with **w** replaced by **u**.

9 NUMERICAL EXPERIMENTS AND COMPARATIVE RESULTS

A number of experiments were conducted to demonstrate the application of the AGE algorithm on parabolic and hyperbolic problems and where appropriate the solutions were compared with that of the GE class of methods either given by Evans and Abdullah (1983) or obtained by Sahimi (1986). In most cases, both the Peaceman-Rachford (PR) and the Douglas-Rachford (DR) variants were employed for the implementation of the AGE scheme and the acceleration parameter *r* was chosen so as to provide the most rapid convergence. Unless otherwise stated, the *convergence criterion* was taken as eps $= 10^{-4}$.

Experiment 1

We considered the following problem taken from Saulev (1964),

$$\frac{\partial U}{\partial t} = \frac{\partial^2 U}{\partial x^2} \qquad 0 \le x \le 1 \tag{9.1a}$$

subject to the initial condition

$$U(x, 0) = 4x(1 - x) \qquad 0 \le x \le 1$$

and the boundary conditions

$$U(0, t) = U(1, t) = 0 \qquad t \geq 0 \tag{9.1b}$$

The exact solution is given by

$$U(x, t) = \frac{32}{\pi^3} \sum_{k=1,(2)}^{\infty} \frac{1}{k^3} e^{-\pi^2 k^2 t} \sin(k\pi x) \tag{9.2}$$

Tables 1–3 provide a comparison of the accuracy of the methods under consideration in terms of the absolute errors at the appropriate grid points for various values of λ. It is very clear that in the AGE class of methods, the Douglas formula (AGE-DG) employing the PR variant is the most accurate in comparison with the fully implicit formula (AGE-IMP) as well as the Crank-Nicolson scheme (AGE-CN). This is to be expected since DG is second-order accurate in space and fourth-order accurate in time whereas CN and IMP have accuracies to the order of $O[(\Delta x)^2 + (\Delta t)^2]$ and $O[(\Delta x)^2 + \Delta t]$, respectively. By the same reasoning, the PR variant which we know is second-order accurate in both space and time is expected to produce a better solution and takes smaller iteration than the DR variant whose truncation error is $O[(\Delta x)^2 + \Delta t]$. It is also apparent that (D)AGE and AGE-CN have comparable accuracies.

To indicate the efficiency of the AGE iterative methods, it is necessary to consider *the computational complexity* for each iteration as well as the number of iterations required for convergence. One way of estimating this computational complexity is to count the number of arithmetic operations performed on each mesh line (time level) where there are m internal points. Thus, using Eqs. (2.4b), (3.15a), (3.15b), (3.18), and (3.19) (with m odd), we find that to complete one iteration of the generalized AGE scheme, the amount of work done is given by Table 4.

In comparison, *the Thomas algorithm* requires approximately $(11m-3)$ multiplications and $(7m-3)$ subtractions to solve the CN scheme directly. We infer from Table 4 that for large m, the number of multiplications incurred in the implementation of the AGE algorithm for CN or DG are only slightly more than that required by the Thomas algorithm. On the average only two iterations are needed by the PR variant for convergence, and it can therefore be concluded that the AGE scheme has merits as an alternative iterative method with respect to stability, accuracy, efficiency, and rate of convergence.

Experiment 2

In this experiment we attempted to solve the following heat conduction problem with periodic boundary conditions,

$$\frac{\partial U}{\partial t} = \frac{\partial^2 U}{\partial t^2} + 10(1 - x)xt \tag{9.3a}$$

subject to the initial condition

$$U(x, 0) = x(1 - x) \tag{9.3b}$$

Table 1 Absolute errors of the numerical solutions to Problem (9.1) (parabolic problem with Dirichlet boundary conditions)

$\lambda = 0.1$, $t = 0.05$, $\Delta t = 0.001$, $\Delta x = 0.1$, $r = 0.5$, eps $= 10^{-4}$

Scheme		0.1	0.2	0.3	0.4	0.5	0.6	0.7	0.8	0.9	Number of iterations
GER		6.0×10^{-5}	1.6×10^{-3}	1.5×10^{-3}	2.3×10^{-3}	2.3×10^{-3}	2.3×10^{-3}	2.6×10^{-3}	1.6×10^{-3}	1.9×10^{-3}	—
GEL		1.9×10^{-3}	1.6×10^{-3}	2.6×10^{-3}	2.3×10^{-3}	2.3×10^{-3}	2.3×10^{-3}	1.5×10^{-3}	1.6×10^{-3}	6.0×10^{-5}	—
(S)AGE		9.0×10^{-4}	1.4×10^{-3}	2.0×10^{-3}	2.0×10^{-3}	2.1×10^{-3}	2.1×10^{-3}	1.9×10^{-3}	1.6×10^{-3}	8.0×10^{-4}	—
(D)AGE		9.0×10^{-4}	1.5×10^{-3}	1.9×10^{-3}	2.1×10^{-3}	2.1×10^{-3}	2.1×10^{-3}	1.9×10^{-3}	1.6×10^{-3}	9.0×10^{-4}	—
AGE-IMP	PR	1.5×10^{-3}	2.6×10^{-3}	3.3×10^{-3}	3.7×10^{-3}	3.8×10^{-3}	3.7×10^{-3}	3.3×10^{-3}	2.6×10^{-3}	1.5×10^{-3}	2
	DR	2.3×10^{-3}	4.2×10^{-3}	5.5×10^{-3}	6.2×10^{-3}	6.4×10^{-3}	6.2×10^{-3}	5.5×10^{-3}	4.2×10^{-3}	2.3×10^{-3}	6
AGE-CN	PR	9.1×10^{-4}	1.6×10^{-3}	2.1×10^{-3}	2.3×10^{-3}	2.4×10^{-3}	2.3×10^{-3}	2.1×10^{-3}	1.6×10^{-3}	9.1×10^{-4}	2
	DR	1.7×10^{-3}	3.2×10^{-3}	4.2×10^{-3}	4.8×10^{-3}	5.0×10^{-3}	4.8×10^{-3}	4.2×10^{-3}	3.2×10^{-3}	1.7×10^{-3}	6
AGE-DG	PR	1.7×10^{-6}	5.5×10^{-6}	1.2×10^{-5}	1.7×10^{-5}	2.0×10^{-5}	1.7×10^{-5}	1.1×10^{-5}	5.5×10^{-6}	1.6×10^{-6}	2
	DR	8.1×10^{-4}	1.5×10^{-3}	2.1×10^{-3}	2.4×10^{-3}	2.6×10^{-3}	2.4×10^{-3}	2.1×10^{-3}	1.5×10^{-3}	8.1×10^{-4}	6
EXACT SOLUTION		0.1950648	0.3707705	0.5098716	0.5989617	0.6296137	0.5989617	0.5098716	0.3707705	0.1950648	—

Table 2 Absolute errors of the numerical solutions to Problem (9.1) (parabolic problem with Dirichlet boundary conditions)

$\lambda = 0.5$, $t = 0.25$, $\Delta t = 0.005$, $\Delta x = 0.1$, $r = 0.5$, eps $= 10^{-4}$

Scheme		x = 0.1	0.2	0.3	0.4	0.5	0.6	0.7	0.8	0.9	Number of iterations
GER		1.0×10^{-4}	1.1×10^{-4}	1.1×10^{-3}	1.7×10^{-3}	1.8×10^{-3}	1.7×10^{-3}	1.9×10^{-3}	1.1×10^{-3}	1.2×10^{-3}	—
GEL		1.2×10^{-3}	1.1×10^{-3}	1.9×10^{-3}	1.7×10^{-3}	1.8×10^{-3}	1.7×10^{-3}	1.1×10^{-3}	1.1×10^{-3}	1.0×10^{-4}	—
(S)AGE		2.0×10^{-4}	1.6×10^{-3}	1.8×10^{-3}	2.5×10^{-3}	2.5×10^{-3}	2.3×10^{-3}	2.1×10^{-3}	1.0×10^{-3}	8.0×10^{-4}	—
(D)AGE		2.0×10^{-4}	7.0×10^{-4}	3.0×10^{-4}	7.0×10^{-4}	7.0×10^{-4}	5.0×10^{-4}	7.0×10^{-4}	3.0×10^{-4}	3.0×10^{-4}	—
AGE-IMP	PR	2.2×10^{-3}	4.2×10^{-3}	5.7×10^{-3}	6.7×10^{-3}	7.0×10^{-3}	6.7×10^{-3}	5.7×10^{-3}	4.1×10^{-3}	2.2×10^{-3}	3
	DR	2.6×10^{-3}	5.0×10^{-3}	6.8×10^{-3}	8.0×10^{-3}	8.4×10^{-3}	7.9×10^{-3}	6.8×10^{-3}	4.9×10^{-3}	2.7×10^{-3}	6
AGE-CN	PR	5.3×10^{-4}	1.0×10^{-3}	1.4×10^{-3}	1.6×10^{-3}	1.7×10^{-3}	1.6×10^{-3}	1.4×10^{-3}	1.0×10^{-3}	5.4×10^{-4}	2
	DR	9.7×10^{-4}	1.9×10^{-3}	2.5×10^{-3}	3.0×10^{-3}	3.1×10^{-3}	3.0×10^{-3}	2.5×10^{-3}	1.8×10^{-3}	9.9×10^{-4}	6
AGE-DG	PR	1.5×10^{-5}	2.5×10^{-5}	3.8×10^{-5}	4.4×10^{-5}	4.7×10^{-5}	4.5×10^{-5}	3.6×10^{-5}	2.8×10^{-5}	1.2×10^{-5}	2
	DR	4.2×10^{-4}	8.0×10^{-4}	1.1×10^{-3}	1.3×10^{-3}	1.3×10^{-3}	1.3×10^{-3}	1.1×10^{-3}	7.9×10^{-4}	4.2×10^{-4}	6
EXACT SOLUTION		0.0270461	0.0514447	0.0708075	0.0832392	0.0875229	0.0832392	0.0708075	0.0514447	0.0270461	—

Table 3 Absolute errors of the numerical solutions to Problem (9.1) (parabolic problem with Dirichlet boundary conditions)

$\lambda = 1.0$, $t = 0.5$, $\Delta t = 0.01$, $\Delta x = 0.1$, $r = 0.5$, eps $= 10^{-4}$

Scheme		0.1	0.2	0.3	0.4	0.5	0.6	0.7	0.8	0.9	Number of iterations
GER		9.0×10^{-6}	2.0×10^{-4}	2.0×10^{-4}	3.0×10^{-4}	3.0×10^{-4}	3.0×10^{-4}	4.0×10^{-4}	2.0×10^{-4}	2.0×10^{-4}	—
GEL		2.2×10^{-4}	2.1×10^{-4}	3.5×10^{-4}	3.3×10^{-4}	3.5×10^{-4}	3.3×10^{-4}	2.1×10^{-4}	2.1×10^{-4}	9.0×10^{-6}	—
(S)AGE		6.0×10^{-4}	1.5×10^{-3}	1.9×10^{-3}	2.4×10^{-3}	2.5×10^{-3}	2.3×10^{-3}	2.0×10^{-3}	1.3×10^{-3}	8.0×10^{-4}	—
(D)AGE		4.0×10^{-4}	3.0×10^{-4}	2.0×10^{-4}	3.0×10^{-4}	3.0×10^{-4}	2.4×10^{-4}	2.8×10^{-4}	7.0×10^{-5}	1.5×10^{-4}	—
AGE-IMP	PR	6.3×10^{-4}	1.3×10^{-3}	1.6×10^{-3}	2.0×10^{-3}	2.0×10^{-3}	1.9×10^{-3}	1.7×10^{-3}	1.2×10^{-3}	7.0×10^{-4}	2
	DR	8.1×10^{-4}	1.6×10^{-3}	2.1×10^{-3}	2.5×10^{-3}	2.6×10^{-3}	2.5×10^{-3}	2.2×10^{-3}	1.5×10^{-3}	9.1×10^{-4}	4
AGE-CN	PR	7.3×10^{-5}	1.5×10^{-4}	1.9×10^{-4}	2.3×10^{-4}	2.4×10^{-4}	2.2×10^{-4}	2.0×10^{-4}	1.4×10^{-4}	8.1×10^{-5}	2
	DR	3.0×10^{-4}	5.8×10^{-4}	7.7×10^{-4}	9.2×10^{-4}	9.6×10^{-4}	9.1×10^{-4}	7.9×10^{-4}	5.6×10^{-4}	3.1×10^{-4}	4
AGE-DG	PR	1.6×10^{-5}	2.7×10^{-5}	4.3×10^{-5}	4.9×10^{-5}	5.3×10^{-5}	5.1×10^{-5}	4.0×10^{-5}	3.1×10^{-5}	1.2×10^{-5}	2
	DR	2.0×10^{-4}	4.0×10^{-4}	5.4×10^{-4}	6.3×10^{-4}	6.6×10^{-4}	6.3×10^{-4}	5.4×10^{-4}	3.9×10^{-4}	2.2×10^{-4}	4
EXACT SOLUTION		0.0022936	0.0043628	0.006048	0.0070591	0.0074224	0.00705091	0.0060048	0.0043628	0.0022936	—

Table 4 Computational complexity of AGE methods

Scheme	Number of multiplications	Number of additions
AGE-IMP	$\dfrac{(19m - 1)}{2} + 31$	$6m + 19$
AGE-CN	$\dfrac{(23m - 1)}{2} + 31$	$8m + 20$
AGE-DG	$\dfrac{(23m - 1)}{2} + 45$	$8m + 23$

and the boundary conditions

$$U(0, t) = U(1, t), \qquad \frac{\partial U}{\partial x}(0, t) = \frac{\partial U}{\partial x}(1, t) \qquad (9.3c)$$

The above problem has the exact solution given by

$$U(x, t) = \frac{(1 + 5t^2)}{6} - \frac{5}{8} \sum_{k=1}^{\infty} \frac{\cos(2k\pi x)}{(k\pi)^6} (4k^2\pi^2 t - 1 + e^{-4k^2\pi^2 t})$$

$$- \sum_{k=1}^{\infty} \frac{e^{-4k^2\pi^2 t}}{k^2\pi^2} \cos(2k\pi x) \qquad (9.4)$$

The AGE solutions at the grid points on selected time rows and for various time steps are displayed in Tables 5 and 6 for both the PR and DR variants. It is observed that except for $\lambda = 0.1$ (at $t = 0.1$), an examination of all the averages of the percentage errors indicates that the AGE solutions are more accurate than the (D)AGE values derived by Abdullah (1983). While the number of iterations for the PR variant remains fairly constant (as well as the percentage errors), the iteration count for the DR variant tends to become large with increasing λ.

Experiment 3

This experiment involved the solution of the diffusion-convection equation,

$$\frac{\partial U}{\partial t} = \epsilon \frac{\partial^2 U}{\partial x^2} - k \frac{\partial U}{\partial x} \qquad (9.5a)$$

with the initial condition

$$U(x, 0) = 0 \qquad 0 < x < 1 \qquad (9.5b)$$

and the Dirichlet boundary conditions

$$\left. \begin{array}{l} U(0, t) = 0 \\ U(1, t) = 1 \end{array} \right\} \quad t \geq 0 \qquad (9.5c)$$

Table 5 Numerical solutions to Problem (9.3) (parabolic problem with periodic boundary conditions)

$\Delta x = 0.1$, $r = 0.5$, eps $= 10^{-4}$

PR variant

Mesh ratio λ	Scheme	x 0.1	0.2	0.3	0.4	0.5	0.6	0.7	0.8	0.9	1.0	Average of percentage errors	Number of iterations
0.1 Δt = 0.001 t = 0.1	(D)AGE	0.1693	0.1718	0.1745	0.1766	0.1774	0.1767	0.1747	0.1720	0.1696	0.1684	≈1%	—
	AGE-IMP	0.1696	0.1720	0.1748	0.1768	0.1776	0.1768	0.1748	0.1720	0.1696	0.1687	1.004%	2
	AGE-CN	0.1697	0.1721	0.1747	0.1767	0.1774	0.1767	0.1747	0.1721	0.1697	0.1686	1.003%	2
	AGE-DG	0.1699	0.1722	0.1746	0.1765	0.1772	0.1765	0.1746	0.1722	0.1699	0.1689	1.001%	2
	EXACT SOL.	0.1718	0.1739	0.1763	0.1781	0.1788	0.1781	0.1763	0.1739	0.1718	0.1709	—	—
0.5 Δt = 0.005 t = 0.5	(D)AGE	0.3556	0.3639	0.3724	0.3784	0.3808	0.3788	0.3730	0.3650	0.3567	0.3517	≈2%	—
	AGE-IMP	0.3607	0.3682	0.3759	0.3814	0.3834	0.3814	0.3759	0.3682	0.3607	0.3568	1.003%	2
	AGE-CN	0.3607	0.3682	0.3759	0.3814	0.3834	0.3814	0.3759	0.3682	0.3607	0.3567	1.004%	2
	AGE-DG	0.3607	0.3682	0.3759	0.3814	0.3834	0.3814	0.3759	0.3682	0.3607	0.3567	1.004%	2
	EXACT SOL.	0.3650	0.3720	0.3793	0.3846	0.3865	0.3846	0.3793	0.3720	0.3650	0.3618	—	—
1.0 Δt = 0.01 t = 1.0	(D)AGE	0.9526	0.9684	0.9841	0.9971	1.0028	0.9989	0.9866	0.9698	0.9525	0.9425	≈3%	—
	AGE-IMP	0.9684	0.9837	0.9994	1.0108	1.0149	1.0108	0.9995	0.9837	0.9684	0.9603	1.002%	2
	AGE-CN	0.9683	0.9837	0.9995	1.0109	1.0150	1.0109	0.9995	0.9837	0.9683	0.9602	1.002%	2
	AGE-DG	0.9683	0.9837	0.9995	1.0109	1.0150	1.0109	0.9995	0.9837	0.9683	0.9602	1.002%	2
	EXACT SOL.	0.9795	0.9937	1.0088	1.0197	1.0237	1.0197	1.0088	0.9937	0.9795	0.9729	—	—
1.5 Δt = 0.015 t = 1.5	(D)AGE	1.9832	1.9999	1.9769	1.9944	2.0403	2.0458	2.0097	1.9685	1.9350	1.9332	≈3%	—
	AGE-IMP	1.9885	2.0117	2.0355	2.0528	2.0590	2.0528	2.0356	2.0117	1.9886	1.9764	1.001%	3
	AGE-CN	1.9884	2.0117	2.0356	2.0529	2.0592	2.0529	2.0356	2.0117	1.9884	1.9762	1.001%	2
	AGE-DG	1.9884	2.0117	2.0356	2.0529	2.0592	2.0529	2.0356	2.0117	1.9884	1.9762	1.001%	2
	EXACT SOL.	2.0107	2.0322	2.0549	2.0715	2.0775	2.0715	2.0549	2.0322	2.0107	2.0007	—	—

Table 6 Numerical solutions to Problem (9.3)

$\Delta x = 0.1$, $r = 0.5$, eps $= 10^{-4}$ DR variant

Mesh ratio	Scheme	x										Average of percentage errors	Number of iterations
		0.1	0.2	0.3	0.4	0.5	0.6	0.7	0.8	0.9	1.0		
0.1 $\Delta t = 0.001$ $t = 0.1$	(D)AGE	0.1693	0.1718	0.1745	0.1766	0.1774	0.1767	0.1747	0.1720	0.1696	0.1684	1%	—
	AGE-IMP	0.1703	0.1738	0.1764	0.1775	0.1764	0.1737	0.1703	0.1672	0.1658	0.1672	1.802%	2
	AGE-CN	0.1703	0.1737	0.1763	0.1772	0.1763	0.1737	0.1703	0.1674	0.1660	0.1674	1.809%	2
	AGE-DG	0.1704	0.1735	0.1760	0.1769	0.1760	0.1735	0.1704	0.1676	0.1663	0.1676	1.828%	2
	EXACT SOL.	0.1718	0.1739	0.1763	0.1781	0.1788	0.1781	0.1763	0.1739	0.1718	0.1709	—	—
0.5 $\Delta t = 0.005$ $t = 0.5$	(D)AGE	0.3556	0.3639	0.3724	0.3784	0.3808	0.3788	0.3730	0.3650	0.3567	0.3517	2%	—
	AGE-IMP	0.3625	0.3701	0.3756	0.3776	0.3756	0.3701	0.3625	0.3550	0.3511	0.3550	2.528%	6
	AGE-CN	0.3625	0.3702	0.3757	0.3777	0.3757	0.3702	0.3625	0.3550	0.3510	0.3550	2.521%	6
	AGE-DG	0.3626	0.3702	0.3758	0.3778	0.3758	0.3702	0.3626	0.3550	0.3511	0.3550	2.508%	6
	EXACT SOL.	0.3650	0.3720	0.3793	0.3846	0.3865	0.3846	0.3793	0.3720	0.3650	0.3618	—	—
1.0 $\Delta t = 0.01$ $t = 1.0$	(D)AGE	0.9526	0.9684	0.9841	0.9971	1.0028	0.9989	0.9866	0.9698	0.9525	0.9425	3%	—
	AGE-IMP	0.9774	0.9931	1.0045	1.0086	1.0045	0.9932	0.9774	0.9621	0.9541	0.9621	1.632%	8
	AGE-CN	0.9773	0.9931	1.0045	1.0087	1.0045	0.9931	0.9773	0.9620	0.9539	0.9620	1.638%	8
	AGE-DG	0.9774	0.9932	1.0046	1.0087	1.0046	0.9932	0.9774	0.9620	0.9539	0.9620	1.632%	8
	EXACT SOL.	0.9795	0.9937	1.0088	1.0197	1.0237	1.0197	1.0088	0.9937	0.9795	0.9729	—	—
1.5 $\Delta t = 0.015$ $t = 1.5$	(D)AGE	1.9832	1.9999	1.9769	1.9944	2.0403	2.0458	2.0097	1.9685	1.9350	1.9332	3%	—
	AGE-IMP	2.0056	2.0294	2.0466	2.0528	2.0466	2.0294	2.0056	1.9824	1.9703	1.9824	1.302%	9
	AGE-CN	2.0052	2.0291	2.0464	2.0526	2.0464	2.0291	2.0052	1.9819	1.9697	1.9819	1.320%	9
	AGE-DG	2.0052	2.0292	2.0465	2.0527	2.0465	2.0292	2.0052	1.9820	1.9697	1.9820	1.316%	9
	EXACT SOL.	2.0107	2.0322	2.0549	2.0715	2.0775	2.0715	2.0549	2.0322	2.0107	2.0007	—	—

The coefficients k and ϵ assumed the same values of 1. The exact solution is given by

$$U(x, t) = \frac{(e^{kx/\epsilon} - 1)}{(e^{k/\epsilon} - 1)}$$

$$+ 2 \sum_{n1}^{\infty} \frac{(-1)^n n\pi}{(n\pi)^2 + (k/2\epsilon)^2} e^{(k/2\epsilon)(x-1)} \sin(n\pi x) e^{-[(n\pi)^2\epsilon + k^2/4\epsilon]t} \quad (9.6)$$

From Abdullah (1983) the solutions of Eq. (9.5a) by means of the Crank-Nicolson scheme with upwinding (CNU) and the (D)AGE scheme were included. These were then compared with the AGE scheme employing the fully implicit formula (AGE-IMP) as well as the methods of Crank-Nicolson (AGE-CN) and Crank-Nicolson with upwinding (AGE-CNU).

The absolute errors of the numerical solutions of these schemes are shown in Tables 7–10. It must be remembered that the CNU scheme is only first-order accurate in space and second-order accurate in time. This explains the poor accuracy of CNU and AGE-CNU even when compared with the fully implicit formula whose truncation error is $O[(\Delta x)^2 + \Delta t]$ since smaller values of Δt were taken in our experiments with $\Delta x = 0.1$. For small λ, the (D)AGE process appears to have a slight edge on AGE-CN. For larger mesh ratios such as $\lambda = 2.0$, however, the AGE-CN method can be very competitive since it exhibits better accuracy and requires only one iteration for convergence and is therefore worthy of recommendation.

Experiment 4

The AGE algorithm was implemented on the same first-order hyperbolic equations of Section 4, i.e., Problem 1 and Problem 2.

(a) Problem 1

$$\frac{\partial U}{\partial t} + \frac{\partial U}{\partial x} = 0 \quad (9.7a)$$

subject to

$$U(x, 0) = \cos x \quad (9.7b)$$

$$U(0, t) = \cos t$$

and

$$U(1, t) = \cos(1 - t) \quad (9.7c)$$

The exact solution is given by

$$U(x, t) = \cos(x - t) \quad (9.8)$$

and

(b) Problem 2

$$\frac{\partial U}{\partial t} + \frac{\partial U}{\partial x} = k(x, t) \quad (9.9a)$$

$$(k(x, t) = -2 \sin(x - t)e^{-2t})$$

Table 7 Absolute errors of the numerical solutions to Problem (9.5) (diffusion-convection problem)

$k = 1.0$, $\epsilon = 1.0$, $\Delta t = 0.001$, $\Delta x = 0.1$, $\lambda = 0.1$, $t = 0.1$, $r = 0.5$, eps $= 5.0 \times 10^{-7}$

Method		0.1	0.2	0.3	0.4	0.5	0.6	0.7	0.8	0.9	Number of iterations
CNU		2.4×10^{-3}	4.9×10^{-3}	7.3×10^{-3}	9.4×10^{-3}	1.1×10^{-2}	1.2×10^{-2}	1.15×10^{-2}	9.4×10^{-3}	5.6×10^{-3}	—
DAGE		1.6×10^{-4}	$.2.7 \times 10^{-4}$	2.8×10^{-4}	1.3×10^{-4}	9.0×10^{-5}	4.0×10^{-4}	6.1×10^{-4}	6.2×10^{-4}	4.3×10^{-4}	—
AGE-IMP	DR	1.3×10^{-4}	1.4×10^{-4}	5.4×10^{-5}	4.8×10^{-4}	1.04×10^{-3}	1.6×10^{-3}	1.9×10^{-3}	1.8×10^{-3}	1.1×10^{-3}	12
	PR	1.3×10^{-4}	1.5×10^{-4}	4.4×10^{-5}	4.7×10^{-4}	1.0×10^{-3}	1.6×10^{-3}	1.9×10^{-3}	1.8×10^{-3}	1.1×10^{-3}	3
AGE-CNU	DR	2.4×10^{-3}	4.9×10^{-3}	7.37×10^{-3}	9.6×10^{-3}	1.1×10^{-2}	1.2×10^{-2}	1.15×10^{-2}	9.4×10^{-3}	5.6×10^{-3}	12
	PR	2.4×10^{-3}	4.9×10^{-3}	7.38×10^{-3}	9.6×10^{-3}	1.1×10^{-2}	1.2×10^{-2}	1.15×10^{-2}	9.4×10^{-3}	5.6×10^{-3}	3
AGE-CN	DR	1.7×10^{-4}	2.8×10^{-4}	2.7×10^{-4}	9.1×10^{-4}	1.8×10^{-4}	5.1×10^{-4}	7.2×10^{-4}	7.2×10^{-4}	4.6×10^{-4}	12
	PR	1.8×10^{-4}	2.9×10^{-4}	2.8×10^{-4}	1.0×10^{-4}	1.6×10^{-4}	4.9×10^{-4}	7.0×10^{-4}	7.1×10^{-4}	4.5×10^{-4}	3
EXACT SOLUTION		0.01895	0.04370	0.07892	0.12982	0.20177	0.29986	0.42794	0.58805	0.77976	—

eps: Convergence criterion.

Table 8 Absolute errors of the numerical solutions to Problem (9.5) (diffusion-convection problem)

$k = 1.0$, $\epsilon = 1.0$, $\Delta t = 0.005$, $\Delta x = 0.1$, $\lambda = 0.5$, $t = 0.5$, $r = 0.5$, eps $= 5.0 \times 10^{-7}$

Method		0.1	0.2	0.3	0.4	0.5	0.6	0.7	0.8	0.9	Number of iterations
CNU		1.7×10^{-3}	3.2×10^{-3}	4.5×10^{-3}	5.4×10^{-3}	6.1×10^{-3}	6.2×10^{-3}	5.8×10^{-3}	4.7×10^{-3}	2.8×10^{-3}	—
DAGE		2.0×10^{-5}	6.0×10^{-5}	5.0×10^{-5}	8.0×10^{-5}	8.0×10^{-5}	8.0×10^{-5}	9.0×10^{-5}	6.0×10^{-5}	6.0×10^{-5}	—
AGE-IMP	DR	1.6×10^{-4}	3.22×10^{-4}	4.6×10^{-4}	5.7×10^{-4}	6.3×10^{-4}	6.3×10^{-4}	5.7×10^{-4}	4.4×10^{-4}	2.4×10^{-4}	9
	PR	1.6×10^{-4}	3.19×10^{-4}	4.6×10^{-4}	5.7×10^{-4}	6.3×10^{-4}	6.3×10^{-4}	5.7×10^{-4}	4.4×10^{-4}	2.5×10^{-4}	4
AGE-CNU	DR	1.7×10^{-3}	3.2×10^{-3}	4.4×10^{-3}	5.4×10^{-3}	6.04×10^{-3}	6.2×10^{-3}	5.8×10^{-3}	4.7×10^{-3}	2.8×10^{-3}	9
	PR	1.7×10^{-3}	3.2×10^{-3}	4.5×10^{-3}	5.4×10^{-3}	6.1×10^{-3}	6.2×10^{-3}	5.8×10^{-3}	4.7×10^{-3}	2.8×10^{-3}	3
AGE-CN	DR	5.8×10^{-5}	1.1×10^{-4}	1.6×10^{-4}	1.98×10^{-4}	2.2×10^{-4}	2.2×10^{-4}	2.0×10^{-4}	1.6×10^{-4}	9.3×10^{-5}	9
	PR	5.6×10^{-5}	1.1×10^{-4}	1.6×10^{-4}	1.9×10^{-4}	2.1×10^{-4}	2.2×10^{-4}	1.97×10^{-4}	1.6×10^{-4}	8.99×10^{-5}	3
EXACT SOLUTION		0.06043	0.12730	0.20136	0.28345	0.37447	0.47539	0.58724	0.71114	0.84830	—

eps: Convergence criterion.

Table 9 Absolute errors of the numerical solutions to Problem (9.5) (diffusion-convection problem)

$k = 1.0$, $\epsilon = 1.0$, $\Delta t = 0.01$, $\Delta x = 0.1$, $\lambda = 1.0$, $t = 1.0$, $r = 0.5$, eps $= 5.0 \times 10^{-7}$

Method		0.1	0.2	0.3	0.4	0.5	0.6	0.7	0.8	0.9	Number of iterations
CNU		1.5×10^{-3}	2.9×10^{-3}	4.1×10^{-3}	4.9×10^{-3}	5.5×10^{-3}	5.7×10^{-3}	5.3×10^{-3}	4.3×10^{-3}	2.6×10^{-5}	—
DAGE		2.7×10^{-5}	5.2×10^{-5}	7.2×10^{-5}	8.8×10^{-5}	9.8×10^{-5}	1.0×10^{-4}	9.4×10^{-5}	7.6×10^{-5}	4.6×10^{-5}	—
AGE-IMP	DR	3.1×10^{-5}	6.0×10^{-5}	8.4×10^{-5}	1.0×10^{-4}	1.1×10^{-4}	1.2×10^{-4}	1.1×10^{-4}	8.8×10^{-5}	5.3×10^{-5}	3
	PR	3.1×10^{-5}	5.9×10^{-5}	8.2×10^{-5}	1.0×10^{-4}	1.1×10^{-4}	1.1×10^{-4}	1.1×10^{-4}	8.6×10^{-5}	5.2×10^{-5}	2
AGE-CNU	DR	1.5×10^{-3}	2.9×10^{-3}	4.1×10^{-3}	4.97×10^{-3}	5.5×10^{-3}	5.7×10^{-3}	5.3×10^{-3}	4.3×10^{-3}	2.6×10^{-3}	3
	PR	1.5×10^{-3}	2.9×10^{-3}	4.1×10^{-3}	4.98×10^{-3}	5.5×10^{-3}	5.7×10^{-3}	5.3×10^{-3}	4.3×10^{-3}	2.6×10^{-3}	2
AGE-CN	DR	2.8×10^{-5}	5.4×10^{-5}	7.6×10^{-5}	9.3×10^{-5}	1.0×10^{-4}	1.1×10^{-4}	9.9×10^{-5}	8.0×10^{-5}	4.8×10^{-5}	3
	PR	2.7×10^{-5}	5.2×10^{-5}	7.3×10^{-5}	8.9×10^{-5}	9.9×10^{-5}	1.0×10^{-4}	9.6×10^{-5}	7.8×10^{-5}	4.7×10^{-5}	2
EXACT SOLUTION		0.06120	0.12884	0.20360	0.28621	0.37752	0.47843	0.58996	0.71322	0.84945	—

eps: Convergence criterion.

Table 10 Absolute errors of the numerical solutions to Problem (9.5) (diffusion-convection problem)

$k = 1.0$, $\epsilon = 1.0$, $\Delta t = 0.020$, $\Delta x = 0.1$, $\lambda = 2.0$, $t = 2.0$, $r = 0.5$, eps $= 5 \times 10^{-7}$

Method		0.1	0.2	0.3	0.4	0.5	0.6	0.7	0.8	0.9	Number of iterations
CNU		1.5×10^{-3}	2.9×10^{-3}	4.1×10^{-3}	4.9×10^{-3}	5.5×10^{-3}	5.7×10^{-3}	5.3×10^{-3}	4.3×10^{-3}	2.6×10^{-3}	—
DAGE TWO-LEVEL		3.0×10^{-5}	5.0×10^{-5}	7.0×10^{-5}	9.0×10^{-5}	1.0×10^{-4}	1.0×10^{-4}	1.0×10^{-4}	8.0×10^{-5}	5.0×10^{-5}	—
AGE-IMP	DR	2.7×10^{-5}	5.1×10^{-5}	7.2×10^{-5}	8.8×10^{-5}	9.8×10^{-5}	10×10^{-4}	9.4×10^{-5}	7.7×10^{-5}	4.6×10^{-5}	1
	PR	2.7×10^{-5}	5.1×10^{-5}	7.2×10^{-5}	8.8×10^{-5}	9.8×10^{-5}	1.0×10^{-4}	9.4×10^{-5}	7.7×10^{-5}	4.6×10^{-5}	1
AGE-CNU	DR	1.5×10^{-3}	2.9×10^{-3}	4.1×10^{-3}	4.97×10^{-3}	5.5×10^{-3}	5.7×10^{-3}	5.3×10^{-3}	4.3×10^{-3}	2.6×10^{-3}	1
	PR	1.5×10^{-3}	2.9×10^{-3}	4.1×10^{-3}	4.97×10^{-3}	5.5×10^{-3}	5.7×10^{-3}	5.3×10^{-3}	4.3×10^{-3}	2.6×10^{-3}	1
AGE-CN	DR	2.7×10^{-5}	5.2×10^{-5}	7.2×10^{-5}	8.8×10^{-5}	9.8×10^{-5}	1.0×10^{-4}	9.4×10^{-5}	7.7×10^{-5}	4.6×10^{-5}	1
	PR	2.7×10^{-5}	5.1×10^{-5}	7.2×10^{-5}	8.8×10^{-5}	9.8×10^{-5}	1.0×10^{-4}	9.4×10^{-5}	7.7×10^{-5}	4.6×10^{-5}	1
EXACT SOLUTION		0.061207	0.128851	0.203610	0.28631	0.377541	0.478454	0.589980	0.713236	0.849455	—

eps: Convergence criterion.

subject to

$$U(x, 0) = \sin x \tag{9.9b}$$

$$U(0, t) = -\sin(t)e^{-2t}$$

and
$$U(1, t) = \sin(1 - t)e^{-2t} \tag{9.9c}$$

The exact solution is given by

$$U(x, t) = \sin(x - t)e^{-2t} \tag{9.10}$$

It is known that the truncation errors of the fully implicit scheme and the Crank-Nicolson type scheme (the centered-in-distance, centered-in-time formula) are $O[(\Delta x)^2 + \Delta t]$ and $O[(\Delta x)^2 + (\Delta t)^2]$, respectively. The accuracies of the AGE method utilizing these schemes (AGE-IMP and AGE-CN) are depicted in Tables 11–14 for both problems using $\lambda = 0.5$ and $\lambda = 1$. They are then compared with the results derived from the GE class of methods and other well-known schemes given in Sahimi (1986). It is interesting to note that the CN method (using the Thomas algorithm) and the AGE-CN scheme (employing the PR variant) exhibit the same order of accuracy although the latter with 3 to 4 iterations requires more computational load than the former. Furthermore, for Problem 1, these schemes evidently emerge as the next most accurate method of solution after the Lax-Wendroff formula. For Problem 2, however, the (D)AGE process appears to be more favorable. As expected, the DR variant of the AGE class of methods produces a slightly less accurate solution and entails two to three times more iterations than the corresponding PR formula.

Experiment 5

This experiment dealt with the following problems involving the one-dimensional wave equation:

(a) *Problem 1 (with Dirichlet boundary conditions)*

$$\frac{\partial^2 U}{\partial x^2} = \frac{\partial^2 U}{\partial t^2} \tag{9.11a}$$

subject to the initial conditions

$$U(x, 0) = \frac{1}{8}\sin(\pi x)$$

$$\frac{\partial U}{\partial t}(x, 0) = 0 \tag{9.11b}$$

and the boundary conditions

$$U(0, t) = U(1, t) = 0 \tag{9.11c}$$

Table 11 Absolute errors of the AGE solutions to hyperbolic Problem 1 (9.7)

(a) $t = 0.4$, $\lambda = 0.5$, $\Delta t = 0.05$, $\Delta x = 0.1$, $r = 0.5$, eps $= 10^{-4}$

Scheme		0.1	0.2	0.3	0.4	0.5	0.6	0.7	0.8	0.9	Average of all absolute errors	No. of iterations
AGE-IMP	PR	2.46×10^{-3}	4.94×10^{-3}	7.16×10^{-3}	8.18×10^{-3}	9.79×10^{-3}	8.39×10^{-3}	1.18×10^{-2}	5.53×10^{-3}	1.53×10^{-2}	8.17×10^{-3}	3
	DR	2.31×10^{-3}	4.92×10^{-3}	7.04×10^{-3}	8.24×10^{-3}	9.93×10^{-3}	8.53×10^{-3}	1.24×10^{-2}	5.67×10^{-3}	1.61×10^{-2}	8.35×10^{-3}	9
AGE-CN	PR	5.81×10^{-5}	7.19×10^{-5}	4.16×10^{-5}	5.25×10^{-6}	9.75×10^{-5}	9.82×10^{-5}	3.44×10^{-4}	5.77×10^{-5}	7.15×10^{-4}	1.65×10^{-4}	3
	DR	1.52×10^{-4}	1.29×10^{-4}	9.54×10^{-5}	4.6×10^{-5}	2.23×10^{-4}	2.42×10^{-4}	7.57×10^{-4}	1.75×10^{-4}	1.51×10^{-3}	3.7×10^{-4}	9
EXACT SOLUTION		0.9553365	0.9800666	0.9950042	1.0	0.9950042	0.9800666	0.9553365	0.9210610	0.8775826	—	—

(b) $t = 1.0$, $\lambda = 0.5$, $\Delta t = 0.05$, $\Delta x = 0.1$, $r = 0.5$, eps $= 10^{-4}$

Scheme		0.1	0.2	0.3	0.4	0.5	0.6	0.7	0.8	0.9	Average of all absolute errors	No. of iterations
AGE-IMP	PR	5.63×10^{-3}	5.39×10^{-4}	1.23×10^{-2}	6.04×10^{-4}	2.21×10^{-2}	2.51×10^{-4}	3.3×10^{-2}	1.59×10^{-4}	4.19×10^{-2}	1.29×10^{-2}	4
	DR	5.43×10^{-3}	1.08×10^{-3}	1.19×10^{-3}	1.64×10^{-3}	2.18×10^{-3}	1.46×10^{-3}	3.28×10^{-2}	6.82×10^{-4}	4.19×10^{-2}	1.32×10^{-2}	9
AGE-CN	PR	1.16×10^{-6}	4.58×10^{-6}	4.99×10^{-5}	8.29×10^{-4}	7.27×10^{-6}	9.04×10^{-4}	4.42×10^{-6}	5.75×10^{-4}	3.49×10^{-5}	3.18×10^{-4}	3
	DR	7.29×10^{-5}	9.64×10^{-4}	2.28×10^{-4}	1.9×10^{-3}	1.29×10^{-4}	2.19×10^{-3}	4.4×10^{-5}	1.45×10^{-3}	1.9×10^{-4}	7.96×10^{-4}	9
EXACT SOLUTION		0.6216100	0.6967067	0.7648422	0.8253356	0.8775826	0.9210610	0.9553365	0.9800666	0.9950042	—	—

Table 12 Absolute errors of the AGE solutions to hyperbolic Problem 1 (9.7)

(a) $t = 0.8$, $\lambda = 1.0$, $\Delta t = 0.1$, $\Delta x = 0.1$, $r = 0.5$, eps $= 10^{-4}$

Scheme		0.1	0.2	0.3	0.4	0.5	0.6	0.7	0.8	0.9	Average of all absolute errors	No. of iterations
AGE-IMP	PR	8.23×10^{-3}	4.36×10^{-3}	2.06×10^{-2}	8.89×10^{-3}	3.67×10^{-2}	1.02×10^{-2}	5.35×10^{-2}	6.73×10^{-3}	6.77×10^{-2}	2.41×10^{-2}	5
	DR	7.92×10^{-3}	4.27×10^{-3}	2.03×10^{-2}	8.68×10^{-3}	3.65×10^{-2}	9.88×10^{-3}	5.34×10^{-2}	6.5×10^{-3}	6.78×10^{-2}	2.39×10^{-2}	12
AGE-CN	PR	1.1×10^{-4}	3.63×10^{-4}	2.11×10^{-4}	6.66×10^{-4}	4.013×10^{-5}	8.18×10^{-4}	4.81×10^{-4}	5.72×10^{-4}	7.29×10^{-4}	4.43×10^{-4}	4
	DR	2.97×10^{-4}	5.47×10^{-4}	4.73×10^{-4}	1.01×10^{-3}	5.85×10^{-5}	1.19×10^{-3}	6.94×10^{-4}	8.31×10^{-4}	1.2×10^{-3}	7.0×10^{-4}	10
EXACT SOLUTION		0.7648422	0.8253356	0.8775826	0.921061	0.9553365	0.9800666	0.9950042	1.0	0.9950042	—	—

(b) $t = 2.0$, $\lambda = 1.0$, $\Delta t = 0.1$, $\Delta x = 0.1$, $r = 0.5$, eps $= 10^{-4}$

Scheme		0.1	0.2	0.3	0.4	0.5	0.6	0.7	0.8	0.9	Average of all absolute errors	No. of iterations
AGE-IMP	PR	7.57×10^{-2}	1.18×10^{-2}	7.09×10^{-2}	1.81×10^{-2}	6.67×10^{-2}	1.79×10^{-2}	6.47×10^{-2}	1.15×10^{-2}	6.61×10^{-2}	4.48×10^{-2}	6
	DR	7.51×10^{-2}	1.15×10^{-2}	7.01×10^{-2}	1.76×10^{-2}	6.57×10^{-2}	1.75×10^{-2}	6.35×10^{-2}	1.12×10^{-2}	6.49×10^{-2}	4.41×10^{-2}	14
AGE-CN	PR	1.6×10^{-4}	4.24×10^{-4}	8.72×10^{-4}	5.26×10^{-4}	1.92×10^{-3}	3.88×10^{-4}	2.99×10^{-3}	1.28×10^{-4}	3.75×10^{-3}	1.24×10^{-3}	4
	DR	1.41×10^{-4}	7.73×10^{-4}	1.25×10^{-3}	1.01×10^{-3}	2.78×10^{-3}	7.92×10^{-4}	4.18×10^{-3}	3.53×10^{-4}	5.07×10^{-3}	1.82×10^{-3}	11
EXACT SOLUTION		-0.3232896	-0.2272021	-0.1288445	-0.2919952	0.0707372	0.1699671	0.2674988	0.3623578	0.4535961	—	—

Table 13 Absolute errors of the AGE solutions to hyperbolic Problem 2 (9.9)

(a) $t = 0.4$, $\lambda = 0.5$, $\Delta t = 0.05$, $\Delta x = 0.1$, $r = 0.5$, eps $= 10^{-4}$

Scheme		0.1	0.2	0.3	0.4	0.5	0.6	0.7	0.8	0.9	Average of all absolute errors	No. of iterations
AGE-IMP	PR	4.42×10^{-3}	1.04×10^{-2}	1.78×10^{-2}	2.1×10^{-2}	2.97×10^{-2}	2.3×10^{-2}	4.27×10^{-2}	1.38×10^{-2}	6.04×10^{-2}	2.48×10^{-2}	4
	DR	4.36×10^{-3}	1.05×10^{-2}	1.8×10^{-2}	2.13×10^{-2}	3.01×10^{-2}	2.33×10^{-2}	4.34×10^{-2}	1.41×10^{-2}	6.13×10^{-2}	2.52×10^{-2}	10
AGE-CN	PR	6.89×10^{-5}	2.08×10^{-4}	3.58×10^{-4}	4.56×10^{-4}	5.62×10^{-4}	5.04×10^{-4}	7.86×10^{-4}	2.9×10^{-4}	1.17×10^{-3}	4.9×10^{-4}	3
	DR	8.27×10^{-5}	3.32×10^{-4}	5.5×10^{-4}	7.39×10^{-4}	9.26×10^{-4}	8.56×10^{-4}	1.36×10^{-3}	5.56×10^{-4}	2.04×10^{-3}	8.26×10^{-4}	9
EXACT SOLUTION		-0.1327158	-0.0890597	-0.044858	0	0.044858	0.0892679	0.1327858	0.1749769	0.2154198	—	—

(b) $t = 1.0$, $\lambda = 0.5$, $\Delta t = 0.05$, $\Delta x = 0.1$, $r = 0.5$, eps $= 10^{-4}$

Scheme		0.1	0.2	0.3	0.4	0.5	0.6	0.7	0.8	0.9	Average of all absolute errors	No. of iterations
AGE-IMP	PR	1.85×10^{-2}	1.45×10^{-2}	2.82×10^{-2}	2.6×10^{-2}	4.37×10^{-2}	2.73×10^{-2}	5.94×10^{-2}	1.65×10^{-2}	7.04×10^{-2}	3.38×10^{-2}	3
	DR	1.83×10^{-2}	1.47×10^{-2}	2.81×10^{-2}	2.64×10^{-2}	4.43×10^{-2}	2.77×10^{-2}	6.08×10^{-2}	1.68×10^{-2}	7.23×10^{-2}	3.44×10^{-2}	7
AGE-CN	PR	2.22×10^{-4}	2.63×10^{-4}	4.99×10^{-4}	5.97×10^{-4}	8.46×10^{-4}	6.47×10^{-4}	1.17×10^{-3}	3.34×10^{-3}	1.44×10^{-3}	6.69×10^{-3}	3
	DR	1.89×10^{-4}	6.38×10^{-4}	6.11×10^{-4}	1.2×10^{-3}	1.43×10^{-3}	1.25×10^{-3}	2.35×10^{-3}	6.76×10^{-4}	3.14×10^{-3}	1.28×10^{-3}	7
EXACT SOLUTION		-0.1060118	-0.0970836	-0.0871854	-0.0764160	-0.0648832	-0.052702	-0.03999431	-0.026887	-0.013511	—	—

Table 14 Absolute errors of the AGE solutions to hyperbolic Problem 2 (9.9)

(a) $t = 0.8$, $\lambda = 1.0$, $\Delta t = 0.1$, $\Delta x = 0.1$, $r = 0.5$, eps $= 10^{-4}$

Scheme		x = 0.1	0.2	0.3	0.4	0.5	0.6	0.7	0.8	0.9	Average of all absolute errors	No. of iterations
AGE-IMP	PR	2.1×10^{-2}	1.1×10^{-2}	4.1×10^{-2}	1.62×10^{-2}	7.85×10^{-2}	1.71×10^{-2}	1.2×10^{-1}	1.27×10^{-2}	1.48×10^{-1}	5.17×10^{-2}	5
	DR	2.07×10^{-2}	1.1×10^{-2}	4.1×10^{-2}	1.62×10^{-2}	7.89×10^{-2}	1.71×10^{-2}	1.2×10^{-1}	1.27×10^{-2}	1.49×10^{-1}	5.19×10^{-2}	10
AGE-CN	PR	9.97×10^{-5}	2.97×10^{-4}	3.27×10^{-4}	7.44×10^{-4}	1.26×10^{-3}	1.04×10^{-3}	2.35×10^{-3}	6.98×10^{-4}	2.74×10^{-3}	1.06×10^{-3}	3
	DR	3.41×10^{-5}	3.81×10^{-4}	3.52×10^{-4}	8.43×10^{-4}	1.55×10^{-3}	1.09×10^{-3}	2.97×10^{-3}	7.22×10^{-4}	3.59×10^{-3}	1.28×10^{-3}	9
EXACT SOLUTION		-0.1300653	-0.113993	-0.0967943	-0.0786222	-0.0596645	-0.0401106	-0.0201560	0	0.020156	—	—

(b) $t = 2.0$, $\lambda = 1.0$, $\Delta t = 0.1$, $\Delta x = 0.1$, $r = 0.5$, eps $= 10^{-4}$

Scheme		x = 0.1	0.2	0.3	0.4	0.5	0.6	0.7	0.8	0.9	Average of all absolute errors	No. of iterations
AGE-IMP	PR	1.19×10^{-1}	1.18×10^{-2}	1.06×10^{-1}	1.88×10^{-2}	8.62×10^{-2}	1.85×10^{-2}	6.64×10^{-2}	1.14×10^{-2}	5.43×10^{-2}	5.48×10^{-2}	4
	DR	1.2×10^{-1}	1.16×10^{-2}	1.07×10^{-1}	1.85×10^{-2}	8.59×10^{-2}	1.83×10^{-2}	6.56×10^{-2}	1.12×10^{-2}	5.32×10^{-2}	5.46×10^{-2}	9
AGE-CN	PR	2.05×10^{-3}	1.05×10^{-3}	2.01×10^{-3}	1.52×10^{-3}	1.28×10^{-3}	1.33×10^{-3}	4.64×10^{-4}	6.61×10^{-4}	2.11×10^{-4}	1.18×10^{-3}	3
	DR	2.89×10^{-3}	9.6×10^{-4}	2.45×10^{-3}	1.37×10^{-3}	1.11×10^{-3}	1.2×10^{-3}	3.03×10^{-4}	5.95×10^{-4}	9.0×10^{-4}	1.31×10^{-3}	6
EXACT SOLUTION		-0.0173321	-0.0178366	-0.018163	-0.0183078	-0.0182698	-0.0180491	-0.0176482	-0.0170709	-0.016323	—	—

The exact solution is given by

$$U(x, t) = \frac{1}{8} \sin(\pi x) \cos(\pi t) \tag{9.12}$$

and

(b) Problem 2 (with derivative boundary conditions)

$$\frac{\partial^2 U}{\partial x^2} = \frac{\partial^2 U}{\partial t^2} \tag{9.13a}$$

subject to the initial conditions

$$U(x, 0) = 100 \, x^2$$

$$\frac{\partial U}{\partial t}(x, 0) = 200 \, x \tag{9.13b}$$

and the boundary conditions

$$\frac{\partial U}{\partial x}(0, t) = 200 \, t$$

and
$$U(1, t) = 100(1 + t)^2 \tag{9.13c}$$

The exact solution is of the form

$$U(x, t) = 100(x + t)^2 \tag{9.14}$$

In applying the AGE algorithm to the general equations (7.3a), we chose $\alpha = 1/4$ to correspond with the implicit method which as we know from Section 2 has second-order accuracy in both space and time. In Table 15 is shown the absolute errors of the AGE solutions for Problem 1 for $\lambda = 0.5$ and $\lambda = 1.0$. As solutions by means of the GE class of methods are available, the results of this particular experiment are compared (Sahimi, 1986). It is immediately evident that while the (S)AGE-LW combination is favored in *applying the GE technique* on the wave equation it is, however, slightly less accurate than the AGE scheme which unlike the former has the additional advantage of being unconditionally stable. The iteration number necessary for convergence is also found to be considerably small. In Table 16 is presented the percentage errors of the AGE solutions to the wave equation for Problem 2 for the values of λ progressing from 0.1 to 2.0. Unfortunately, no solutions from other schemes have been worked out to provide a comparison for this particular example.

Experiment 6

This experiment was concerned with the solution of the nonlinear problem,

$$\frac{\partial U}{\partial t} = \frac{\partial^2 U^2}{\partial x^2} \tag{9.15}$$

Table 15 Absolute errors of the AGE solutions to the wave equation for Problem 1 (9.11)

(a) $t = 1.0$, $\lambda = 0.5$, $\Delta t = 0.05$, $\Delta x = 0.1$, $r = 0.5$, eps $= 10^{-4}$

Method		0.1	0.2	0.3	0.4	0.5	0.6	0.7	0.8	0.9	Average of all absolute errors	No. of iterations
AGE	PR	6.43×10^{-5}	1.22×10^{-4}	1.69×10^{-4}	1.98×10^{-4}	2.08×10^{-4}	1.98×10^{-4}	1.69×10^{-4}	1.22×10^{-4}	6.44×10^{-5}	1.46×10^{-4}	2
	DR	1.86×10^{-3}	3.54×10^{-3}	4.87×10^{-3}	5.72×10^{-3}	6.02×10^{-3}	5.72×10^{-3}	4.87×10^{-3}	3.54×10^{-3}	1.86×10^{-3}	4.22×10^{-3}	6
EXACT SOLUTION		-0.0386272	-0.0734732	-0.1011271	-0.1188821	-0.125000	-0.1188827	-0.1011271	-0.0734732	-0.0386272	—	—

(b) $t = 2.0$, $\lambda = 1.0$, $\Delta t = 0.1$, $\Delta x = 0.1$, $r = 0.5$, eps $= 10^{-4}$

Method		0.1	0.2	0.3	0.4	0.5	0.6	0.7	0.8	0.9	Average of all absolute errors	No. of iterations
AGE	PR	5.6×10^{-4}	1.06×10^{-3}	1.47×10^{-3}	1.73×10^{-3}	1.81×10^{-3}	1.73×10^{-3}	1.47×10^{-3}	1.07×10^{-3}	5.59×10^{-4}	1.27×10^{-3}	3
	DR	1.47×10^{-3}	2.8×10^{-3}	3.83×10^{-3}	4.52×10^{-3}	4.74×10^{-3}	4.5×10^{-3}	3.85×10^{-3}	2.78×10^{-3}	1.47×10^{-3}	3.33×10^{-3}	8
EXACT SOLUTION		0.0386272	0.0734732	0.1011271	0.1188821	0.125000	0.1188821	0.1011271	0.0734732	0.0386272	—	—

Table 16 Percentage errors of the AGE solutions to the wave equation for Problem 2 (9.13)

Method	x	0	0.12	0.24	0.36	0.48	0.6	0.72	0.84	0.96	Average of all percentage errors	No. of iterations

(a) $t = 0.4$, $\lambda = 0.1$, $\Delta t = 0.004$, $\Delta x = 0.04$, $r = 0.5$, eps $= 10^{-4}$

Method		0	0.12	0.24	0.36	0.48	0.6	0.72	0.84	0.96	Average	No. of iterations
AGE	PR	9.99×10^{-1}	5.92×10^{-1}	3.91×10^{-1}	2.77×10^{-1}	2.07×10^{-1}	1.55×10^{-1}	8.87×10^{-2}	4.15×10^{-2}	8.57×10^{-3}	2.86×10^{-1}	2
	DR	1.49×10	9.17×10^{-1}	6.64×10^{-1}	5.29×10^{-1}	4.42×10^{-1}	3.73×10^{-1}	2.64×10^{-1}	1.47×10^{-1}	3.57×10^{-2}	5.16×10^{-1}	14
EXACT SOLUTION		16	27.04	40.96	57.76	77.44	100	125.44	153.76	184.96	—	—

(b) $t = 2.0$, $\lambda = 0.5$, $\Delta t = 0.02$, $\Delta x = 0.04$, $r = 0.5$, eps $= 10^{-4}$

Method		0	0.12	0.24	0.36	0.48	0.6	0.72	0.84	0.96	Average	No. of iterations
AGE	PR	1.46×10^{-3}	7.53×10^{-4}	5.58×10^{-4}	1.32×10^{-4}	1.34×10^{-3}	2.05×10^{-3}	2.61×10^{-3}	3.79×10^{-3}	2.93×10^{-3}	1.51×10^{-3}	4
	DR	2.6×10^{-2}	2.35×10^{-2}	2.17×10^{-2}	1.83×10^{-2}	1.38×10^{-2}	1.39×10^{-2}	5.82×10^{-3}	8.64×10^{-3}	4.14×10^{-3}	1.53×10^{-2}	17
EXACT SOLUTION		400	449.44	501.76	556.96	615.04	676	739.84	806.56	876.16	—	—

(c) $t = 8.0$, $\lambda = 2.0$, $\Delta t = 0.08$, $\Delta x = 0.04$, $r = 0.5$, eps $= 10^{-4}$

Method		0	0.12	0.24	0.36	0.48	0.6	0.72	0.84	0.96	Average	No. of iterations
AGE	PR	3.12×10^{-3}	9.96×10^{-4}	2.97×10^{-3}	3.33×10^{-3}	1.75×10^{-3}	7.36×10^{-3}	8.77×10^{-3}	4.91×10^{-3}	8.35×10^{-4}	3.53×10^{-3}	14
	DR	3.11×10^{-3}	1.01×10^{-3}	2.97×10^{-3}	3.33×10^{-3}	1.75×10^{-3}	7.37×10^{-3}	8.77×10^{-3}	4.91×10^{-3}	8.35×10^{-4}	3.53×10^{-3}	33
EXACT SOLUTION		6400	6593.44	6789.76	6988.96	7191.04	7396	7603.84	7814.56	8028.16	—	—

with the exact solution (Abdullah, 1983)

$$2U - 3 + \ln\left(U - \frac{1}{2}\right) = 2(2t - x) \qquad (9.16)$$

The appropriate boundary data were given to satisfy the above exact solution.

The analytical solution was obtained iteratively by means of the Newton-Raphson method through the formula

$$U^{(p+1)} = \frac{1}{2} + \left[\frac{3}{2} - \frac{1}{2}\ln\left(U^{(p)} - \frac{1}{2}\right) - x + 2t\right]$$

$$\times \left(1 - \frac{1}{2U^{(p)}}\right) \qquad p = 1, 2, 3, \ldots \qquad (9.17a)$$

and the initial guesses were taken as

$$U^{(0)} = 1 + \frac{4 - e^{2(x-2t)}}{4 + 2e^{2(x-2t)}} \qquad (9.17b)$$

The AGE solutions using the fully implicit, Crank-Nicolson and Douglas formulas which were linearized using the Richtmyer's method [AGE-IMP(RCHM), AGE-CN(RCHM) and AGE-DG(RCHM)] are shown in Tables 17 and 18. They are compared with the results obtained from the linearized schemes of Crank-Nicolson (using the Richtmyer's method) and Lee and as well as from the (D)AGE formulas. The linearized schemes of Crank-Nicolson (CN) and Lee employed the Thomas algorithm as a method of solution to the resulting tridiagonal system of equations. As the (D)AGE scheme had to be solved iteratively, convergence of the iterative process was considered for every group of 2 points and the figures in brackets in the tables indicate the number of iterations required.

It is observed that in the AGE class of methods, the AGE-LEE scheme provides the most accurate solutions for both $\lambda = 0.05$ and $\lambda = 0.1$. The approximate amount of arithmetic involved in the computation of the AGE solutions per iteration is given in Table 19.

To reduce the storage requirement to a minimum, the entries of the coefficient matrix A and the right-hand side vector $\mathbf{f}$ of Eq. (8.6) were generated rather than stored at each of the $(p + 1/2)$ and $(p + 1)$ iterates. Consequently as is expected for a nonlinear problem, the computational complexity of the AGE schemes can be quite substantial.

It is also apparent that the (D)AGE, CN and LEE schemes possess solutions close to the analytical ones. However, it must be mentioned that the (D)AGE scheme requires 3 iterations for *every group of 2 points* while the AGE methods require only 1 or 2 iterations over the *whole mesh line*.

Experiment 7

The following Burger's equation was considered:

$$\frac{\partial U}{\partial t} = \epsilon \frac{\partial^2 U}{\partial x^2} - U \frac{\partial U}{\partial x} \qquad (9.18)$$

Table 17 Numerical solutions to nonlinear problem (9.15)

$t = 0.05$, $\Delta x = 0.1$, $\lambda = 0.05$, eps $= 10^{-4}$

Scheme		0.1	0.2	0.3	0.4	0.5	0.6	0.7	0.8	0.9	Average of all absolute errors	No. of iterations
(D) AGE		1.5000 (3)	1.43409	1.36976 (3)	1.30713	1.24630 (3)	1.18740	1.13056 (3)	1.07590	1.02356 (3)	NA	—
AGE-IMP (RCHM)	PR	1.49914	1.43226	1.36766	1.30452	1.24381	1.18481	1.12845	1.07411	1.02267	1.92×10^{-3}	1
	DR	1.49830	1.43049	1.36561	1.30200	1.24141	1.18233	1.12643	1.07240	1.02183	3.77×10^{-3}	1
AGE-CN (RCHM)	PR	1.49957	1.43318	1.36868	1.30580	1.24502	1.18608	1.12947	1.07498	1.02309	9.82×10^{-4}	1
	DR	1.49914	1.43227	1.36762	1.30450	1.24375	1.18478	1.12839	1.07408	1.02264	1.95×10^{-3}	1
CN		1.50000	1.43409	1.36976	1.30713	1.24630	1.18740	1.13056	1.07590	1.02355	NA	—
AGE-DG (RCHM)	PR	1.50098	1.43618	1.37241	1.31033	1.24954	1.19069	1.13337	1.07820	1.02478	2.42×10^{-3}	1
	DR	1.50204	1.43841	1.37527	1.31381	1.25308	1.19430	1.13647	1.08075	1.02613	5.06×10^{-3}	2
AGE-LEE	PR	1.500098	1.43430	1.37011	1.30753	1.24677	1.18785	1.13099	1.07619	1.02370	3.14×10^{-4}	2
	DR	1.49909	1.43254	1.36779	1.30485	1.24394	1.18505	1.12845	1.07415	1.02247	1.82×10^{-3}	3
LEE		1.50000	1.43409	1.36976	1.30713	1.24630	1.18740	1.13056	1.07590	1.02355	NA	—
EXACT SOLUTION		1.50000	1.43409	1.36974	1.30713	1.24631	1.18741	1.13057	1.07590	1.02355	—	—

Table 18 Numerical solutions to nonlinear problem (9.15)

$t = 0.1$, $\Delta x = 0.1$, $\lambda = 0.1$, eps $= 10^{-4}$

Scheme		0.1	0.2	0.3	0.4	0.5	0.6	0.7	0.8	0.9	Average of all absolute errors	No. of iterations
(D) AGE		1.56738 (3)	1.50001	1.43407 (3)	1.36976	1.30710 (3)	1.24630	1.18738 (3)	1.13056	1.07588 (3)	NA	—
AGE-IMP (RCHM)	PR	1.56514	1.49530	1.42859	1.36286	1.30041	1.23928	1.18162	1.12569	1.07351	5.13×10^{-3}	1
	DR	1.56300	1.49078	1.42334	1.35629	1.29402	1.23263	1.17614	1.12110	1.07125	9.99×10^{-3}	1
AGE-CN (RCHM)	PR	1.56627	1.49766	1.43126	1.36625	1.30364	1.24270	1.18438	1.12805	1.07462	2.63×10^{-3}	1
	DR	1.56517	1.49536	1.42848	1.36280	1.30021	1.23915	1.18141	1.12558	1.07337	5.22×10^{-3}	1
CN		1.56739	1.5000	1.43408	1.36976	1.30712	1.24630	1.18740	1.13056	1.07590	NA	—
AGE-DG (RCHM)	PR	1.56809	1.50148	1.43599	1.37210	1.30953	1.24874	1.18950	1.13227	1.07680	1.77×10^{-3}	1
	DR	1.56882	1.50302	1.43797	1.37452	1.31201	1.25127	1.19168	1.13404	1.07774	3.62×10^{-3}	2
AGE-LEE	PR	1.5675	1.50025	1.43450	1.37024	1.30769	1.24684	1.18792	1.13092	1.07608	3.78×10^{-4}	2
	DR	1.56691	1.49920	1.43309	1.36859	1.30591	1.24507	1.18631	1.12964	1.07531	9.45×10^{-4}	4
LEE		1.56739	1.50000	1.43408	1.36976	1.30712	1.24630	1.18740	1.13056	1.07590	NA	—
EXACT SOLUTION		1.56739	1.50000	1.43409	1.36974	1.30713	1.24631	1.18741	1.13057	1.07590	—	—

Table 19 Computational complexity of AGE methods

Method	Operation	Number of multiplications	Number of additions
AGE-IMP/CN/DG		$41m - 16$	$24m + 1$
AGE-LEE		$50m - 21$	$45m - 20$

and the initial and boundary conditions were prescribed so as to satisfy the exact solution. Two problems were solved using the AGE algorithm.

Problem 1

This problem has the exact solution (Madsen and Sincovec, 1976),

$$U(x, t) = \frac{0.1e^{-A} + 0.5e^{-B} + e^{-C}}{e^{-A} + e^{-B} + e^{-C}} \qquad 0 \le x \le 1 \qquad t \ge 0 \qquad (9.19a)$$

where

$$A = \frac{0.05}{\epsilon}(x - 0.5 + 4.95t)$$

$$B = \frac{0.25}{\epsilon}(x - 0.5 + 0.75t) \qquad\qquad (9.19b)$$

and

$$C = \frac{0.5}{\epsilon}(x - 0.375)$$

Problem 2

The exact solution to this problem is given by (Cole, 1951),

$$U(x, t) = \frac{2\pi \sum_{k=1}^{\infty} kA_k \sin(k\pi x)e^{(-\epsilon k^2 \pi^2 t)}}{1/\epsilon[A_0 + \sum_{k=1}^{\infty} A_k \cos(k\pi x)e^{(-\epsilon k^2 \pi^2 t)}]} \qquad (9.20a)$$

where

$$A_k = 2 \int_0^1 \cos(k\pi x) \exp\left[-\frac{1}{2\epsilon} \int_0^x F(x')dx'\right]dx \qquad (k = 1, 2, 3, \ldots)$$

$$A_0 = \int_0^1 \exp\left[-\frac{1}{2\epsilon} \int_0^x F(x')dx'\right]dx \qquad\qquad (9.20b)$$

and

$$F(x) = 4x(1 - x) \qquad\qquad (9.20c)$$

$$F(x) = \sin(\pi x) \qquad\qquad (9.20d)$$

Comparative results for Problem 1 using the (D)AGE and AGE schemes for $\lambda = 1.0$ and $\epsilon = 0.1$ and $\epsilon = 1.0$ at $t = 1.0$ are given in Table 20. The figures

Table 20 Absolute errors of the numerical solutions to Burger's equation for Problem 1 (9.19)

$\epsilon = 1.0$, $\lambda = 1.0$, $\Delta t = 0.01$, $\Delta x = 0.1$, $t = 1.0$, eps $= 10^{-6}$

Method		0.1	0.2	0.3	0.4	0.5	0.6	0.7	0.8	0.9	No. of iterations
DAGE		7.0×10^{-8} (2)	2.5×10^{-7}	3.7×10^{-7} (2)	2.9×10^{-7}	6.5×10^{-7} (2)	3.5×10^{-7}	6.1×10^{-7} (2)	9.6×10^{-7}	1.22×10^{-6} (2)	—
AGE-IMP	PR	8.4×10^{-4}	1.5×10^{-3}	1.99×10^{-3}	2.3×10^{-3}	2.4×10^{-3}	2.3×10^{-3}	2.0×10^{-3}	1.6×10^{-3}	8.8×10^{-4}	5
	DR	8.5×10^{-4}	1.5×10^{-3}	1.99×10^{-3}	2.3×10^{-3}	2.4×10^{-3}	2.3×10^{-3}	2.0×10^{-3}	1.6×10^{-3}	8.8×10^{-4}	10
AGE-CN	PR	8.2×10^{-8}	1.1×10^{-7}	1.8×10^{-7}	1.4×10^{-7}	2.2×10^{-7}	1.1×10^{-7}	2.9×10^{-7}	1.1×10^{-7}	2.7×10^{-7}	4
	DR	1.8×10^{-6}	2.9×10^{-6}	3.9×10^{-6}	4.3×10^{-6}	4.5×10^{-6}	4.2×10^{-6}	3.8×10^{-6}	2.5×10^{-6}	2.4×10^{-6}	10
EXACT SOLUTION		0.58918851	0.58233803	0.5747863	0.56861393	0.56174757	0.55488319	0.54802440	0.54117482	0.53433803	—

$\epsilon = 0.1$, $\lambda = 1.0$, $\Delta t = 0.01$, $\Delta x = 0.1$, $t = 1.0$, eps $= 10^{-6}$

Method		0.1	0.2	0.3	0.4	0.5	0.6	0.7	0.8	0.9	No. of iterations
DAGE		1.83×10^{-4} (3)	3.26×10^{-4}	3.88×10^{-4} (3)	2.39×10^{-4}	9.6×10^{-5} (3)	6.49×10^{-4}	1.19×10^{-3} (3)	1.51×10^{-3}	1.22×10^{-3} (3)	—
AGE-IMP	PR	2.9×10^{-2}	6.1×10^{-2}	9.4×10^{-2}	1.236×10^{-1}	1.4498×10^{-1}	1.542×10^{-1}	1.4769×10^{-1}	1.2229×10^{-1}	7.485×10^{-2}	5
	DR	2.9×10^{-2}	6.1×10^{-2}	9.4×10^{-2}	1.236×10^{-1}	1.450×10^{-1}	1.543×10^{-1}	1.4777×10^{-1}	1.2235×10^{-1}	7.489×10^{-2}	14
AGE-CN	PR	1.62×10^{-4}	2.78×10^{-4}	2.83×10^{-4}	1.07×10^{-4}	2.8×10^{-4}	8.3×10^{-4}	1.38×10^{-3}	1.64×10^{-3}	1.30×10^{-3}	3
	DR	1.60×10^{-4}	2.66×10^{-4}	2.59×10^{-4}	7.17×10^{-5}	3.3×10^{-4}	8.94×10^{-4}	1.45×10^{-3}	1.71×10^{-3}	1.34×10^{-3}	13
EXACT SOLUTION		0.932745	0.911271	0.883314	0.847514	0.802758	0.748601	0.685736	0.616304	0.543775	—

in brackets for the (D)AGE scheme indicate the number of iterations required for convergence at every group of two points. The numerical solutions to Burger's equation for the same problem obtained by means of the various difference schemes at different time levels for smaller values of ϵ are presented in Tables 21 and 22. The solutions to Problem 2 are shown in Tables 23 and 24 where only the PR variant was employed for the generalized AGE method.

It can be inferred from Table 20 that the AGE-CN and (D)AGE processes exhibit comparable accuracies at the grid point. The number of iterations required by the AGE-CN (PR) scheme (about 3–4 iterations) can be further reduced by relaxing the convergence criterion which at 10^{-6} may be regarded as quite stringent. Similar features of accuracy of the (D)AGE and AGE-CN schemes are also observed in Tables 21 and 22 for a much smaller value $\epsilon = 0.003$ although now the solutions are at slight variance with the exact ones. However, a *large number of iterations* were required to achieve convergence of the (D)AGE procedure (Evans and Abdullah, 1984). The AGE-CN (PR) method, on the other hand, required only 2 iterations over the whole mesh line at $t = 0.1$ and $t = 0.5$. Hence, it can be said that the AGE-CN scheme is just as competitive, if not better, than the (D)AGE method.

For Problem 2 with $F(x)$ given by Eq. (9.20c), we find from Table 23 that while the (D)AGE and AGE-CN schemes have about the same order of accuracy,

Table 21 Numerical solutions to Burger's equation for Problem 1 (9.19)
$\epsilon = 0.003$, $\lambda = 1.0$, $\Delta x = 0.01$, $\Delta t = 0.0001$, $t = 0.1$, eps $= 10^{-6}$

Method		PR		DR		
x	DAGE	AGE-IMP	AGE-CN	AGE-IMP	AGE-CN	EXACT SOLUTION
0.05	1.000000	1.000000	1.000000	0.999999	0.999999	1.000000
0.10	1.000000	0.999999	0.999999	0.999999	0.999999	1.000000
0.15	1.000000	0.999999	0.999999	0.999999	0.999999	1.000000
0.20	0.999993	0.999976	0.999994	0.999976	0.999994	0.999985
0.25	0.999572	0.998482	0.999572	0.998475	0.999569	0.999037
0.30	0.952565	0.930082	0.952562	0.929798	0.952191	0.944636
0.35	0.557343	0.593433	0.557344	0.593049	0.557074	0.555361
0.40	0.501467	0.503739	0.501468	0.503713	0.501458	0.500894
0.45	0.498438	0.499422	0.498438	0.499421	0.498435	0.498093
0.50	0.454613	0.484010	0.454613	0.483994	0.454527	0.452319
0.55	0.181832	0.325894	0.181833	0.325743	0.181710	0.183443
0.60	0.103999	0.129626	0.103999	0.129561	0.103992	0.103726
0.65	0.100148	0.101435	0.100148	0.101430	0.100148	0.100134
0.70	0.100005	0.100053	0.100005	0.100053	0.100005	0.100004
0.75	0.100000	0.100002	0.100000	0.100002	0.100000	0.100000
0.80	0.100000	0.100000	0.100000	0.100000	0.100000	0.100000
0.85	0.100000	0.100000	0.100000	0.100000	0.100000	0.100000
0.90	0.100000	0.100000	0.100000	0.100000	0.100000	0.100000
0.95	0.100000	0.100000	0.100000	0.100000	0.099999	0.100000
Number of iterations	—	2	2	9	10	—

Table 22 Numerical solution to Burger's equation for Problem 1 (9.19)
$\epsilon = 0.003$, $\lambda = 1.0$, $\Delta x = 0.01$, $\Delta t = 0.0001$, $t = 0.5$, eps = 10^{-6}

| Method | PR | | | DR | | EXACT |
x	DAGE	AGE-IMP	AGE-CN	AGE-IMP	AGE-CN	SOLUTION
0.05	1.0000	1.000000	1.000000	1.000000	1.000000	1.0000
0.10	1.0000	0.999999	1.000000	0.999999	1.000000	1.0000
0.15	1.0000	0.999999	1.000000	0.999999	1.000000	1.0000
0.20	1.0000	0.999999	1.000000	0.999999	0.999999	1.0000
0.25	1.0000	0.999999	1.000000	0.999999	0.999999	1.0000
0.30	1.0000	0.999995	0.999999	0.999995	0.999999	1.0000
0.35	1.0000	0.999755	0.999999	0.999746	0.999999	1.0000
0.40	1.0000	0.992646	0.999999	0.992410	0.999999	1.0000
0.45	1.0000	0.894929	0.999999	0.892547	0.999999	1.0000
0.50	1.0000	0.620463	1.000001	0.618006	1.000001	0.9999
0.55	1.0000	0.492275	1.000020	0.491687	1.000028	0.9999
0.60	0.9552	0.360375	0.955208	0.359512	0.953063	0.9413
0.65	0.3362	0.173939	0.336164	0.173327	0.334005	0.3410
0.70	0.1145	0.109650	0.114505	0.109531	0.114373	0.1138
0.75	0.1006	0.100808	0.100642	0.100795	0.100636	0.1005
0.80	0.1000	0.100049	0.100027	0.100048	0.100026	0.1000
0.85	0.1000	0.100002	0.100001	0.100002	0.100001	0.1000
0.90	0.1000	0.100000	0.100000	0.100000	0.100000	0.1000
0.95	0.1000	0.100000	0.100000	0.100000	0.100000	0.1000
Number of iterations	—	2	2	8	10	—

the AGE-IMP method appears to be more accurate. The same conclusion can be drawn from Table 24 with $F(x)$ given by Eq. (9.20d). In this particular case the explicit scheme (Caldwell and Smith, 1982) demonstrates that it provides a better solution. In general, it is not expected of these two schemes to perform better. However, the inaccuracies in the solutions for small values of ϵ seems to be a difficulty experienced by most finite difference methods.

Finally, as an indication of efficiency of the AGE algorithm, we present in Table 25 estimates of the computational complexity per iteration of the relevant difference schemes employed to solve the above Burger's equation, where we have assumed that m is odd.

Experiment 8

This experiment dealt with the calculation of the propagation of a one-dimensional wave governed by the reaction-diffusion equation,

$$\frac{\partial U}{\partial t} = \frac{\partial^2 U}{\partial x^2} + S \qquad -\infty < x < \infty \qquad t \geq 0 \qquad (9.21a)$$

where $\qquad S = U^2(1 - U)$ $\qquad\qquad\qquad\qquad\qquad\qquad (9.21b)$

Table 23 Numerical solution to Burger's equation for Problem 2 (9.20)

Case (i) $F(x) = 4x(1 - x)$ with $\epsilon = 0.01$, $\Delta t = 0.01$, $\Delta x = 0.25$, $r = 0.5$, eps $= 10^{-6}$

$x = 0.25$						PR
t	Exact	Implicit	Explicit	DAGE	AGE-IMP	AGE-CN
0.01	0.7492	0.7346	0.7342	0.7342	0.7417	0.7344
0.05	0.7460	0.6766	0.6748	0.6755	0.7098	0.6757
0.10	0.7420	0.6122	0.6087	0.6104	0.6724	0.6104
0.15	0.7380	0.5562	0.5512	0.5537	0.6376	0.5537
0.20	0.7340	NA	0.5016	0.5046	0.6054	0.5047
0.25	0.7300	NA	NA	0.4652	0.5755	0.4625

$x = 0.5$						PR
t	Exact	Implicit	Explicit	DAGE	AGE-IMP	AGE-CN
0.01	0.9992	0.9986	0.9992	0.9989	0.9991	0.9989
0.05	0.9960	0.9873	0.9901	0.9887	0.9938	0.9887
0.10	0.9920	0.9611	0.9662	0.9636	0.9841	0.9636
0.15	0.9880	0.9233	0.9299	0.9263	0.9712	0.9265
0.20	0.9840	NA	0.8835	0.8796	0.9554	0.8797
0.25	0.9800	NA	NA	0.8256	0.9370	0.8254

$x = 0.75$						PR
t	Exact	Implicit	Explicit	DAGE	AGE-IMP	AGE-CN
0.01	0.7492	0.7644	0.7642	0.7645	0.7567	0.7643
0.05	0.7460	0.8241	0.8232	0.8240	0.7837	0.8237
0.10	0.7420	0.9027	0.9012	0.9020	0.8181	0.9019
0.15	0.7380	0.9843	0.9828	0.9832	0.8531	0.9836
0.20	0.7340	NA	1.0065	1.0671	0.8886	1.0670
0.25	0.7300	NA	NA	1.1513	0.9242	1.1507

	Method	Number of Iterations	
t		AGE-IMP	AGE-CN
0.01		2	3
0.05		2	3
0.10		2	3
0.15		2	3
0.20		2	3
0.25		2	3

As was explained in Section 8(iii), for practical purposes the above domain was truncated to $-50 \leq x \leq 400$, $t \geq 0$ and the following initial and boundary conditions were prescribed,

$$U(x, 0) = \frac{1}{1 + e^{Vx}} \qquad (9.21c)$$

$$U(-50, t) = 1$$

and $\qquad U(400, t) = 0 \qquad (9.21d)$

Table 24 Numerical solutions to Burger's equation for Problem 2 (9.20)

Case (ii) $F(x) = \sin(\pi x)$ with $\epsilon = 1.0$, $\Delta t = 0.01$, $\Delta x = 0.25$, $r = 0.5$, eps $= 10^{-6}$

$x = 0.25$						PR
t	Exact	Implicit	Explicit	DAGE	AGE-IMP	AGE-CN
0.01	0.6290	0.6377	0.6267	0.6259	0.6416	0.6327
0.05	0.4131	0.4339	0.4099	0.4168	0.4419	0.4224
0.10	0.2536	0.2768	0.2525	0.2639	0.2821	0.2648
0.15	0.1566	0.1784	0.1565	0.1681	0.1811	0.1676
0.20	0.9064	NA	0.0967	0.1047	0.1163	0.1059
0.25	0.0592	NA	NA	0.0651	0.0746	0.0668

$x = 0.5$						PR
t	Exact	Implicit	Explicit	DAGE	AGE-IMP	AGE-CN
0.01	0.9057	0.9141	0.9063	0.9082	0.9142	0.9103
0.05	0.6096	0.6380	0.6100	0.6222	0.6386	0.6246
0.10	0.3716	0.4075	0.3729	0.3876	0.4080	0.3906
0.15	0.2268	0.2604	0.2281	0.2417	0.2607	0.2445
0.20	0.1385	NA	0.1395	0.1509	0.1666	0.1530
0.25	0.0845	NA	NA	0.0942	0.1064	0.0957

$x = 0.75$						PR
t	Exact	Implicit	Explicit	DAGE	AGE-IMP	AGE-CN
0.01	0.6524	0.6556	0.6550	0.6612	0.6514	0.6550
0.05	0.4502	0.4702	0.4556	0.4668	0.4617	0.4631
0.10	0.2726	0.3007	0.2762	0.2871	0.2952	0.2888
0.15	0.1644	0.1904	0.1663	0.1753	0.1877	0.1786
0.20	0.0994	NA	0.1006	0.1094	0.1193	0.1105
0.25	0.0603	NA	NA	0.0684	0.0759	0.0686

	Method	Number of Iterations	
t		AGE-IMP	AGE-CN
0.01		4	4
0.05		4	4
0.10		4	4
0.15		4	3
0.20		4	3
0.25		4	3

Table 25 Computational complexity of AGE methods

Schemes	Operations	Number of multiplications	Number of additions
AGE-IMP		$24m$	$\dfrac{39m - 1}{2} - 3$
AGE-CN		$41m - 16$	$37m - 19$

where $V = 1/\sqrt{2}$, the steady-state wave speed. The exact traveling wave solution is given by

$$U(x, t) = \frac{1}{1 + e^{V(x - Vt)}} \tag{9.22}$$

The numerical solutions to the reaction-diffusion problem were obtained using the GE-PC, AGE-PC and AGE-TL schemes and comparisons among these methods were presented in terms of the absolute errors of the solutions. The mesh ratio was taken as $\lambda = 0.2$ and the calculations were performed with different spatial grids and time steps.

An examination of the absolute errors in Tables 26–29 for solutions at different time levels with $\Delta x = 1.0$ and $\Delta t = 0.2$ shows that for small t, the AGE-4TL method is the most accurate followed by the AGE-PC, (D)AGE-PC, AGE-2TL, (S)AGE-PC, AGE-1TL, AGE-3TL, (S)AGE-EXP and (D)AGE-EXP schemes, while for large t we have in decreasing order of accuracy, the AGE-4TL, AGE-2TL, AGE-PC, (D)AGE-PC, (S)AGE-PC, AGE-1TL, AGE-3TL, (D)AGE-EXP and (S)AGE-EXP schemes. The same conclusion can be drawn for finer grids with $\Delta x = 0.5$ and $\Delta t = 0.05$ (Tables 30 and 31). However, as t progresses, AGE-3TL becomes more accurate than AGE-1TL. One possible explanation to the above observation is that the accuracy of the methods is very much dependent on *the time step* employed in the calculation. For the TL methods, for example, the nonlinear reaction terms are expanded in Taylor series about the previous time and the 1TL, 2TL, 3TL and 4TL formulas are known to have accuracies to the order of $O[(\Delta x)^2 + \Delta t]$, $O[(\Delta x)^2 + (\Delta t)^2]$, $O(\Delta x)^4 + \Delta t]$ and $O[(\Delta x)^4 + (\Delta t)^2]$, respectively. The results presented in Tables 26–29, 30, and 31 therefore suggest that *temporal approximations play* a more dominant role than *spatial approximations* in determining the accuracy of the methods. A similar reasoning can also be applied to other methods employing the PC technique. For each grid point along the mesh line (time row), the Runge-Kutta method of $O[(\Delta t)^5]$ accuracy is used which is then corrected by utilizing the Crank-Nicolson formula which in turn is second-order accurate in both space and time. Thus, in the (S)AGE-PC and (D)AGE-PC methods, the utilization of these high-order formulas coupled with the cancelation of errors resulting from the alternate use of the constituent GER and GEL schemes lead to accurate solutions of these schemes.

In Tables 32 and 33, we compare the accuracy of the various methods in terms of the computed wave speed V. For a steady state wave propagation problem, the following condition applies:

$$\frac{\partial U}{\partial t} + V \frac{\partial U}{\partial x} = 0 \tag{9.23}$$

By substituting $\partial U/\partial t$ into Eq. (9.21) and then integrating the result using Eq. (9.21c), V can be obtained as

$$V = \int_{-50}^{400} U^2 (1 - U) dx \tag{9.24}$$

Table 26 Absolute errors of the numerical solutions to nonlinear problem (9.21) (the reaction-diffusion equation)

$t = 3$, $\Delta x = 1.0$, $\Delta t = 0.2$, $\lambda = 0.2$, eps $= 10^{-4}$

Method		-49	-33	-17	-1	15	207	339	Average of all absolute errors	Number of iterations
(S) AGE-EXP		7.11×10^{-14}	8.17×10^{-13}	7.96×10^{-8}	3.31×10^{-3}	2.45×10^{-6}	3.37×10^{-70}	3.01×10^{-123}	1.06×10^{-4}	—
(D) AGE-EXP		4.26×10^{-14}	6.47×10^{-13}	6.58×10^{-8}	1.82×10^{-3}	5.4×10^{-6}	7.34×10^{-70}	2.58×10^{-123}	1.12×10^{-4}	—
(S) AGE-PC		8.53×10^{-14}	1.67×10^{-12}	1.23×10^{-7}	1.64×10^{-3}	2.47×10^{-6}	3.37×10^{-70}	3.01×10^{-123}	3.65×10^{-5}	—
(D) AGE-PC		5.68×10^{-14}	1.88×10^{-12}	1.41×10^{-7}	2.06×10^{-4}	5.43×10^{-6}	7.34×10^{-70}	2.58×10^{-123}	3.62×10^{-5}	—
AGE-PC	PR	1.42×10^{-13}	8.81×10^{-13}	7.91×10^{-8}	9.2×10^{-4}	7.25×10^{-6}	9.84×10^{-70}	2.47×10^{-123}	3.52×10^{-5}	2
	DR	0	9.09×10^{-13}	8.75×10^{-8}	3.95×10^{-4}	6.58×10^{-6}	8.95×10^{-70}	2.53×10^{-123}	4.21×10^{-5}	8
AGE-1TL	PR	0	2.3×10^{-12}	1.84×10^{-7}	7.46×10^{-4}	1.76×10^{-5}	2.4×10^{-69}	1.5×10^{-123}	5.92×10^{-5}	3
	DR	5.68×10^{-14}	2.57×10^{-12}	1.94×10^{-7}	1.26×10^{-3}	1.66×10^{-5}	2.27×10^{-69}	1.6×10^{-123}	4.85×10^{-5}	8
AGE-2TL	PR	5.68×10^{-14}	1.11×10^{-12}	8.66×10^{-8}	1.08×10^{-3}	7.25×10^{-6}	9.84×10^{-70}	2.47×10^{-123}	3.64×10^{-5}	2
	DR	1.14×10^{-13}	1.53×10^{-12}	9.51×10^{-8}	5.57×10^{-4}	6.58×10^{-6}	8.95×10^{-70}	2.53×10^{-123}	4.17×10^{-5}	8
AGE-3TL	PR	2.42×10^{-13}	1.68×10^{-12}	9.6×10^{-8}	1.64×10^{-3}	8.95×10^{-6}	1.2×10^{-69}	2.32×10^{-123}	7.82×10^{-5}	5
	DR	3.27×10^{-13}	1.68×10^{-12}	9.98×10^{-8}	2.02×10^{-3}	8.27×10^{-6}	1.11×10^{-69}	2.38×10^{-123}	6.95×10^{-5}	11
AGE-4TL	PR	1.56×10^{-13}	1.85×10^{-13}	3.73×10^{-9}	1.9×10^{-4}	2.67×10^{-8}	4.18×10^{-72}	3.14×10^{-123}	3.17×10^{-6}	2
	DR	1.43×10^{-13}	5.26×10^{-13}	4.69×10^{-9}	3.43×10^{-4}	6.44×10^{-7}	8.65×10^{-71}	3.2×10^{-123}	1.63×10^{-5}	8
EXACT SOLUTION		1.0000	1.0000	0.9999987	0.9008854	1.11×10^{-4}	1.48×10^{-68}	1.32×10^{-122}	—	—

Table 27 Absolute errors of the numerical solutions to nonlinear problem (9.21) (the reaction-diffusion equation)

$t = 12$, $\Delta x = 1.0$, $\Delta t = 0.2$, $\lambda = 0.2$, eps $= 10^{-4}$

Method		-49	-33	-17	-1	15	207	339	Average of all absolute errors	Number of iterations
(S) AGE-EXP		5.68×10^{-14}	1.28×10^{-13}	2.43×10^{-9}	6.77×10^{-7}	5.99×10^{-4}	1.8×10^{-67}	1.68×10^{-121}	4.28×10^{-4}	—
(D) AGE-EXP		8.53×10^{-14}	1.07×10^{-13}	3.03×10^{-9}	2.63×10^{-5}	6.43×10^{-4}	1.94×10^{-67}	1.65×10^{-121}	4.29×10^{-4}	—
(S) AGE-PC		7.11×10^{-14}	2.63×10^{-13}	5.24×10^{-9}	1.11×10^{-4}	9.91×10^{-5}	1.8×10^{-67}	1.68×10^{-121}	1.0×10^{-4}	—
(D) AGE-PC		8.53×10^{-14}	2.63×10^{-13}	4.49×10^{-9}	8.67×10^{-5}	4.74×10^{-5}	1.94×10^{-67}	1.65×10^{-121}	1.01×10^{-4}	—
AGE-PC	PR	1.42×10^{-14}	7.11×10^{-14}	2.78×10^{-9}	6.58×10^{-5}	5.48×10^{-4}	3.92×10^{-67}	2.43×10^{-122}	1.03×10^{-4}	2
	DR	0	2.84×10^{-14}	3.21×10^{-9}	9.6×10^{-5}	3.24×10^{-4}	3.53×10^{-67}	5.07×10^{-121}	1.7×10^{-4}	8
AGE-1TL	PR	0	2.56×10^{-13}	8.04×10^{-13}	1.22×10^{-4}	2.87×10^{-3}	1.1×10^{-66}	4.55×10^{-121}	2.92×10^{-4}	3
	DR	5.68×10^{-14}	3.55×10^{-13}	8.65×10^{-9}	1.54×10^{-4}	2.57×10^{-3}	1.03×10^{-66}	4.06×10^{-121}	2.22×10^{-4}	8
AGE-2TL	PR	5.68×10^{-14}	1.56×10^{-13}	3.05×10^{-9}	6.69×10^{-5}	5.53×10^{-4}	3.92×10^{-67}	2.43×10^{-122}	9.6×10^{-5}	2
	DR	1.14×10^{-13}	8.38×10^{-13}	3.48×10^{-9}	9.72×10^{-5}	3.29×10^{-4}	3.53×10^{-67}	5.07×10^{-122}	1.63×10^{-4}	8
AGE-3TL	PR	2.42×10^{-13}	5.83×10^{-13}	4.27×10^{-9}	5.68×10^{-5}	2.33×10^{-3}	4.86×10^{-67}	3.97×10^{-122}	3.67×10^{-4}	5
	DR	3.27×10^{-13}	9.66×10^{-13}	4.49×10^{-9}	7.81×10^{-5}	2.08×10^{-3}	4.47×10^{-67}	1.29×10^{-122}	3.05×10^{-4}	11
AGE-4TL	PR	1.14×10^{-13}	4.83×10^{-13}	1.49×10^{-10}	5.33×10^{-6}	1.22×10^{-5}	1.51×10^{-69}	2.91×10^{-121}	3.61×10^{-6}	2
	DR	1.42×10^{-13}	5.26×10^{-13}	4.69×10^{-9}	3.43×10^{-4}	6.44×10^{-7}	8.65×10^{-71}	3.2×10^{-123}	1.63×10^{-5}	8
EXACT SOLUTION		1.0000	1.0000	0.9999999	0.9987793	0.0098869	1.33×10^{-66}	1.19×10^{-120}	—	—

Table 28 Absolute errors of the numerical solutions to nonlinear problem (9.21) (the reaction-diffusion equation)

$t = 21$, $\Delta x = 1.0$, $\Delta t = 0.2$, $\lambda = 0.2$, eps $= 10^{-4}$

Method		-49	-33	-17	-1	15	207	339	Average of all absolute errors	Number of iterations
(S) AGE-EXP		7.11×10^{-14}	1.85×10^{-13}	2.66×10^{-11}	3.17×10^{-7}	6.97×10^{-2}	2.66×10^{-65}	9.04×10^{-120}	7.88×10^{-4}	—
(D) AGE-EXP		8.53×10^{-14}	1.85×10^{-13}	3.32×10^{-11}	7.13×10^{-9}	7.01×10^{-2}	3.21×10^{-65}	5.65×10^{-120}	7.87×10^{-4}	—
(S) AGE-PC		8.53×10^{-14}	1.85×10^{-13}	7.96×10^{-11}	2.73×10^{-6}	1.68×10^{-2}	2.66×10^{-65}	9.04×10^{-120}	1.88×10^{-4}	—
(D) AGE-PC		9.95×10^{-14}	1.85×10^{-13}	7.05×10^{-11}	2.43×10^{-6}	1.71×10^{-2}	3.21×10^{-65}	5.65×10^{-120}	1.87×10^{-4}	2
AGE-PC	PR	1.42×10^{-14}	2.84×10^{-14}	4.5×10^{-11}	1.79×10^{-6}	1.56×10^{-2}	6.84×10^{-65}	2.02×10^{-119}	1.64×10^{-4}	8
	DR	0	2.84×10^{-14}	5.39×10^{-11}	2.42×10^{-6}	2.56×10^{-2}	6.1×10^{-65}	1.52×10^{-118}	2.85×10^{-4}	8
AGE-1TL	PR	0	1.14×10^{-13}	1.26×10^{-10}	2.57×10^{-6}	4.41×10^{-2}	2.24×10^{-64}	1.25×10^{-118}	5.88×10^{-4}	3
	DR	5.68×10^{-14}	1.56×10^{-13}	1.4×10^{-10}	3.27×10^{-6}	3.39×10^{-2}	2.06×10^{-64}	1.14×10^{-118}	4.58×10^{-4}	8
AGE-2TL	PR	5.68×10^{-14}	2.7×10^{-13}	4.93×10^{-11}	1.87×10^{-6}	1.48×10^{-2}	6.84×10^{-65}	2.02×10^{-119}	1.54×10^{-4}	2
	DR	1.14×10^{-13}	3.98×10^{-13}	5.85×10^{-11}	2.51×10^{-6}	2.49×10^{-2}	6.1×10^{-65}	1.52×10^{-119}	2.76×10^{-4}	8
AGE-3TL	PR	2.42×10^{-13}	3.55×10^{-13}	6.37×10^{-11}	7.57×10^{-7}	5.64×10^{-2}	8.68×10^{-65}	3.27×10^{-119}	7.19×10^{-4}	5
	DR	3.27×10^{-13}	7.39×10^{-13}	6.91×10^{-11}	1.18×10^{-6}	4.76×10^{-2}	7.9×10^{-65}	2.74×10^{-119}	6.07×10^{-4}	11
AGE-4TL	PR	2.13×10^{-13}	4.97×10^{-13}	2.03×10^{-12}	9.6×10^{-8}	7.99×10^{-5}	2.37×10^{-67}	2.62×10^{-119}	3.54×10^{-6}	2
	DR	1.42×10^{-13}	9.95×10^{-13}	5.22×10^{-12}	4.75×10^{-7}	1.01×10^{-2}	4.82×10^{-66}	2.93×10^{-119}	1.25×10^{-4}	8
EXACT SOLUTION		1.0000	1.0000	0.9999999	0.9999864	0.4733711	1.198×10^{-64}	1.07×10^{-118}	—	—

Table 29 Absolute errors of the numerical solutions to nonlinear problem (9.21) (the reaction-diffusion equation)

$t = 30$, $\Delta x = 1.0$, $\Delta t = 0.2$, $\lambda = 0.2$, eps $= 10^{-4}$

Method		-49	-33	-17	-1	15	207	339	Average of all absolute errors	Number of iterations
(S) AGE-EXP		5.68×10^{-14}	1.85×10^{-13}	5.68×10^{-14}	1.24×10^{-8}	3.84×10^{-3}	3.84×10^{-63}	2.57×10^{-118}	1.15×10^{-3}	—
(D) AGE-EXP		7.11×10^{-14}	1.85×10^{-13}	1.28×10^{-13}	7.85×10^{-9}	3.8×10^{-3}	4.7×10^{-63}	8.2×10^{-118}	1.14×10^{-3}	—
(S) AGE-PC		7.11×10^{-14}	1.85×10^{-13}	1.42×10^{-12}	5.48×10^{-8}	1.3×10^{-3}	3.84×10^{-63}	2.57×10^{-118}	2.75×10^{-4}	—
(D) AGE-PC		7.11×10^{-14}	1.85×10^{-13}	1.3×10^{-12}	5.03×10^{-8}	1.25×10^{-3}	4.7×10^{-63}	8.2×10^{-118}	2.71×10^{-4}	—
AGE-PC	PR	1.42×10^{-14}	2.84×10^{-14}	5.68×10^{-13}	3.2×10^{-8}	9.05×10^{-4}	9.78×10^{-63}	4.27×10^{-117}	2.23×10^{-4}	2
	DR	0	2.84×10^{-14}	7.11×10^{-13}	4.29×10^{-8}	1.66×10^{-3}	8.63×10^{-63}	3.5×10^{-117}	3.99×10^{-4}	8
AGE-1TL	PR	0	1.14×10^{-13}	1.79×10^{-12}	4.2×10^{-8}	1.82×10^{-3}	3.78×10^{-62}	2.33×10^{-116}	8.94×10^{-4}	3
	DR	5.68×10^{-14}	1.42×10^{-13}	2.3×10^{-12}	5.42×10^{-8}	1.21×10^{-3}	3.43×10^{-62}	2.09×10^{-116}	7.08×10^{-4}	8
AGE-2TL	PR	5.68×10^{-14}	2.7×10^{-13}	9.24×10^{-13}	3.38×10^{-8}	8.54×10^{-4}	9.78×10^{-63}	4.27×10^{-117}	2.1×10^{-4}	2
	DR	1.14×10^{-13}	3.98×10^{-13}	1.28×10^{-12}	4.48×10^{-8}	1.61×10^{-3}	8.63×10^{-63}	3.5×10^{-117}	3.86×10^{-4}	8
AGE-3TL	PR	2.42×10^{-13}	3.84×10^{-13}	1.12×10^{-12}	9.21×10^{-9}	2.45×10^{-3}	1.27×10^{-62}	6.26×10^{-117}	1.07×10^{-3}	5
	DR	3.27×10^{-13}	7.39×10^{-13}	1.52×10^{-12}	1.6×10^{-8}	1.98×10^{-3}	1.15×10^{-62}	5.41×10^{-117}	9.14×10^{-4}	11
AGE-4TL	PR	1.42×10^{-13}	4.97×10^{-13}	2.98×10^{-13}	1.48×10^{-9}	3.54×10^{-5}	3.05×10^{-65}	2.37×10^{-117}	3.41×10^{-6}	2
	DR	2.56×10^{-13}	9.95×10^{-13}	6.54×10^{-13}	7.63×10^{-9}	6.8×10^{-4}	6.15×10^{-64}	2.76×10^{-117}	1.79×10^{-4}	8
EXACT SOLUTION		1.0000	1.0000	1.0000	0.9999985	0.9877919	1.08×10^{-62}	9.65×10^{-117}	—	—

Table 30 Absolute errors of the numerical solutions to nonlinear problem (9.21) (the reaction-diffusion equation)

$\Delta x = 0.5,\ \Delta t = 0.05,\ \lambda = 0.2,\ eps = 10^{-4},\ t = 3$

Method		-49	-33	-17	-1	15	207	339	Average of all absolute errors	Number of iterations
AGE-PC	PR	2.84×10^{-14}	2.27×10^{-13}	2.04×10^{-8}	2.27×10^{-4}	1.74×10^{-6}	2.34×10^{-70}	2.97×10^{-123}	8.79×10^{-6}	2
	DR	2.84×10^{-14}	2.13×10^{-13}	5.3×10^{-8}	1.89×10^{-3}	8.79×10^{-7}	1.15×10^{-70}	3.21×10^{-123}	7.69×10^{-5}	6
AGE-1TL	PR	2.27×10^{-13}	1.07×10^{-12}	4.7×10^{-8}	2.6×10^{-4}	3.87×10^{-6}	5.2×10^{-70}	2.78×10^{-123}	1.46×10^{-5}	2
	DR	3.84×10^{-13}	2.06×10^{-12}	8.0×10^{-8}	2.38×10^{-3}	1.16×10^{-6}	1.59×10^{-70}	3.02×10^{-123}	6.57×10^{-5}	6
AGE-2TL	PR	4.0×10^{-13}	4.97×10^{-13}	2.12×10^{-8}	2.35×10^{-4}	1.74×10^{-6}	2.34×10^{-70}	2.97×10^{-123}	8.76×10^{-6}	2
	DR	4.41×10^{-13}	1.49×10^{-12}	5.38×10^{-8}	1.88×10^{-3}	8.79×10^{-7}	1.15×10^{-70}	3.21×10^{-123}	7.67×10^{-5}	6
AGE-3TL	PR	4.26×10^{-13}	1.99×10^{-12}	2.7×10^{-8}	6.67×10^{-4}	1.83×10^{-6}	2.45×10^{-70}	2.97×10^{-123}	1.72×10^{-5}	4
	DR	3.27×10^{-13}	2.57×10^{-12}	5.19×10^{-8}	2.44×10^{-3}	4.97×10^{-7}	6.59×10^{-71}	3.18×10^{-123}	6.03×10^{-5}	8
AGE-4TL	PR	1.42×10^{-12}	8.1×10^{-13}	2.45×10^{-10}	1.25×10^{-5}	1.8×10^{-9}	2.83×10^{-73}	3.13×10^{-123}	2.05×10^{-7}	2
	DR	8.53×10^{-14}	2.07×10^{-12}	3.2×10^{-8}	2.1×10^{-3}	2.54×10^{-6}	3.39×10^{-70}	3.36×10^{-123}	7.34×10^{-5}	6
EXACT SOLUTION		1.0000	1.0000	0.9999987	0.9008854	1.11×10^{-4}	1.48×10^{-68}	1.32×10^{-122}	—	—

$\Delta x = 0.5,\ \Delta t = 0.05,\ \lambda = 0.2,\ eps = 10^{-4},\ t = 12$

Method		-49	-33	-17	-1	15	207	339	Average of all absolute errors	Number of iterations
AGE-PC	PR	2.84×10^{-14}	1.71×10^{-13}	7.12×10^{-10}	1.73×10^{-5}	1.39×10^{-4}	8.63×10^{-68}	2.31×10^{-121}	2.55×10^{-5}	2
	DR	2.84×10^{-14}	2.84×10^{-14}	2.27×10^{-9}	1.39×10^{-4}	7.36×10^{-4}	4.1×10^{-68}	3.17×10^{-121}	3.16×10^{-4}	6
AGE-1TL	PR	2.7×10^{-13}	1.04×10^{-12}	1.83×10^{-9}	3.26×10^{-4}	6.79×10^{-4}	1.97×10^{-67}	1.56×10^{-121}	7.0×10^{-5}	2
	DR	3.84×10^{-13}	2.09×10^{-12}	3.49×10^{-9}	1.55×10^{-4}	2.47×10^{-4}	5.76×10^{-68}	2.51×10^{-121}	2.27×10^{-4}	6
AGE-2TL	PR	3.98×10^{-13}	1.85×10^{-13}	7.4×10^{-10}	1.75×10^{-5}	1.41×10^{-4}	8.63×10^{-68}	2.31×10^{-121}	2.46×10^{-4}	2
	DR	4.41×10^{-13}	2.4×10^{-12}	2.3×10^{-9}	1.39×10^{-4}	7.35×10^{-4}	4.1×10^{-68}	3.17×10^{-121}	3.15×10^{-4}	6
AGE-3TL	PR	4.26×10^{-13}	3.17×10^{-12}	1.12×10^{-9}	2.5×10^{-5}	4.47×10^{-4}	9.01×10^{-68}	2.29×10^{-121}	6.14×10^{-5}	4
	DR	3.27×10^{-13}	3.95×10^{-12}	2.35×10^{-12}	1.28×10^{-4}	3.7×10^{-4}	2.45×10^{-68}	3.06×10^{-121}	2.01×10^{-4}	8
AGE-4TL	PR	9.52×10^{-13}	1.17×10^{-12}	8.63×10^{-12}	3.62×10^{-7}	8.32×10^{-7}	1.02×10^{-70}	2.9×10^{-121}	2.35×10^{-7}	2
	DR	8.53×10^{-14}	3.5×10^{-12}	1.48×10^{-9}	1.2×10^{-4}	8.69×10^{-4}	1.18×10^{-67}	3.69×10^{-121}	2.93×10^{-4}	6
EXACT SOLUTION		1.0000	1.0000	0.9999999	0.9987793	0.0098869	1.33×10^{-66}	1.19×10^{-120}	—	—

Table 31 Absolute errors of the numerical solutions to nonlinear problem (9.21) (the reaction-diffusion equation)

$\Delta x = 0.5,\ \Delta t = 0.05,\ \lambda = 0.2,\ eps = 10^{-4},\ t = 21$

Method		x = -49	-33	-17	-1	15	207	339	Average of all absolute errors	Number of iterations
AGE-PC	PR	2.84×10^{-14}	0	1.07×10^{-11}	4.57×10^{-7}	3.84×10^{-3}	1.39×10^{-65}	1.66×10^{-119}	4.1×10^{-5}	2
	DR	2.84×10^{-14}	1.28×10^{-13}	4.24×10^{-11}	2.96×10^{-6}	4.43×10^{-2}	6.38×10^{-66}	3.04×10^{-120}	5.52×10^{-4}	6
AGE-1TL	PR	2.7×10^{-13}	1.21×10^{-12}	2.73×10^{-11}	6.61×10^{-7}	1.0×10^{-2}	3.27×10^{-65}	3.93×10^{-120}	1.41×10^{-4}	2
	DR	3.84×10^{-13}	2.26×10^{-12}	6.26×10^{-11}	3.20×10^{-6}	3.0×10^{-2}	9.21×10^{-66}	1.98×10^{-119}	3.76×10^{-4}	6
AGE-2TL	PR	3.98×10^{-13}	3.55×10^{-13}	1.2×10^{-11}	4.66×10^{-6}	3.71×10^{-3}	1.39×10^{-65}	1.66×10^{-119}	3.96×10^{-5}	2
	DR	4.41×10^{-13}	2.57×10^{-12}	4.44×10^{-11}	2.97×10^{-6}	4.41×10^{-2}	6.38×10^{-66}	3.04×10^{-119}	5.51×10^{-4}	6
AGE-3TL	PR	4.26×10^{-13}	3.34×10^{-12}	1.82×10^{-11}	3.88×10^{-7}	9.98×10^{-3}	1.46×10^{-65}	1.62×10^{-119}	1.25×10^{-4}	4
	DR	3.27×10^{-13}	4.12×10^{-12}	4.4×10^{-11}	2.47×10^{-6}	2.6×10^{-2}	3.85×10^{-66}	2.87×10^{-119}	3.33×10^{-4}	8
AGE-4TL	PR	9.52×10^{-13}	1.34×10^{-12}	8.95×10^{-13}	6.54×10^{-9}	1.6×10^{-6}	1.6×10^{-68}	2.6×10^{-119}	2.42×10^{-7}	2
	DR	8.53×10^{-14}	3.67×10^{-12}	3.14×10^{-11}	2.42×10^{-6}	4.04×10^{-2}	1.79×10^{-65}	3.82×10^{-119}	5.13×10^{-4}	6
EXACT SOLUTION		1.0000	1.0000	0.9999999	0.9999864	0.4733711	1.198×10^{-64}	1.07×10^{-118}	—	—

$\Delta x = 0.5,\ \Delta t = 0.05,\ \lambda = 0.2,\ eps = 10^{-4},\ t = 30$

Method		x = -49	-33	-17	-1	15	207	339	Average of all absolute errors	Number of iterations
AGE-PC	PR	2.84×10^{-14}	0	2.7×10^{-13}	7.99×10^{-9}	2.26×10^{-4}	1.83×10^{-63}	1.1×10^{-117}	5.59×10^{-5}	2
	DR	2.84×10^{-14}	1.28×10^{-13}	3.69×10^{-13}	4.99×10^{-8}	3.43×10^{-3}	8.11×10^{-64}	2.9×10^{-117}	7.87×10^{-4}	6
AGE-1TL	PR	2.7×10^{-13}	1.21×10^{-12}	9.09×10^{-13}	1.05×10^{-8}	4.59×10^{-4}	4.44×10^{-63}	6.59×10^{-118}	2.15×10^{-4}	2
	DR	3.84×10^{-13}	2.26×10^{-12}	2.53×10^{-13}	5.31×10^{-8}	2.58×10^{-3}	1.20×10^{-63}	1.53×10^{-117}	5.23×10^{-4}	6
AGE-2TL	PR	3.98×10^{-13}	3.55×10^{-13}	7.82×10^{-13}	8.15×10^{-9}	2.2×10^{-4}	1.83×10^{-63}	1.1×10^{-117}	5.38×10^{-5}	2
	DR	4.41×10^{-13}	2.57×10^{-12}	2.47×10^{-12}	5.01×10^{-8}	3.43×10^{-3}	8.11×10^{-64}	2.9×10^{-117}	7.85×10^{-4}	6
AGE-3TL	PR	4.26×10^{-13}	3.34×10^{-12}	2.81×10^{-12}	5.4×10^{-9}	3.97×10^{-4}	1.92×10^{-63}	1.05×10^{-117}	1.9×10^{-4}	4
	DR	3.27×10^{-13}	4.12×10^{-12}	3.5×10^{-12}	4.0×10^{-8}	2.25×10^{-3}	4.92×10^{-64}	2.68×10^{-117}	4.65×10^{-4}	8
AGE-4TL	PR	9.52×10^{-13}	1.34×10^{-12}	1.41×10^{-12}	9.99×10^{-11}	2.69×10^{-6}	2.06×10^{-66}	2.35×10^{-117}	2.51×10^{-7}	2
	DR	8.53×10^{-4}	3.67×10^{-12}	2.7×10^{-12}	3.98×10^{-8}	3.16×10^{-3}	2.23×10^{-63}	3.86×10^{-117}	7.33×10^{-4}	6
EXACT SOLUTION		1.0000	1.0000	1.0000	0.9999985	0.9877919	1.08×10^{-62}	9.65×10^{-117}	—	—

Table 32 Computed wave speeds and their percentage errors

$\Delta x = 1, \lambda = 0.2, \Delta t = 0.2, \text{eps} = 10^{-4}$

t	Method	(S) AGE-EXP	(D) AGE-EXP	(S) AGE-PC	(D) AGE-PC	AGE-PC		AGE-1TL		AGE-2TL		AGE-3TL		AGE-4TL	
						PR	DR	PR	DR	PR	DR	PR	DR	PR	DR
3	CV	0.6908973	0.6887024	0.7038429	0.7036834	0.7032153	0.7032711	0.7144151	0.714478	0.7030501	0.7031086	0.7178107	0.7175869	0.7066162	0.7066676
	CV-R	(0.6916288)		(0.7031970)		(0.7031970)		(0.714411)		(0.7030459)		(0.7178736)		(0.7066156)	
	PCV-R (%)	(2.19)		(0.553)		(0.533)		(1.033)		(0.574)		(1.523)		(0.069)	
	PCV (%)	2.292	2.603	0.461	0.484	0.55	0.542	1.034	1.043	0.574	0.565	1.514	1.482	0.0692	0.062
12	CV	0.6889379	0.6890496	0.7026014	0.7027083	0.7035437	0.703606	0.7203569	0.7203964	0.7032528	0.7033159	0.7234951	0.7232126	0.7065187	0.7065909
	CV-R	(0.6870319)		(0.7035245)		(0.7035245)		(0.7203591)		(0.7032573)		(0.7235846)		(0.7065183)	
	PCV-R (%)	(2.84)		(0.506)		(0.506)		(1.874)		(0.544)		(2.330)		(0.083)	
	PCV (%)	2.569	2.554	0.637	0.622	0.504	0.495	1.874	1.88	0.545	0.536	2.318	2.278	0.083	0.073
21	CV	0.6892023	0.6892408	0.7027116	0.7027591	0.7038907	0.7039569	0.7211176	0.7211574	0.7035924	0.7036592	0.7239050	0.7236222	0.7065232	0.7065970
	CV-R	(0.6869790)		(0.7038702)		(0.7038702)		(0.7211216)		(0.7035991)		(0.7239955)		(0.7065230)	
	PCV-R (%)	(2.846)		(0.458)		(0.458)		(1.982)		(0.496)		(2.389)		(0.0824)	
	PCV (%)	2.532	2.526	0.621	0.615	0.455	0.445	1.982	1.987	0.497	0.487	2.376	2.336	0.0824	0.072
30	CV	0.6892778	0.689422	0.7027429	0.7028983	0.7039567	0.7040239	0.7212402	0.7212807	0.7036572	0.7037251	0.7239644	0.7236825	0.7065240	0.7065980
	CV-R	(0.6869822)		(0.7039361)		(0.7039361)		(0.7212447)		(0.7036646)		(0.7240550)		(0.7065237)	
	PCV-R (%)	(2.846)		(0.448)		(0.448)		(2.00)		(0.487)		(2.397)		(0.0823)	
	PCV (%)	2.521	2.501	0.617	0.595	0.445	0.436	1.999	2.005	0.488	0.478	2.384	12.344	0.0823	0.0718

CV: computed wave speed; CV-R: computed wave speed from Ramos; PCV: percentage error in CV; PCV-R: percentage error in CV-R; Exact $V = 1/\sqrt{2} = 0.7071058$.

Table 33 Computed wave speeds and their percentage errors

$\Delta x = 0.5$, $\Delta t = 0.05$, $\lambda = 0.2$, eps $= 10^{-4}$

t	Method	AGE-PC		AGE-1TL		AGE-2TL		AGE-3TL		AGE-4TL	
		PR	DR	PR	DR	PR	DR	PR	DR	PR	DR
3	CV	0.7061664	0.7062216	0.7090783	0.7091131	0.7061763	0.7062321	0.7098810	0.7096623	0.7070767	0.7071058
	CV-R	(0.7061863)		(0.7090694)		(0.7061753)		(0.7099601)		(0.7070766)	
	PCV-R (%)	(0.13)		(2.78)		(0.132)		(0.404)		(0.00413)	
	PCV (%)	0.133	0.125	0.279	0.284	0.131	0.124	0.392	0.362	0.00412	0.00246
12	CV	0.7062323	0.7062913	0.7105140	0.7105653	0.7062357	0.706295	0.7112576	0.7109959	0.7070700	0.7071058
	CV-R	(0.7062578)		(0.7105382)		(0.7062386)		(0.7113624)		(0.7070711)	
	PCV-R (%)	(0.120)		(0.485)		(0.123)		(0.602)		(0.0049)	
	PCV (%)	0.124	0.115	0.482	0.489	0.123	0.115	0.587	0.55	0.00505	0.00458
21	CV	0.7063150	0.706377	0.7106940	0.7107515	0.7063178	0.7063801	0.7113578	0.7110972	0.7070703	0.7071058
	CV-R	(0.7063419)		(0.7107250)		(0.7063221)		(0.7114640)		(0.7070724)	
	PCV-R (%)	(0.108)		(0.512)		(0.111)		(0.616)		(0.00472)	
	PCV (%)	0.112	0.103	0.507	0.516	0.111	0.103	0.601	0.564	0.00502	0.00484
30	CV	0.7063307	0.7063937	0.7107242	0.7107833	0.7063334	0.7063966	0.7113730	0.7111128	0.7071058	0.7071058
	CV-R	(0.7063589)		(0.7107582)		(0.7063389)		(0.7114793)		(0.7070734)	
	PCV-R (%)	(0.106)		(0.517)		(0.108)		(0.619)		(0.00458)	
	PCV (%)	0.11	0.101	0.512	0.52	0.109	0.1	0.603	0.567	0.00501	0.00488

See Table 32 for abbreviations.

In our experiment the solutions worked out from each of the numerical methods were used along the whole mesh line and then the composite trapezium rule was employed to compute V which was then compared with the exact value $1/\sqrt{2} = 0.7071058$. It must be mentioned that Ramos (1985) employed the Thomas algorithm to solve tridiagonal systems of equations arising from the application of the relevant finite difference formulas. Since the values of V from the solutions of these methods are available, we are then able to compare the accuracy of the solutions of the (S)AGE, (D)AGE and AGE class of methods with those obtained from the application of the Thomas algorithm. In general, we find that the computed values of V from the AGE(PR) solutions (CV) are in close agreement with those of Ramos (CV-R). This can also be read from a comparison of the appropriate percentage errors in the computed values (PCV and PCV-R) as presented in Tables 32 and 33 for both coarse and fine grids.

An estimate of the computational effort involved by all the methods to execute the calculations per time row (in the case of the GE schemes) or per iteration (in the case of the AGE schemes) is given in Table 34. The estimate of the amount of work done is based on an odd number (m odd) of internal points on the time row.

From the preceding discussion, we conclude that among all the methods worthy of recommendation, the AGE-4TL(PR) procedure provides the most accurate solution to the reaction-diffusion problem. As a high-order method, it is expected to yield more accurate results for a wide range of spatial and temporal step sizes. By contrast, the 3TL technique is the least competitive in the AGE class because it is not only less accurate but also requires more arithmetic work and iterations. Although the GE-PC and AGE-PC procedures result in highly accurate solutions

Table 34 Computational complexity

Schemes	Operations	Number of multiplications	Number of additions
GE-EXP		$12m + 10$	$7m + 1$
	P	$38m$	29
GE-PC	C	$18m + 10$	$9m + 1$
	Total	$56m + 10$	$9m + 30$
	P	$38m$	29
AGE-PC	C	$22m - 23$	$\dfrac{(33m - 1)}{2} + 6$
	Total	$60m - 23$	$\dfrac{33m - 1}{2} + 35$
AGE-1TL		$\dfrac{(73m - 1)}{2} + 34$	$31m - 13$
AGE-2TL		$38m - 12$	$\dfrac{63m - 1}{2} - 11$
AGE-3TL		$76m - 12$	$52m - 13$
AGE-4TL		$75m - 10$	$55m - 11$

for our particular example, the *explicit evaluations* of the diffusion terms at the predictor stage may lead to some stability restrictions on the time step size.

Experiment 9

The following nonlinear first order hyperbolic (convection) equation was considered:

$$\frac{\partial U}{\partial t} = -U \frac{\partial U}{\partial x} \tag{9.25}$$

and two problems were solved using the AGE algorithm.

Problem 1

Equation (9.25) was solved subject to the auxiliary conditions (Casulli et al., 1984),

$$U(x, 0) = 1 - x \qquad 0 \leq x \leq 1 \tag{9.26a}$$

$$U(0, t) = 1 \qquad 0 < t \leq 1$$

and
$$U(1, t) = \begin{cases} 1 & \text{for } 1 \leq t \\ 0 & \text{for } t < 1 \end{cases} \tag{9.26b}$$

The exact solution is given by

$$U(x, t) = \begin{cases} 1 & \text{for } x \leq t \\ \dfrac{x - 1}{t - 1} & \text{for } t < x \leq 1 \end{cases} \tag{9.27}$$

Problem 2

For this problem the initial and boundary conditions take the form (Ames, 1977),

$$U(x, 0) = x \qquad 0 \leq x \leq 1 \tag{9.28a}$$

$$U(0, t) = 0 \qquad t \geq 0$$

and
$$U(1, t) = \frac{1}{1 + t} \qquad t \geq 0 \tag{9.28b}$$

This problem has the exact solution given by

$$U(x, t) = \frac{x}{1 + t} \tag{9.29}$$

The solutions to both problems were obtained using the AGE algorithm which was implemented on the CDBT and CDCT formulas given by the linearized tridiagonal systems (8.77) and (8.79), respectively. A number of computer runs

Table 35 Absolute errors of the numerical solutions to Eq. (9.25) for Problem 1 [nonlinear first order hyperbolic (convection) equation]

$t = 1.0$, $\Delta x = 0.05$, $\Delta t = 0.025$, $\lambda = 0.5$, eps $= 10^{-6}$

Method		0.1	0.2	0.3	0.4	0.5	0.6	0.7	0.8	0.9	Average of all absolute errors	No. of iterations
AGE-CDBT	PR	1.21×10^{-6}	5.25×10^{-6}	3.66×10^{-5}	1.2×10^{-4}	2.19×10^{-4}	1.55×10^{-4}	2.15×10^{-3}	7.97×10^{-4}	3.04×10^{-2}	1.21×10^{-2}	5
	DR	1.19×10^{-6}	5.24×10^{-6}	3.65×10^{-5}	1.21×10^{-4}	2.18×10^{-4}	1.54×10^{-4}	2.15×10^{-3}	7.93×10^{-4}	3.04×10^{-2}	1.21×10^{-2}	16
AGE-CDCT	PR	9.9×10^{-4}	1.89×10^{-3}	1.88×10^{-3}	8.25×10^{-4}	6.55×10^{-3}	7.06×10^{-3}	1.22×10^{-2}	1.63×10^{-2}	1.3×10^{-1}	1.64×10^{-2}	4
	DR	9.9×10^{-4}	1.89×10^{-3}	1.88×10^{-3}	8.23×10^{-4}	6.55×10^{-3}	7.07×10^{-3}	1.22×10^{-2}	1.64×10^{-2}	1.3×10^{-1}	1.64×10^{-2}	15
EXACT SOLUTION		1.00	1.00	1.00	1.00	1.00	1.00	1.00	1.00	1.00	—	—

$t = 1.0$, $\Delta x = 0.05$, $\Delta t = 0.05$, $\lambda = 1.0$, eps $= 10^{-6}$

Method		0.1	0.2	0.3	0.4	0.5	0.6	0.7	0.8	0.9	Average of all absolute errors	No. of iterations
AGE-CDBT	PR	4.62×10^{-7}	2.43×10^{-7}	5.49×10^{-6}	2.22×10^{-5}	1.55×10^{-4}	2.27×10^{-4}	3.44×10^{-3}	2.1×10^{-2}	6.77×10^{-2}	3.25×10^{-2}	8
	DR	1.84×10^{-7}	8.19×10^{-7}	5.82×10^{-6}	2.28×10^{-5}	1.53×10^{-4}	2.24×10^{-4}	3.45×10^{-3}	2.11×10^{-2}	6.77×10^{-2}	3.25×10^{-2}	20
AGE-CDCT	PR	7.42×10^{-4}	2.22×10^{-3}	4.22×10^{-3}	3.69×10^{-3}	4.28×10^{-3}	1.27×10^{-2}	8.78×10^{-3}	3.7×10^{-2}	2.2×10^{-1}	2.93×10^{-2}	6
	DR	7.42×10^{-4}	2.22×10^{-3}	4.22×10^{-3}	3.69×10^{-3}	4.28×10^{-3}	1.27×10^{-2}	8.77×10^{-3}	3.7×10^{-2}	2.2×10^{-1}	2.93×10^{-2}	18
EXACT SOLUTION		1.00	1.00	1.00	1.00	1.00	1.00	1.00	1.00	1.00	—	—

$t = 1.0$, $\Delta x = 0.05$, $\Delta t = 0.1$, $\lambda = 2.0$, eps $= 10^{-6}$

Method		0.1	0.2	0.3	0.4	0.5	0.6	0.7	0.8	0.9	Average of all absolute errors	No. of iterations
AGE-CDBT	PR	3.07×10^{-5}	8.7×10^{-5}	2.61×10^{-4}	5.14×10^{-4}	1.81×10^{-4}	4.91×10^{-3}	1.98×10^{-2}	5.51×10^{-2}	1.21×10^{-1}	6.22×10^{-2}	17
	DR	3.08×10^{-5}	8.91×10^{-5}	2.65×10^{-4}	5.18×10^{-4}	1.77×10^{-4}	4.91×10^{-3}	1.98×10^{-2}	5.51×10^{-2}	1.21×10^{-1}	6.22×10^{-2}	34
AGE-CDCT	PR	2.11×10^{-3}	5.16×10^{-3}	4.95×10^{-3}	6.32×10^{-3}	1.38×10^{-2}	1.48×10^{-2}	2.47×10^{-2}	1.27×10^{-1}	3.75×10^{-1}	6.03×10^{-2}	10
	DR	2.11×10^{-3}	5.16×10^{-3}	4.95×10^{-3}	6.32×10^{-3}	1.38×10^{-2}	1.48×10^{-2}	2.47×10^{-2}	1.27×10^{-1}	3.75×10^{-1}	6.03×10^{-2}	24
EXACT SOLUTION		1.00	1.00	1.00	1.00	1.00	1.00	1.00	1.00	1.00	—	—

385

Table 36 Absolute errors of the numerical solutions to Eq. (9.25) for Problem 2 [nonlinear first order hyperbolic (convection) equation]

$t = 1.0$, $\Delta x = 0.05$, $\Delta t = 0.025$, $\lambda = 0.5$, eps $= 10^{-6}$

Method		0.1	0.2	0.3	0.4	0.5	0.6	0.7	0.8	0.9	Average of all absolute errors	No. of iterations
AGE-CDBT	PR	1.07×10^{-8}	1.69×10^{-6}	6.38×10^{-5}	8.59×10^{-4}	4.69×10^{-3}	1.04×10^{-2}	9.73×10^{-3}	7.61×10^{-3}	7.19×10^{-3}	4.41×10^{-3}	3
	DR	1.56×10^{-6}	4.86×10^{-6}	6.87×10^{-5}	8.66×10^{-4}	4.7×10^{-3}	1.04×10^{-2}	9.73×10^{-3}	7.62×10^{-3}	7.19×10^{-3}	4.41×10^{-3}	13
AGE-CDCT	PR	1.12×10^{-11}	2.0×10^{-10}	1.32×10^{-9}	3.63×10^{-9}	7.37×10^{-10}	4.72×10^{-9}	5.56×10^{-9}	2.65×10^{-9}	2.22×10^{-9}	3.3×10^{-9}	3
	DR	1.76×10^{-6}	3.53×10^{-6}	5.33×10^{-6}	7.11×10^{-6}	8.46×10^{-6}	8.17×10^{-6}	6.29×10^{-6}	4.72×10^{-6}	3.29×10^{-6}	9.38×10^{-6}	13
EXACT SOLUTION		0.05	0.1	0.15	0.2	0.25	0.3	0.35	0.4	0.45	—	—

$t = 1.0$, $\Delta x = 0.05$, $\Delta t = 0.05$, $\lambda = 1.0$, eps $= 10^{-6}$

Method		0.1	0.2	0.3	0.4	0.5	0.6	0.7	0.8	0.9	Average of all absolute errors	No. of iterations
AGE-CDBT	PR	1.65×10^{-7}	1.45×10^{-5}	3.17×10^{-4}	2.68×10^{-3}	1.02×10^{-2}	1.91×10^{-2}	1.94×10^{-2}	1.6×10^{-2}	1.44×10^{-2}	8.87×10^{-3}	4
	DR	5.82×10^{-7}	1.54×10^{-5}	3.19×10^{-4}	2.68×10^{-3}	1.02×10^{-2}	1.91×10^{-2}	1.94×10^{-2}	1.6×10^{-2}	1.44×10^{-2}	8.87×10^{-3}	15
AGE-CDCT	PR	3.09×10^{-11}	7.17×10^{-10}	3.73×10^{-9}	8.38×10^{-10}	7.65×10^{-9}	9.69×10^{-9}	9.78×10^{-9}	9.4×10^{-9}	3.2×10^{-9}	5.31×10^{-9}	4
	DR	7.46×10^{-7}	1.53×10^{-6}	2.37×10^{-6}	3.23×10^{-6}	3.78×10^{-6}	3.53×10^{-6}	2.82×10^{-6}	2.07×10^{-6}	1.37×10^{-6}	4.26×10^{-6}	14
EXACT SOLUTION		0.05	0.1	0.15	0.2	0.25	0.3	0.35	0.4	0.45	—	—

$t = 1.0$, $\Delta x = 0.05$, $\Delta t = 0.1$, $\lambda = 2.0$, eps $= 10^{-6}$

Method		0.1	0.2	0.3	0.4	0.5	0.6	0.7	0.8	0.9	Average of all absolute errors	No. of iterations
AGE-CDBT	PR	4.78×10^{-6}	1.79×10^{-4}	1.85×10^{-3}	8.52×10^{-3}	2.17×10^{-2}	3.44×10^{-2}	3.78×10^{-2}	3.38×10^{-2}	2.95×10^{-2}	1.8×10^{-2}	6
	DR	4.81×10^{-6}	1.79×10^{-4}	1.85×10^{-3}	8.52×10^{-3}	2.17×10^{-2}	3.44×10^{-2}	3.78×10^{-2}	3.38×10^{-2}	2.95×10^{-2}	1.8×10^{-2}	18
AGE-CDCT	PR	4.68×10^{-11}	5.76×10^{-10}	1.07×10^{-9}	7.76×10^{-9}	9.13×10^{-9}	4.75×10^{-9}	2.11×10^{-8}	3.15×10^{-8}	1.19×10^{-8}	8.92×10^{-9}	5
	DR	1.92×10^{-7}	4.36×10^{-7}	7.78×10^{-7}	1.12×10^{-6}	1.11×10^{-6}	7.7×10^{-7}	6.43×10^{-7}	5.89×10^{-7}	2.98×10^{-7}	1.5×10^{-6}	16
EXACT SOLUTION		0.05	0.1	0.15	0.2	0.25	0.3	0.35	0.4	0.45	—	—

Table 37 Absolute errors of the numerical solutions to Eq. (9.25) for Problem 2 [nonlinear first order hyperbolic (convection) equation]

$t = 5$, $\Delta x = 0.1$, $\Delta t = 0.05$, $\lambda = 0.5$, eps $= 10^{-6}$

Method		0.1	0.2	0.3	0.4	0.5	0.6	0.7	0.8	0.9	Average of all absolute errors	No. of iterations
AGE-CDBT	PR	5.93×10^{-4}	3.34×10^{-3}	6.11×10^{-3}	4.87×10^{-3}	1.31×10^{-3}	2.48×10^{-3}	2.74×10^{-3}	8.61×10^{-4}	1.33×10^{-3}	2.63×10^{-3}	3
	DR	5.89×10^{-4}	3.35×10^{-3}	6.1×10^{-3}	4.89×10^{-3}	1.29×10^{-3}	2.49×10^{-3}	2.7×10^{-3}	8.68×10^{-4}	1.27×10^{-3}	2.62×10^{-3}	11
AGE-CDCT	PR	1.11×10^{-10}	2.18×10^{-10}	4.55×10^{-11}	1.06×10^{-10}	9.55×10^{-10}	6.45×10^{-10}	1.64×10^{-10}	2.1×10^{-9}	3.16×10^{-9}	1.1×10^{-9}	2
	DR	3.26×10^{-6}	5.95×10^{-6}	1.17×10^{-5}	8.13×10^{-6}	2.39×10^{-6}	8.26×10^{-6}	3.8×10^{-5}	5.61×10^{-6}	5.49×10^{-5}	1.77×10^{-5}	11
EXACT SOLUTION		0.0166699	0.0333393	0.05000	0.0666667	0.0833333	0.1000	0.1166667	0.1333333	0.1500	—	—

$t = 10$, $\Delta x = 0.1$, $\Delta t = 0.1$, $\lambda = 1.0$, eps $= 10^{-6}$

Method		0.1	0.2	0.3	0.4	0.5	0.6	0.7	0.8	0.9	Average of all absolute errors	No. of iterations
AGE-CDBT	PR	2.24×10^{-3}	7.02×10^{-3}	3.51×10^{-3}	3.81×10^{-3}	2.34×10^{-3}	1.49×10^{-3}	5.29×10^{-4}	2.09×10^{-3}	2.53×10^{-4}	2.21×10^{-3}	2
	DR	2.23×10^{-3}	7.03×10^{-3}	3.5×10^{-3}	3.9×10^{-3}	2.32×10^{-3}	1.5×10^{-3}	4.92×10^{-4}	2.09×10^{-3}	2.01×10^{-4}	2.2×10^{-3}	10
AGE-CDCT	PR	1.43×10^{-10}	1.48×10^{-10}	5.8×10^{-11}	1.59×10^{-9}	7.64×10^{-10}	2.03×10^{-10}	2.67×10^{-9}	2.71×10^{-9}	3.07×10^{-9}	1.26×10^{-9}	2
	DR	3.15×10^{-6}	5.16×10^{-6}	1.17×10^{-5}	7.57×10^{-6}	2.24×10^{-5}	7.77×10^{-6}	3.55×10^{-5}	5.81×10^{-6}	5.1×10^{-5}	1.67×10^{-5}	10
EXACT SOLUTION		0.0090909	0.0181818	0.0272273	0.0363636	0.0454545	0.0545455	0.0636364	0.0727273	0.0818181	—	—

$t = 20$, $\Delta x = 0.1$, $\Delta t = 0.2$, $\lambda = 2.0$, eps $= 10^{-6}$

Method		0.1	0.2	0.3	0.4	0.5	0.6	0.7	0.8	0.9	Average of all absolute errors	No. of iterations
AGE-CDBT	PR	3.29×10^{-3}	5.18×10^{-3}	1.81×10^{-3}	2.76×10^{-3}	2.4×10^{-4}	1.43×10^{-4}	7.23×10^{-4}	2.26×10^{-4}	6.29×10^{-4}	1.67×10^{-3}	2
	DR	3.28×10^{-3}	5.19×10^{-3}	1.81×10^{-3}	2.77×10^{-3}	2.23×10^{-4}	1.49×10^{-4}	6.91×10^{-4}	2.22×10^{-4}	5.8×10^{-4}	1.66×10^{-3}	9
AGE-CDCT	PR	5.59×10^{-11}	4.11×10^{-10}	6.25×10^{-10}	3.53×10^{-9}	2.79×10^{-9}	3.19×10^{-9}	2.21×10^{-9}	1.77×10^{-9}	3.38×10^{-9}	2.0×10^{-9}	2
	DR	3.08×10^{-6}	4.8×10^{-6}	1.12×10^{-5}	7.3×10^{-6}	2.15×10^{-5}	7.44×10^{-6}	3.42×10^{-5}	5.34×10^{-6}	4.88×10^{-5}	1.6×10^{-5}	9
EXACT SOLUTION		0.0047619	0.0095238	0.0142857	0.0190476	0.0238095	0.0285714	0.0333333	0.0380952	0.0428571	—	—

Table 38 Computational complexity per iteration (m odd)

Schemes	Operations Number of multiplications	Number of additions
AGE-CDBT	$22m - 2$	$\dfrac{29(m - 1)}{2} + 14$
AGE-CDCT	$34m - 16$	$26m - 12$

were carried out using $\Delta x = 0.05$ and with varying time steps to determine the accuracy of the AGE-CDBT and AGE-CDCT schemes. The results given in terms of the absolute errors are shown in Tables 35–37.

It is observed from Table 35 that for small Δt, the average of absolute errors on time level $t = 1.0$ indicates that the AGE-CDBT scheme is more accurate than the AGE-CDCT procedure. When the mesh ratio is progressively increased (Δx being fixed), we find that the accuracy of the AGE-CDCT scheme begins to improve over the AGE-CDBT method. Since the truncation errors of the CDCT formula is $O[(\Delta x)^2 + (\Delta t)^2]$, it is therefore clear that the AGE-CDCT scheme produces a more accurate solution than AGE-CDBT whose truncation error is only $O[(\Delta x)^2 + \Delta t]$. The accuracy of the AGE-CDCT scheme is even more pronounced for the second example. It is immediately evident from Table 37 that despite having to satisfy the stringent convergence requirement of eps $= 10^{-6}$, the AGE-CDCT(PR) scheme demonstrates that it requires only two iterations while at the same time it maintains its high order of accuracy for large time steps. Thus, although the computational complexity of the AGE-CDCT scheme may be slightly large as illustrated in Table 38, it offers great promise as an accurate, stable, and efficient numerical procedure to solve nonlinear hyperbolic problems.

REFERENCES

Peaceman, D. W., and Rachford, H. H.: The numerical solution of parabolic and elliptic differential equations, *J. Soc. Indust. Appl. Maths*. 3, pp. 28–41 (1955).

Wachspress, E. L.: *Iterative solution of elliptic systems*, Englewood Cliffs: Prentice-Hall, Inc. (1966).

Douglas, J., and Rachford, H. H.: On the numerical solution of heat conduction problems in two or three space variables, *Trans. Amer. Math. Soc*. 82, pp. 421–439 (1956).

Mitchell, A. R., and Griffiths, D. F.: *The finite difference methods in partial differential equations*, Chichester: John Wiley and Sons (1980).

Lees, M.: A linear three level difference scheme for quasi-linear parabolic equations, *Maths. Comp*. 20, pp. 516–522 (1966).

Friedman, A.: *Partial differential equations of parabolic type*, Englewood Cliffs: Prentice-Hall Inc. (1964).

Douglas, J., and Jones, B. F.: On predictor-corrector methods for non-linear parabolic differential equations, *J. Soc. Indust. Appl. Math*. 11, pp. 195–204 (1963).

Ramos, J. I.: Numerical solution of reactive-diffusive systems, Part 1: Explicit methods, *Int. Jour. Comp. Maths*. 18, pp. 43–66 (1985), Part 2: Methods of lines and implicit algorithms, *Int. J.*

Comp. Maths. 18, pp. 141–162 (1985), Part 3: Time linearisation and operator-splitting techniques, *Int. J. Comp. Maths.* 18, pp. 289–310 (1985).

Kopal, Z.: *Numerical analysis,* New York: John Wiley and Sons (1961).

Saulev, V. K.: *Integration of equations of parabolic type by the method of nets,* Oxford: Pergramon Press (1964).

Sahimi, M.S.: *Numerical Methods for Solving Hyperbolic and Parabolic Differential Equations,* Ph.D. thesis, Loughborough University of Technology (1986).

Evans, D. J., and Abdullah, A. R. B.: A new explicit method for the solution of $\partial u/\partial t = \partial^2 u/\partial x^2 + \partial^2 u/\partial y^2$, *Int. J. Computer Math.* 14, pp. 325–352 (1983).

Abdullah, A. R.: The study of some numerical methods for solving parabolic partial differential equations, Ph.D. Thesis, Loughborough University of Technology (1983).

Madsen, N. K., and Sincovec, R. F.: *General software for partial differential equations in numerical methods for differential system,* pp. 229–242 (eds. Lapidus, L. and Schiesser, W. E.), New York: Academic Press, 1976.

Caldwell, J., and Smith P.: Solution of Burger's equation with a large Reynolds number, *Appl. Math. Modelling* 6, pp. 381–385 (1982).

Casilli, V., Cheng, R. T., and Bulgarelli, U.: Eulerian-Lagrangian solution of convection dominated diffusion problems, *Numerical methods for non-linear problems,* vol. 2, pp. 962–971, (eds. Taylor, C., Hinton, E., and Owen, D. R. J.), Swansea: Pineridge Press (1984).

Cole, J. D.: On a quasilinear parabolic equation occurring in aero-dynamics, *A. Appl. Maths.* 9, pp. 225–236 (1951).

Evans, D. J., and Abdullah, A. R. B.: The group explicit method for the solution of Burger's equation, *Computing 32*, pp. 239–253 (1984).

INDEX